CAMBRIDGE LIBRARY COLLECTION

Books of enduring scholarly value

Mathematical Sciences

From its pre-historic roots in simple counting to the algorithms powering modern desktop computers, from the genius of Archimedes to the genius of Einstein, advances in mathematical understanding and numerical techniques have been directly responsible for creating the modern world as we know it. This series will provide a library of the most influential publications and writers on mathematics in its broadest sense. As such, it will show not only the deep roots from which modern science and technology have grown, but also the astonishing breadth of application of mathematical techniques in the humanities and social sciences, and in everyday life.

Babbage's Calculating Engines

The famous and prolific nineteenth-century mathematician, engineer and inventor Charles Babbage (1791–1871) was an early pioneer of computing. He planned several calculating machines, but none was built in his lifetime. On his death his youngest son, Henry P. Babbage, was charged with the task of completing an unfinished volume of papers on the machines, which was finally published in 1889 and is reissued here. The papers, by a variety of authors, were collected from journals including *The Philosophical Magazine*, *The Edinburgh Review* and *Scientific Memoirs*. They relate to the construction and potential application of Charles Babbage's calculating engines, notably the Difference Engine and the more complex Analytical Engine, which was to be programmed using punched cards. The book also includes correspondence with members of scientific societies, as well as proceedings, catalogues and drawings. Included is a complete catalogue of the drawings of the Analytical Engine.

Babbage's Calculating Engines

Being a Collection of Papers Relating to Them;
Their History, and Construction

CHARLES BABBAGE
EDITED BY HENRY P. BABBAGE

CAMBRIDGE
UNIVERSITY PRESS

CAMBRIDGE UNIVERSITY PRESS

Cambridge, New York, Melbourne, Madrid, Cape Town, Singapore,
São Paolo, Delhi, Dubai, Tokyo

Published in the United States of America by Cambridge University Press, New York

www.cambridge.org
Information on this title: www.cambridge.org/9781108000963

© in this compilation Cambridge University Press 2010

This edition first published 1889
This digitally printed version 2010

ISBN 978-1-108-00096-3 Paperback

Ever truly yours
C. Babbage

FROM A PHOTOGRAPH TAKEN IN 1860.

BABBAGE'S

CALCULATING ENGINES.

BEING

A COLLECTION OF PAPERS RELATING TO THEM;
THEIR HISTORY, AND CONSTRUCTION.

LONDON:
E. AND F. N. SPON, 125, STRAND.
1889.

PREFACE.

THIS volume is intended to bring together, as much as possible, the information scattered in various places, regarding the Calculating Machines of Charles Babbage, and to make it available for those interested in the subject, who may wish to pursue it further.

He has never himself written a full description of any of his machines. He preferred, while his energies lasted, to devote them to the actual progress and development of the design. As regards the Analytical Engine, he considered the sketch at the commencement of this volume by the Count Menabrea, and the translation of it with notes by Lady Lovelace, as entirely disposing of the mathematical aspect; proving " that the whole of the developments and operations of Analysis are now capable of being executed by machinery."

In the " *Passages from the Life of a Philosopher*," published in 1864, and written with the object of interesting the public in these machines, he mentions on the last page a forthcoming " History of the Analytical Engine." As materials towards it he had an Article from the " Philosophical Magazine " of September 1853, and Article XXIX. from Vol. III. of the " Scientific Memoirs," containing the translation and notes by Lady Lovelace above mentioned, reprinted; and added several chapters of " The Passages " and other papers. He also supplied a complete list of the drawings, notations and note books of the Analytical Engine. Thus the first 294 pages of this book were printed in his lifetime; but owing to the weariness and infirmities of age, he did nothing more, and the historical and descriptive part was never written.

A great deal has been written about the Difference Engine, but it is still to be regretted that he did not himself contribute more information about the expanded uses which he saw in the possible future for such an engine. As to his own contrivance, he was quite content to let it be judged by the beautiful fragment put together in 1833, which will for ever remain to answer all detractors. He died at his house No. 1, Dorset Street, London, on 18th October, 1871. By his Will he left the Calculating Machines and all that belonged to them at my absolute disposal, and so the materials of this unfinished volume came to me.

It is to this, and not to any special fitness for the task, that it has fallen to me to complete it. I need not enter into the reasons which have impelled me to do the best I can unaided rather than to devolve the work on another.

Pages 295 to 322 have been added agreeably to a memorandum found among his papers.

The report of the Committee of the British Association 1878 has been reprinted by permission from the Proceedings of the Association 1878 (pp. 323 to 330). A Paper read by me at the recent meeting of the Association at Bath, slightly modified to suit its position in this book, has been added (pp. 331 to 338). Lithographed copies of the general plan No. 25, dated 6th August, 1840, of the Analytical Engine, and impressions from the Plates mentioned at page 6, have been supplied, and I have added a few concluding remarks.

The arrangement of the book is unavoidably faulty, but in what I have written I have given references to the preceding pages, which will, I hope, be of some use, and if ever a fresh edition is wanted, the whole will, of course, be recast.

HENRY P. BABBAGE.

October, 1888.

CONTENTS.

CONTENTS.

1 Plate of Portion of Difference Engine.

13 Plates (see page 6).

Plan No. 25, dated 6 August, 1840, of the Analytical Engine.

List of Mr. Babbage's Printed Papers.

HISTORY

OF

MR. BABBAGE'S CALCULATING ENGINES.

Statement of the circumstances attending the Invention and Construction of Mr. Babbage's Calculating Engines.

[From the Philosophical Magazine, Sept. 1843, p. 235.]

MUCH misapprehension having arisen as to the circumstances attending the invention and construction of Mr. Babbage's Calculating Engines, it is necessary to state *from authority* the facts relating to them.

In 1823, Mr. Babbage, who had previously invented an Engine for calculating and printing tables by means of *differences*, undertook, at the desire of the Government, to superintend the construction of such an Engine. He bestowed his whole time upon the subject for many years, refusing for that purpose other avocations which would have been attended with considerable pecuniary advantage. During this period about £17,000 had been expended by the Government in the construction of the Difference Engine. A considerable part of this sum had from time to time been advanced by Mr. Babbage for the payment of the workmen, and was of course repaid; but it was never contemplated by either party that any portion of this sum should be appropriated to Mr. Babbage himself, and in truth not one single shilling of the money was in any shape whatever received by Mr. Babbage for his invention, his time, or his services, a fact which Sir Robert Peel admitted in the House of Commons in March 1843.

Early in 1833 the construction of this Engine was suspended on account of some dissatisfaction with the workmen, which it is now unnecessary to detail. It was expected that the interruption, which arose from circumstances over which Mr. Babbage had no control, would be only temporary. About twelve months after the progress of the Difference Engine had been thus suspended, Mr. Babbage discovered a principle of an entirely new order, the power of which over the most complicated arithmetical operations seemed nearly unbounded. The invention of simpler mechanical means for executing the elementary operations of that Engine, now acquired far greater importance than it had hitherto possessed.

In the Engine for calculating by differences, such simplifications affected only about a hundred and twenty similar parts, while in the new, or Analytical Engine, they might affect several thousand. The Difference Engine might be constructed with more or less advantage, by employing various mechanical modes for the operation of addition. The Analytical Engine could not exist without inventing for it a method of mechanical addition possessed of the utmost simplicity. In fact it was not until upwards of twenty different modes for performing the operation of addition had been designed and drawn, that the necessary degree of simplicity required for the Analytical Engine was ultimately attained.

B

These new views acquired great additional importance from their bearings upon the Difference Engine already partly executed for the Government; for if such simplifications should be discovered, it might happen that the Analytical Engine would execute with greater rapidity the calculations for which the Difference Engine was intended; or that the Difference Engine would itself be superseded by a far simpler mode of construction.

Though these views might, perhaps, at that period, have appeared visionary, they have subsequently been completely realized.

To have allowed the construction of the Difference Engine to be resumed while these new views were withheld from the Government, would have been improper; yet the state of uncertainty in which those views were then necessarily involved, rendered any written communication respecting their probable bearing on that engine a matter of very great difficulty. It therefore appeared to Mr. Babbage that the most straightforward course was to ask for an interview with the head of the Government, and to communicate to him the exact state of the case. Various circumstances occurred to delay, and ultimately to prevent that interview.

From the year 1833 to the close of 1842, Mr. Babbage repeatedly applied to the Government for its decision upon the subject. These applications were unavailing. Years of delay and anxiety followed each other, impairing those energies which were now directed to the invention of the Analytical Engine. This state of uncertainty had many injurious effects. It prevented Mr. Babbage from entering into any engagement with other Governments respecting the Analytical Engine, by which he might have been enabled to employ a greater number of assistants, and thus to have applied his faculties only to the highest departments of the subject, instead of exhausting them on inferior objects, that might have been executed with less fatigue by other heads. It also became necessary, from motives of prudence, that the heavy expense incurred for this purpose should be spread over a period of many years. This consideration naturally caused a new source of anxiety and risk, arising from the uncertain tenure of human life and of human faculties,—a reflection ever present to distract and torment the mind, and itself calculated to cause the fulfilment of its own forebodings.

Amidst such distractions the author of the Analytical Engine has steadily pursued his single purpose. The numberless misrepresentations of the facts connected with both Engines have not induced him to withdraw his attention from the new Invention; and the circumstance of his not having printed a description of either Engine has arisen entirely from his determination never to employ his mind upon the *description* of those Machines so long as a single difficulty remained which might *limit* the *power* of the Analytical Engine. The drawings, however, and the notations have been freely shown; and the great principles on which the Analytical Engine is founded have been explained and discussed with some of the first philosophers of the present day. Copies of the engravings were sent to the libraries of several public institutions, and the effect of the publicity thus given to the subject is fully proved by its having enabled a distinguished Italian Geometer to draw up from these sources an excellent account of that Engine.

Throughout the whole of these labours connected with the Analytical Engine, neither the Science, nor the Institutions, nor the Government of his Country have ever afforded him the slightest encouragement. When the Invention was noticed in the House of Commons, one single voice* alone was raised in its favour.

---

* That of Mr. Hawes, Member for Lambeth.

During nearly the whole of a period of upwards of twenty years, Mr. Babbage had maintained, in his own house, and at his own expense, an establishment for aiding him in carrying out his views, and in making experiments, which most materially assisted in improving the Difference Engine. When that work was suspended he still continued his own inquiries, and having discovered principles of far wider extent, he ultimately embodied them in the Analytical Engine.

The establishment necessary in the former part of this period for the actual construction of the Difference Engine, and of the extensive drawings which it demanded, as well as for the formation of those tools which were contrived to overcome the novel difficulties of the case, and in the latter part of the same period by the drawings and notations of the Analytical Engine, and the experiments relating to its construction, gave occupation to a considerable number of workmen of the greatest skill. During the many years in which this work proceeded, the workmen were continually changing, who carried into the various workshops in which they were afterwards employed the practical knowledge acquired in the construction of these machines.

To render the drawings of the Difference Engine intelligible, Mr. Babbage had invented a compact and comprehensive language (the Mechanical Notation), by which every contemporaneous or successive movement of this Machine became known. Another addition to mechanical science was subsequently made in establishing principles for the *lettering* of drawings; one consequence of which is, that although many parts of a machine may be projected upon any plan, it will be easily seen, by the nature of the letter attached to each working point, to which of those parts it really belongs.

By the means of this system, combined with the Mechanical Notations, it is now possible to express the forms and actions of the most complicated machine in language which is at once condensed, precise and universal.

At length, in November 1842, Mr. Babbage received a letter from the Chancellor of the Exchequer, stating that Sir Robert Peel and himself had jointly and reluctantly come to the conclusion that it was the duty of the Government, on the ground of expense, to abandon the further construction of the Difference Engine. The same letter contained a proposal to Mr. Babbage, on the part of Government, that he should accept the whole of the drawings, together with the part of the Engine already completed, as well as the materials in a state of preparation. This proposition he declined.

The object of the ANALYTICAL ENGINE (the drawings and the experiments for which have been wholly carried on at Mr. Babbage's expense, by his own draftsmen, workmen and assistants) is to convert into numbers all the formulæ of analysis, and to work out the algebraical development of all formulæ whose laws are known.

The present state of the Analytical Engine is as follows:—

All the great principles on which the discovery rests have been explained, and drawings of mechanical structures have been made, by which each may be carried into operation.

Simpler mechanisms, as well as more extensive principles than were required for the Difference Engine, have been discovered for all the elementary portions of the Analytical Engine, and numerous drawings of these successive simplifications exist.

The mode of combining the various sections of which the Engine is formed has been examined with unceasing anxiety, for the purpose of reducing the whole combination to the greatest possible simplicity. Drawings of almost all the plans thus discussed have been made, and the latest of the drawings (bearing the number 28) shows how many have been superseded, and also, from its extreme comparative simplicity, that little further advance can be expected in that direction.

Mechanical Notations have been made both of the actions of detached parts and of the general action of the whole, which cover about four or five hundred large folio sheets of paper.

The original rough sketches are contained in about five volumes.

There are upwards of one hundred large drawings.

No part of the construction of the Analytical Engine has yet been commenced. A long series of experiments have, however, been made upon the art of shaping metals; and the tools to be employed for that purpose have been discussed, and many drawings of them prepared. The great object of these inquiries and experiments is, on the one hand, by simplifying as much as possible the construction, and on the other, by contriving new and cheaper means of execution, at length to reduce the expense within those limits which a private individual may command.

Article XXIX.

[From the Scientific Memoirs, vol. iii. p. 666.]

[BEFORE submitting to our readers the translation of M. Menabrea's memoir 'On the Mathematical Principles of the ANALYTICAL ENGINE' invented by Mr. Babbage, we shall present to them a list of the printed papers connected with the subject, and also of those relating to the Difference Engine by which it was preceded.

For information on Mr. Babbage's "*Difference* Engine," which is but slightly alluded to by M. Menabrea, we refer the reader to the following sources:—

1. Letter to Sir Humphry Davy, Bart., P.R.S., on the Application of Machinery to Calculate and Print Mathematical Tables. By Charles Babbage, Esq., F.R.S. London, July 1822. Reprinted, with a Report of the Council of the Royal Society, by order of the House of Commons, May 1823.

2. On the Application of Machinery to the Calculation of Astronomical and Mathematical Tables. By Charles Babbage, Esq.—Memoirs of the Astronomical Society, vol. i. part 2. London, 1822.

3. Address to the Astronomical Society by Henry Thomas Colebrooke, Esq., F.R.S., President. on presenting the first Gold Medal of the Society to Charles Babbage, Esq., for the invention of the Calculating Engine.—Memoirs of the Astronomical Society. London, 1822.

4. On the Determination of the General Term of a New Class of Infinite Series. By Charles Babbage, Esq.—Transactions of the Cambridge Philosophical Society.

5. On Mr. Babbage's New Machine for Calculating and Printing Mathematical Tables.—Letter from Francis Baily, Esq., F.R.S., to M. Schumacher. No. 46, Astronomische Nachrichten. Reprinted in the Philosophical Magazine, May 1824.

6. On a Method of expressing by Signs the Action of Machinery. By Charles Babbage, Esq.—Philosophical Transactions. London, 1826.

7. On Errors common to many Tables of Logarithms. By Charles Babbage, Esq.—Memoirs of the Astronomical Society. London, 1827.

8. Report of the Committee appointed by the Council of the Royal Society to consider the subject referred to in a communication received by them from the Treasury respecting Mr. Babbage's Calculating Engine, and to report thereon. London, 1829.

9. Economy of Manufactures, chap. xx. 8vo. London, 1832.

10. Article on Babbage's Calculating Engine.—Edinburgh Review, July 1834. No. 120. vol. lix.

The present state of the Difference Engine, which has always been the property of Government, is as follows:—The drawings are nearly finished, and the mechanical notation of the whole, recording every motion of which it is susceptible, is completed. A part of that Engine, comprising sixteen figures, arranged in three orders of differences, has been put together, and has frequently been used during the last eight years. It performs its work with absolute precision. This portion of the Difference Engine, together with all the drawings, are at present deposited in the Museum of King's College, London.

Of the ANALYTICAL ENGINE, which forms the principal object of the present memoir, we are not aware that any notice has hitherto appeared, except a Letter from the Inventor to M. Quetelet, Secretary to the Royal Academy of Sciences at Brussels, by whom it was communicated to that body. We subjoin a translation of this Letter, which was itself a translation of the original, and was not intended for publication by its author.

Royal Academy of Sciences at Brussels. General Meeting of the 7th and 8th of May, 1835.

"A Letter from Mr. Babbage announces that he has for six months been engaged in making the drawings of a new calculating machine of far greater power than the first.

"'I am myself astonished,' says Mr. Babbage, 'at the power I have been enabled to give to this machine; a year ago I should not have believed this result possible. This machine is intended to contain a hundred variables (or numbers susceptible of changing); each of these numbers may consist of twenty-five figures, $v_1, v_2, \ldots v_n$ being any numbers whatever, n being less than a hundred; if $f(v_1, v_2, v_3, \ldots v_n)$ be any given function which can be formed by addition, subtraction, multiplication, division, extraction of roots, or elevation to powers, the machine will calculate its numerical value; it will afterwards substitute this value in the place of v, or of any other variable, and will calculate this second function with respect to v. It will reduce to tables almost all equations of finite differences. Let us suppose that we have observed a thousand values of a, b, c, d, and that we wish to calculate them by the formula $p = \sqrt{\dfrac{a+b}{c\,d}}$, the machine must be set to calculate the formula; the first series of the values of a, b, c, d must be adjusted to it; it will then calculate them, print them, and reduce them to zero; lastly, it will ring a bell to give notice that a new set of constants must be inserted. When there exists a relation between any number of successive coefficients of a series, provided it can be expressed as has already been said, the machine will calculate them and make their terms known in succession; and it may afterwards be disposed so as to find the value of the series for all the values of the variable.'

"Mr. Babbage announces, in conclusion, 'that the greatest difficulties of the invention have already been surmounted, and that the plans will be finished in a few months.'"

In the Ninth Bridgewater Treatise, Mr. Babbage has employed several arguments deduced from the Analytical Engine, which afford some idea of its powers. See Ninth Bridgewater Treatise, 8vo, second edition. London, 1834.

Some of the numerous drawings of the Analytical Engine have been engraved on wooden blocks, and from these (by a mode contrived by Mr. Babbage) various stereotype plates have been taken. They comprise—

1. Plan of the figure wheels for one method of adding numbers.
2. Elevation of the wheels and axis of ditto.
3. Elevation of framing only of ditto.
4. Section of adding wheels and framing together.
5. Section of the adding wheels, sign wheels and framing complete.
6. Impression from the original wooden block.
7. Impressions from a stereotype cast of No. 6, with the letters and signs inserted. Nos. 2, 3, 4 and 5 were stereotypes taken from this.
8. Plan of adding wheels and of long and short pinions, by means of which *stepping* is accomplished.

N.B. This process performs the operation of multiplying or dividing a number by any power of ten.

9. Elevation of long pinions in the position for addition.
10. Elevation of long pinions in the position for stepping.
11. Plan of mechanism for carrying the tens (by anticipation), connected with long pinions.
12. Section of the chain of wires for anticipating carriage.
13. Sections of the elevation of parts of the preceding carriage.

All these were executed about five years ago. At a later period (August 1840) Mr. Babbage caused one of his general plans (No. 25) of the whole Analytical Engine to be lithographed at Paris.

Although these illustrations have not been published, on account of the time which would be required to describe them, and the rapid succession of improvements made subsequently, yet copies have been freely given to many of Mr. Babbage's friends, and were in August 1838 presented at Newcastle to the British Association for the Advancement of Science, and in August 1840 to the Institute of France through M. Arago, as well as to the Royal Academy of Turin through M. Plana. —EDITOR.]

Sketch of the Analytical Engine invented by Charles Babbage, *Esq. By* L. F. MENABREA, *of Turin, Officer of the Military Engineers.*

[From the *Bibliothèque Universelle de Genève*, No. 82. October 1842.]

THOSE labours which belong to the various branches of the mathematical sciences, although on first consideration they seem to be the exclusive province of intellect, may, nevertheless, be divided into two distinct sections; one of which may be called the mechanical, because it is subjected to precise and invariable laws, that are capable of being expressed by means of the operations of matter; while the other, demanding the intervention of reasoning, belongs more specially to the domain of the understanding. This admitted, we may propose to execute, by means of machinery, the mechanical branch of these labours, reserving for pure intellect that which depends on the reasoning faculties. Thus the rigid exactness of those laws which regulate numerical calculations must frequently have suggested the employment of material instruments, either for executing the whole of such calculations or for abridging them; and thence have arisen several inventions having this object in view, but which have in general but partially attained it. For instance, the much-admired machine of Pascal is now simply an object of curiosity, which, whilst it displays the powerful intellect of its inventor, is yet of little utility in itself. Its powers extended no further than the execution of the first four* operations of arithmetic, and indeed were in reality confined to that of the first two, since multiplication and division were the result of a series of additions and subtractions. The chief drawback hitherto on most of such machines is, that they require the continual intervention of a human agent to regulate their movements, and thence arises a source of errors; so that, if their use has not become general for large numerical calculations, it is because they have not in fact resolved the double problem which the question presents, that of *correctness* in the results, united with *economy* of time.

Struck with similar reflections, Mr. Babbage has devoted some years to the realization of a gigantic idea. He proposed to himself nothing less than the construction of a machine capable of executing not merely arithmetical calculations, but even all those of analysis, if their laws are known. The imagination is at first astounded at the idea of such an undertaking; but the more calm reflection we bestow on it, the less impossible does success appear, and it is felt that it may depend on the discovery of some principle so general, that, if applied to machinery, the latter may be capable of mechanically translating the operations which may be indicated to it by algebraical notation. The illustrious inventor having been kind enough to communicate to me some of his views on this subject during a visit he made at Turin, I have, with his approbation, thrown together the impressions they have left on my mind. But the reader must not expect to find a description of Mr. Babbage's engine; the comprehension of this would entail studies of much length; and I shall endeavour merely to give an insight into the end proposed, and to develope the principles on which its attainment depends.

I must first premise that this engine is entirely different from that of which there is a notice in the 'Treatise on the Economy of Machinery,' by the same author. But as the latter gave rise† to the idea of the engine in question, I consider it will be a useful preliminary briefly to recall what were Mr. Babbage's first essays, and also the circumstances in which they originated.

It is well known that the French government, wishing to promote the extension of the decimal

---

* This remark seems to require further comment, since it is in some degree calculated to strike the mind as being at variance with the subsequent passage (page 10), where it is explained that *an engine which can effect these four operations can in fact effect every species of calculation.* The apparent discrepancy is stronger too in the translation than in the original, owing to its being impossible to render precisely into the English tongue all the niceties of distinction which the French idiom happens to admit of in the phrases used for the two passages we refer to. The explanation lies in this: that in the one case the execution of these four operations is the *fundamental starting-point,* and the object proposed for attainment by the machine is the *subsequent combination of these* in every possible variety; whereas in the other case the execution of some *one* of these four operations, selected at pleasure, is the *ultimatum,* the sole and utmost result that can be proposed for attainment by the machine referred to, and which result it cannot any further combine or work upon. The one *begins* where the other *ends.* Should this distinction not now appear perfectly clear, it will become so on perusing the rest of the Memoir, and the Notes that are appended to it.—Note by Translator.

† The idea that the one engine is the offspring and has grown out of the other, is an exceedingly natural and plausible supposition, until reflection reminds us that no *necessary* sequence and connexion need exist between two such inventions, and that they *may* be wholly independent. M. Menabrea has shared this idea in common with persons who have not his profound and accurate insight into the nature of either engine. In Note A. (see the Notes at the end of the Memoir) it will be found sufficiently explained, however, that this supposition is unfounded. M. Menabrea's opportunities were by no means such as could be adequate to afford him information on a point like this, which would be naturally and almost unconsciously *assumed,* and would scarcely suggest any inquiry with reference to it.—Note by Translator.

system, had ordered the construction of logarithmical and trigonometrical tables of enormous extent. M. de Prony, who had been entrusted with the direction of this undertaking, divided it into three sections, to each of which was appointed a special class of persons. In the first section the formulæ were so combined as to render them subservient to the purposes of numerical calculation; in the second, these same formulæ were calculated for values of the variable, selected at certain successive distances; and under the third section, comprising about eighty individuals, who were most of them only acquainted with the first two rules of arithmetic, the values which were intermediate to those calculated by the second section were interpolated by means of simple additions and subtractions.

An undertaking similar to that just mentioned having been entered upon in England, Mr. Babbage conceived that the operations performed under the third section might be executed by a machine; and this idea he realized by means of mechanism, which has been in part put together, and to which the name Difference Engine is applicable, on account of the principle upon which its construction is founded. To give some notion of this, it will suffice to consider the series of whole square numbers, 1, 4, 9, 16, 25, 36, 49, 64, &c. By subtracting each of these from the succeeding one, we obtain a new series, which we will name the Series of First Differences, consisting of the numbers 3, 5, 7, 9, 11, 13, 15, &c. On subtracting from each of these the preceding one, we obtain the Second Differences, which are all constant and equal to 2. We may represent this succession of operations, and their results, in the following table :—

A. Column o Square Numbers.	B. First Differ- ences.	C. Second Differ- ences.
1		
....	3	
4		2 *b*
....	5	
9		2 *d*
....	7	
16		2
....	9	
25		2
....	11	
36		

From the mode in which the last two columns B and C have been formed, it is easy to see, that if, for instance, we desire to pass from the number 5 to the succeeding one 7, we must add to the former the constant difference 2; similarly, if from the square number 9 we would pass to the following one 16, we must add to the former the difference 7, which difference is in other words the preceding difference 5, plus the constant difference 2; or again, which comes to the same thing, to obtain 16 we have only to add together the three numbers 2, 5, 9, placed obliquely in the direction *a b*. Similarly, we obtain the number 25 by summing up the three numbers placed in the oblique direction *d c*: commencing by the addition 2 + 7, we have the first difference 9 consecutively to 7; adding 16 to the 9 we have the square 25. We see then that the three numbers 2, 5, 9 being given, the whole series of successive square numbers, and that of their first differences likewise, may be obtained by means of simple additions.

Now, to conceive how these operations may be reproduced by a machine, suppose the latter to have three dials, designated as A, B, C, on each of which are traced, say a thousand divisions, by way of example, over which a needle shall pass. The two dials, C, B, shall have in addition a registering hammer, which is to give a number of strokes equal to that of the divisions indicated by the needle. For each stroke of the registering hammer of the dial C, the needle B shall advance one division; similarly, the needle A shall advance one division for every stroke of the registering hammer of the dial B. Such is the general disposition of the mechanism.

This being understood, let us, at the beginning of the series of operations we wish to execute, place the needle C on the division 2, the needle B on the division 5, and the needle A on the division 9. Let us allow the hammer of the dial C to strike; it will strike twice, and at the same time the needle B will pass over two divisions. The latter will then indicate the number 7, which succeeds the number 5 in the column of first differences. If we now permit the hammer of the dial B to strike in its turn, it will strike seven times, during which the needle A will advance seven divisions; these added to the nine already marked by it will give the number 16, which is the square number consecutive to 9. If we now recommence these operations, beginning with the needle C, which is always to be left on the division 2, we shall perceive that by repeating them indefinitely, we may successively reproduce the series of whole square numbers by means of a very simple mechanism.

The theorem on which is based the construction of the machine we have just been describing, is a particular case of the following more general theorem: that if in any polynomial whatever, the highest power of whose variable is m, this same variable be increased by equal degrees; the corresponding values of the polynomial then calculated, and the first, second, third, &c. differences of these be taken (as for the preceding series of squares); the mth differences will all be equal to each other. So that, in order to reproduce the series of values of the polynomial by means of a machine analogous to the one above described, it is sufficient that there be $(m + 1)$ dials, having the mutual relations we have indicated. As the differences may be either positive or negative, the machine will have a contrivance for either advancing or retrograding each needle, according as the number to be algebraically added may have the sign *plus* or *minus*.

If from a polynomial we pass to a series having an infinite number of terms, arranged according to the ascending powers of the variable, it would at first appear, that in order to apply the machine to the calculation of the function represented by such a series, the mechanism must include an infinite number of dials, which would in fact render the thing impossible. But in many cases the difficulty will disappear, if we observe that for a great number of functions the series which represent them may be rendered convergent; so that, according to the degree of approximation desired, we may limit ourselves to the calculation of a certain number of terms of the series, neglecting the rest. By this method the question is reduced to the primitive case of a finite polynomial. It is thus that we can calculate the succession of the logarithms of numbers. But since, in this particular instance, the terms which had been originally neglected receive increments in a ratio so continually increasing for equal increments of the variable, that the degree of approximation required would ultimately be affected, it is necessary, at certain intervals, to calculate the value of the function by different methods, and then respectively to use the results thus obtained, as data whence to deduce, by means of the machine, the other intermediate values. We see that the machine here performs the office of the third section of calculators mentioned in describing the tables computed by order of the French government, and that the end originally proposed is thus fulfilled by it.

Such is the nature of the first machine which Mr. Babbage conceived. We see that its use is confined to cases where the numbers required are such as can be obtained by means of simple additions or subtractions; that the machine is, so to speak, merely the expression of one* particular theorem of analysis; and that, in short, its operations cannot be extended so as to embrace the solution of an infinity of other questions included within the domain of mathematical analysis. It was while contemplating the vast field which yet remained to be traversed, that Mr. Babbage, renouncing his original essays, conceived the plan of another system of mechanism whose operations should themselves possess all the generality of algebraical notation, and which, on this account, he denominates the *Analytical Engine*.

Having now explained the state of the question, it is time for me to develope the principle on

* See Note A.

c

which is based the construction of this latter machine. When analysis is employed for the solution of any problem, there are usually two classes of operations to execute: first, the numerical calculation of the various coefficients; and secondly, their distribution in relation to the quantities affected by them. If, for example, we have to obtain the product of two binomials $(a + bx) (m + nx)$, the result will be represented by $am + (an + bm)\, x + bnx^2$, in which expression we must first calculate am, an, bm, bn; then take the sum of $an + bm$; and lastly, respectively distribute the coefficients thus obtained amongst the powers of the variable. In order to reproduce these operations by means of a machine, the latter must therefore possess two distinct sets of powers: first, that of executing numerical calculations; secondly, that of rightly distributing the values so obtained.

But if human intervention were necessary for directing each of these partial operations, nothing would be gained under the heads of correctness and economy of time; the machine must therefore have the additional requisite of executing by itself all the successive operations required for the solution of a problem proposed to it, when once the *primitive numerical data* for this same problem have been introduced. Therefore, since, from the moment that the nature of the calculation to be executed or of the problem to be resolved have been indicated to it, the machine is, by its own intrinsic power, of itself to go through all the intermediate operations which lead to the proposed result, it must exclude all methods of trial and guess-work, and can only admit the direct processes of calculation*.

It is necessarily thus; for the machine is not a thinking being, but simply an automaton which acts according to the laws imposed upon it. This being fundamental, one of the earliest researches its author had to undertake, was that of finding means for effecting the division of one number by another without using the method of guessing indicated by the usual rules of arithmetic. The difficulties of effecting this combination were far from being among the least; but upon it depended the success of every other. Under the impossibility of my here explaining the process through which this end is attained, we must limit ourselves to admitting that the first four operations of arithmetic, that is addition, subtraction, multiplication and division, can be performed in a direct manner through the intervention of the machine. This granted, the machine is thence capable of performing every species of numerical calculation, for all such calculations ultimately resolve themselves into the four operations we have just named. To conceive how the machine can now go through its functions according to the laws laid down, we will begin by giving an idea of the manner in which it materially represents numbers.

Let us conceive a pile or vertical column consisting of an indefinite number of circular discs, all pierced through their centres by a common axis, around which each of them can take an independent rotatory movement. If round the edge of each of these discs are written the ten figures which constitute our numerical alphabet, we may then, by arranging a series of these figures in the same vertical line, express in this manner any number whatever. It is sufficient for this purpose that the first disc represent units, the second tens, the third hundreds, and so on. When two numbers have been thus written on two distinct columns, we may propose to combine them arithmetically with

---

* This must not be understood in too unqualified a manner. The engine is capable, under certain circumstances, of feeling about to discover which of two or more possible contingencies has occurred, and of then shaping its future course accordingly.—NOTE BY TRANSLATOR.

each other, and to obtain the result on a third column. In general, if we have a series of columns *
consisting of discs, which columns we will designate as V_0, V_1, V_2, V_3, V_4, &c., we may require, for
instance, to divide the number written on the column V_1 by that on the column V_4, and to obtain
the result on the column V_7. To effect this operation, we must impart to the machine two distinct
arrangements; through the first it is prepared for executing *a division*, and through the second the
columns it is to operate on are indicated to it, and also the column on which the result is to be
represented. If this division is to be followed, for example, by the addition of two numbers taken on
other columns, the two original arrangements of the machine must be simultaneously altered. If,
on the contrary, a series of operations of the same nature is to be gone through, then the first of
the original arrangements will remain, and the second alone must be altered. Therefore, the arrangements that may be communicated to the various parts of the machine may be distinguished
into two principal classes:

First, that relative to the *Operations*.

Secondly, that relative to the *Variables*.

By this latter we mean that which indicates the columns to be operated on. As for the operations
themselves, they are executed by a special apparatus, which is designated by the name of *mill*, and
which itself contains a certain number of columns, similar to those of the Variables. When two
numbers are to be combined together, the machine commences by effacing them from the columns
where they are written, that is, it places *zero* † on every disc of the two vertical lines on which the
numbers were represented; and it transfers the numbers to the mill. There, the apparatus having
been disposed suitably for the required operation, this latter is effected, and, when completed, the
result itself is transferred to the column of Variables which shall have been indicated. Thus the
mill is that portion of the machine which works, and the columns of Variables constitute that where
the results are represented and arranged. After the preceding explanations, we may perceive that
all fractional and irrational results will be represented in decimal fractions. Supposing each column
to have forty discs, this extension will be sufficient for all degrees of approximation generally required.

It will now be inquired how the machine can of itself, and without having recourse to the hand
of man, assume the successive dispositions suited to the operations. The solution of this problem
has been taken from Jacquard's apparatus ‡, used for the manufacture of brocaded stuffs, in the
following manner:—

Two species of threads are usually distinguished in woven stuffs; one is the *warp* or longitudinal
thread, the other the *woof* or transverse thread, which is conveyed by the instrument called the
shuttle, and which crosses the longitudinal thread or warp. When a brocaded stuff is required,
it is necessary in turn to prevent certain threads from crossing the woof, and this according to a
succession which is determined by the nature of the design that is to be reproduced. Formerly this
process was lengthy and difficult, and it was requisite that the workman, by attending to the design
which he was to copy, should himself regulate the movements the threads were to take. Thence
arose the high price of this description of stuffs, especially if threads of various colours entered into

---

* See Note B.

† Zero is not *always* substituted when a number is transferred to the mill. This is explained further on in the memoir,
and still more fully in Note D.—NOTE BY TRANSLATOR.

‡ See Note C.

the fabric. To simplify this manufacture, Jacquard devised the plan of connecting each group of threads that were to act together, with a distinct lever belonging exclusively to that group. All these levers terminate in rods, which are united together in one bundle, having usually the form of a parallelopiped with a rectangular base. The rods are cylindrical, and are separated from each other by small intervals. The process of raising the threads is thus resolved into that of moving these various lever-arms in the requisite order. To effect this, a rectangular sheet of pasteboard is taken, somewhat larger in size than a section of the bundle of lever-arms. If this sheet be applied to the base of the bundle, and an advancing motion be then communicated to the pasteboard, this latter will move with it all the rods of the bundle, and consequently the threads that are connected with each of them. But if the pasteboard, instead of being plain, were pierced with holes corresponding to the extremities of the levers which meet it, then, since each of the levers would pass through the pasteboard during the motion of the latter, they would all remain in their places. We thus see that it is easy so to determine the position of the holes in the pasteboard, that, at any given moment, there shall be a certain number of levers, and consequently of parcels of threads, raised, while the rest remain where they were. Supposing this process is successively repeated according to a law indicated by the pattern to be executed, we perceive that this pattern may be reproduced on the stuff. For this purpose we need merely compose a series of cards according to the law required, and arrange them in suitable order one after the other; then, by causing them to pass over a polygonal beam which is so connected as to turn a new face for every stroke of the shuttle, which face shall then be impelled parallelly to itself against the bundle of lever-arms, the operation of raising the threads will be regularly performed. Thus we see that brocaded tissues may be manufactured with a precision and rapidity formerly difficult to obtain.

Arrangements analogous to those just described have been introduced into the Analytical Engine. It contains two principal species of cards: first, Operation cards, by means of which the parts of the machine are so disposed as to execute any determinate series of operations, such as additions, subtractions, multiplications, and divisions; secondly, cards of the Variables, which indicate to the machine the columns on which the results are to be represented. The cards, when put in motion, successively arrange the various portions of the machine according to the nature of the processes that are to be effected, and the machine at the same time executes these processes by means of the various pieces of mechanism of which it is constituted.

In order more perfectly to conceive the thing, let us select as an example the resolution of two equations of the first degree with two unknown quantities. Let the following be the two equations, in which x and y are the unknown quantities :—

$$\begin{cases} m\,x + n\,y = d \\ m'x + n'y = d'. \end{cases}$$

We deduce $x = \dfrac{d\,n' - d'\,n}{n'm - n\,m'}$, and for y an analogous expression. Let us continue to represent by V_0, V_1, V_2, &c. the different columns which contain the numbers, and let us suppose that the first eight columns have been chosen for expressing on them the numbers represented by m, n, d, m', n', d', n and n', which implies that $V_0 = m$, $V_1 = n$, $V_2 = d$, $V_3 = m'$, $V_4 = n'$, $V_5 = d'$, $V_6 = n$, $V_7 = n'$.

The series of operations commanded by the cards, and the results obtained, may be represented in the following table :—

Number of the operations.	Symbols indicating the nature of the operations.	Columns on which operations are to be performed.	Columns which receive results of operations.	Progress of the operations.
1	×	$V_2 \times V_4 =$	V_8	$= d\,n'$
2	×	$V_5 \times V_1 =$	V_9	$= d'\,n$
3	×	$V_4 \times V_0 =$	V_{10}	$= n'\,m$
4	×	$V_1 \times V_3 =$	V_{11}	$= n\,m'$
5	—	$V_8 - V_9 =$	V_{12}	$= d\,n' - d'\,n$
6	—	$V_{10} - V_{11} =$	V_{13}	$= n'\,m - n\,m'$
7	÷	$\dfrac{V_{12}}{V_{13}} =$	V_{14}	$= x = \dfrac{d\,n' - d'\,n}{n'\,m - n\,m'}$

Since the cards do nothing but indicate in what manner and on what columns the machine shall act, it is clear that we must still, in every particular case, introduce the numerical data for the calculation. Thus, in the example we have selected, we must previously inscribe the numerical values of m, n, d, m', n', d', in the order and on the columns indicated, after which the machine when put in action will give the value of the unknown quantity x for this particular case. To obtain the value of y, another series of operations analogous to the preceding must be performed. But we see that they will be only four in number, since the denominator of the expression for y, excepting the sign, is the same as that for x, and equal to $n'm - nm'$. In the preceding table it will be remarked that the column for operations indicates four successive *multiplications*, two *subtractions*, and one *division*. Therefore, if desired, we need only use three operation-cards; to manage which, it is sufficient to introduce into the machine an apparatus which shall, after the first multiplication, for instance, retain the card which relates to this operation, and not allow it to advance so as to be replaced by another one, until after this same operation shall have been four times repeated. In the preceding example we have seen, that to find the value of x we must begin by writing the coefficients m, n, d, m', n', d', upon eight columns, thus repeating n and n' twice. According to the same method, if it were required to calculate y likewise, these coefficients must be written on twelve different columns. But it is possible to simplify this process, and thus to diminish the chances of errors, which chances are greater, the larger the number of the quantities that have to be inscribed previous to setting the machine in action. To understand this simplification, we must remember that every number written on a column must, in order to be arithmetically combined with another number, be effaced from the column on which it is, and transferred to the *mill*. Thus, in the example we have discussed, we will take the two coefficients m and n', which are each of them to enter into *two* different products, that is m into mn' and md', n' into mn' and $n'd$. These coefficients will be inscribed on the columns V_0 and V_4. If we commence the series of operations by the product of m into n', these numbers will be effaced from the columns V_0 and V_4, that they may be transferred to the mill, which will multiply them into each other, and will then command the machine to represent the result, say on the column V_6. But as these numbers are each to be used again in another operation, they must again be inscribed somewhere; therefore, while the mill is working out their product, the machine will inscribe them anew on any two columns that may be indicated to it through the cards; and as, in the actual case, there is no reason why they should not resume their former places, we will suppose them again inscribed on V_0 and V_4, whence in short they would not finally disappear, to be reproduced no more, until they should have gone through all the combinations in which they might have to be used.

We see, then, that the whole assemblage of operations requisite for resolving the two* above equations of the first degree may be definitively represented in the following table:—

Columns on which are inscribed the primitive data.	Number of the operations.	Cards of the operations.		Variable cards.			Statement of results.
		Number of the Operation-cards.	Nature of each operation.	Columns acted on by each operation.	Columns that receive the result of each operation.	Indication of change of value on any column.	
$^1V_0 = m$	1	1	×	$^1V_0 \times {}^1V_4 = {}^1V_6$		$\left\{ \begin{array}{l} ^1V_0 = {}^1V_0 \\ ^1V_4 = {}^1V_4 \end{array} \right\}$	$^1V_6 = mn'$
$^1V_1 = n$	2	,,	×	$^1V_3 \times {}^1V_1 = {}^1V_7$		$\left\{ \begin{array}{l} ^1V_3 = {}^1V_3 \\ ^1V_1 = {}^1V_1 \end{array} \right\}$	$^1V_7 = m'n$
$^1V_2 = d$	3	,,	×	$^1V_2 \times {}^1V_4 = {}^1V_8$		$\left\{ \begin{array}{l} ^1V_2 = {}^1V_2 \\ ^1V_4 = {}^0V_4 \end{array} \right\}$	$^1V_8 = dn'$
$^1V_3 = m'$	4	,,	×	$^1V_5 \times {}^1V_1 = {}^1V_9$		$\left\{ \begin{array}{l} ^1V_5 = {}^1V_5 \\ ^1V_1 = {}^0V_1 \end{array} \right\}$	$^1V_9 = d'n$
$^1V_4 = n'$	5	,,	×	$^1V_0 \times {}^1V_5 = {}^1V_{10}$		$\left\{ \begin{array}{l} ^1V_0 = {}^0V_0 \\ ^1V_5 = {}^0V_5 \end{array} \right\}$	$^1V_{10} = d'm$
$^1V_5 = d'$	6	,,	×	$^1V_2 \times {}^1V_3 = {}^1V_{11}$		$\left\{ \begin{array}{l} ^1V_2 = {}^0V_2 \\ ^1V_3 = {}^0V_3 \end{array} \right\}$	$^1V_{11} = dm'$
	7	2	−	$^1V_6 - {}^1V_7 = {}^1V_{12}$		$\left\{ \begin{array}{l} ^1V_6 = {}^0V_6 \\ ^1V_7 = {}^0V_7 \end{array} \right\}$	$^1V_{12} = mn' - m'n$
	8	,,	−	$^1V_8 - {}^1V_9 = {}^1V_{13}$		$\left\{ \begin{array}{l} ^1V_8 = {}^0V_8 \\ ^1V_9 = {}^0V_9 \end{array} \right\}$	$^1V_{13} = dn' - d'n$
	9	,,	−	$^1V_{10} - {}^1V_{11} = {}^1V_{14}$		$\left\{ \begin{array}{l} ^1V_{10} = {}^0V_{10} \\ ^1V_{11} = {}^0V_{11} \end{array} \right\}$	$^1V_{14} = d'm - dm'$
	10	3	÷	$^1V_{13} \div {}^1V_{12} = {}^1V_{15}$		$\left\{ \begin{array}{l} ^1V_{13} = {}^0V_{13} \\ ^1V_{12} = {}^1V_{12} \end{array} \right\}$	$^1V_{15} = \dfrac{dn' - d'n}{mn' - m'n} = x$
	11	,,	÷	$^1V_{14} \div {}^1V_{12} = {}^1V_{16}$		$\left\{ \begin{array}{l} ^1V_{14} = {}^0V_{14} \\ ^1V_{12} = {}^0V_{12} \end{array} \right\}$	$^1V_{16} = \dfrac{d'm - dm'}{mn' - m'n} = y$
1	2	3	4	5	6	7	8

In order to diminish to the utmost the chances of error in inscribing the numerical data of the problem, they are successively placed on one of the columns of the mill; then, by means of cards arranged for this purpose, these same numbers are caused to arrange themselves on the requisite columns, without the operator having to give his attention to it; so that his undivided mind may be applied to the simple inscription of these same numbers.

According to what has now been explained, we see that the collection of columns of Variables may be regarded as a *store* of numbers, accumulated there by the mill, and which, obeying the orders transmitted to the machine by means of the cards, pass alternately from the mill to the store and from the store to the mill, that they may undergo the transformations demanded by the nature of the calculation to be performed.

Hitherto no mention has been made of the *signs* in the results, and the machine would be far from perfect were it incapable of expressing and combining amongst each other positive and negative quantities. To accomplish this end, there is, above every column, both of the mill and of the store, a disc, similar to the discs of which the columns themselves consist. According as the digit on this disc is even or uneven, the number inscribed on the corresponding column below it will be considered as positive or negative. This granted, we may, in the following manner, conceive how the signs can be algebraically combined in the machine. When a number is to be transferred from the store to the mill, and *vice versâ*, it will always be transferred with its sign, which will be effected by

---

* See Note D.

means of the cards, as has been explained in what precedes. Let any two numbers then, on which we are to operate arithmetically, be placed in the mill with their respective signs. Suppose that we are first to add them together; the operation-cards will command the addition: if the two numbers be of the same sign, one of the two will be entirely effaced from where it was inscribed, and will go to add itself on the column which contains the other number; the machine will, during this operation, be able, by means of a certain apparatus, to prevent any movement in the disc of signs which belongs to the column on which the addition is made, and thus the result will remain with the sign which the two given numbers originally had. When two numbers have two different signs, the addition commanded by the card will be changed into a subtraction through the intervention of mechanisms which are brought into play by this very difference of sign. Since the subtraction can only be effected on the larger of the two numbers, it must be arranged that the disc of signs of the larger number shall not move while the smaller of the two numbers is being effaced from its column and subtracted from the other, whence the result will have the sign of this latter, just as in fact it ought to be. The combinations to which algebraical subtraction give rise, are analogous to the preceding. Let us pass on to multiplication. When two numbers to be multiplied are of the same sign, the result is positive; if the signs are different, the product must be negative. In order that the machine may act conformably to this law, we have but to conceive that on the column containing the product of the two given numbers, the digit which indicates the sign of that product has been formed by the mutual addition of the two digits that respectively indicated the signs of the two given numbers; it is then obvious that if the digits of the signs are both even, or both odd, their sum will be an even number, and consequently will express a positive number; but that if, on the contrary, the two digits of the signs are one even and the other odd, their sum will be an odd number, and will consequently express a negative number. In the case of division, instead of adding the digits of the discs, they must be subtracted one from the other, which will produce results analogous to the preceding; that is to say, that if these figures are both even or both uneven, the remainder of this subtraction will be even; and it will be uneven in the contrary case. When I speak of mutually adding or subtracting the numbers expressed by the digits of the signs, I merely mean that one of the sign-discs is made to advance or retrograde a number of divisions equal to that which is expressed by the digit on the other sign-disc. We see, then, from the preceding explanation, that it is possible mechanically to combine the signs of quantities so as to obtain results conformable to those indicated by algebra*.

The machine is not only capable of executing those numerical calculations which depend on a given algebraical formula, but it is also fitted for analytical calculations in which there are one or several variables to be considered. It must be assumed that the analytical expression to be operated on can be developed according to powers of the variable, or according to determinate functions of this same variable, such as circular functions, for instance; and similarly for the result that is to be attained. If we then suppose that above the columns of the store, we have inscribed the powers or the functions of the variable, arranged according to whatever is the prescribed law of development,

---

* Not having had leisure to discuss with Mr. Babbage the manner of introducing into his machine the combination of algebraical signs, I do not pretend here to expose the method he uses for this purpose; but I considered that I ought myself to supply the deficiency, conceiving that this paper would have been imperfect if I had omitted to point out one means that might be employed for resolving this essential part of the problem in question.

the coefficients of these several terms may be respectively placed on the corresponding column below each. In this manner we shall have a representation of an analytical development; and, supposing the position of the several terms composing it to be invariable, the problem will be reduced to that of calculating their coefficients according to the laws demanded by the nature of the question. In order to make this more clear, we shall take the following* very simple example, in which we are to multiply $(a + b\,x^1)$ by $(A + B\cos^1 x)$. We shall begin by writing $x^0, x^1, \cos^0 x, \cos^1 x$, above the columns V_0, V_1, V_2, V_3; then since, from the form of the two functions to be combined, the terms which are to compose the products will be of the following nature, $x^0.\cos^0 x$, $x^0.\cos^1 x$, $x^1.\cos^0 x$, $x^1.\cos^1 x$, these will be inscribed above the columns V_4, V_5, V_6, V_7. The coefficients of $x^0, x^1, \cos^0 x$, $\cos^1 x$ being given, they will, by means of the mill, be passed to the columns V_0, V_1, V_2 and V_3. Such are the primitive data of the problem. It is now the business of the machine to work out its solution, that is, to find the coefficients which are to be inscribed on V_4, V_5, V_6, V_7. To attain this object, the law of formation of these same coefficients being known, the machine will act through the intervention of the cards, in the manner indicated by the following table:—

† Columns above which are written the functions of the variable.	Coefficients.		Cards of the operations.		Cards of the variables.			
	Given.	To be formed.	Number of the operations.	Nature of the operation.	Columns on which operations are to be performed.	Columns on which are to be inscribed the results of the operations.	Indication of change of value on any column submitted to an operation.	Results of the operations.
$x^0 \ldots\ldots {}^1V_0$	a	,,	,,	,,	,,	,,	,,	,, ,,
$x^1 \ldots\ldots {}^1V_1$	b	,,	,,	,,	,,	,,	,,	,, ,,
$\cos^0 x \ldots {}^1V_2$	A	,,	,,	,,	,,	,,	,,	,, ,,
$\cos^1 x \ldots {}^1V_3$	B	,,	,,	,,	,,	,,	,,	,, ,,
$x^0\cos^0 x ..{}^0V_4$	...	$a\,A$	1	$\times$	${}^1V_0 \times {}^1V_2 =$	${}^1V_4 \ldots\ldots$	$\left\{\begin{array}{l}{}^1V_0 = {}^1V_0\\{}^1V_2 = {}^1V_2\end{array}\right\}$	${}^1V_4 = a\,A$ coefficients of $x^0\cos^0 x$
$x^0\cos^1 x ..{}^0V_5$	...	$a\,B$	2	$\times$	${}^1V_0 \times {}^1V_3 =$	${}^1V_5 \ldots\ldots$	$\left\{\begin{array}{l}{}^1V_0 = {}^0V_0\\{}^1V_3 = {}^1V_3\end{array}\right\}$	${}^1V_5 = a\,B$ ... ... $x^0\cos^1 x$
$x^1\cos^0 x ..{}^0V_6$	...	$b\,A$	3	$\times$	${}^1V_1 \times {}^1V_2 =$	${}^1V_6 \ldots\ldots$	$\left\{\begin{array}{l}{}^1V_1 = {}^1V_1\\{}^1V_2 = {}^0V_2\end{array}\right\}$	${}^1V_6 = b\,A$ ... ... $x^1\cos^0 x$
$x^1\cos^1 x ..{}^0V_7$	...	$b\,B$	4	$\times$	${}^1V_1 \times {}^1V_3 =$	${}^1V_7 \ldots\ldots$	$\left\{\begin{array}{l}{}^1V_1 = {}^0V_1\\{}^1V_3 = {}^0V_3\end{array}\right\}$	${}^1V_7 = b\,B$ ... ... $x^1\cos^1 x$

It will now be perceived that a general application may be made of the principle developed in the preceding example, to every species of process which it may be proposed to effect on series submitted to calculation. It is sufficient that the law of formation of the coefficients be known, and that this law be inscribed on the cards of the machine, which will then of itself execute all the calculations requisite for arriving at the proposed result. If, for instance, a recurring series were proposed, the law of formation of the coefficients being here uniform, the same operations which must be performed for one of them will be repeated for all the others; there will merely be a change in the locality of the operation, that is, it will be performed with different columns. Generally, since every analytical expression is susceptible of being expressed in a series ordered according to certain functions of the variable, we perceive that the machine will include all analytical calculations which can be definitively reduced to the formation of coefficients according to certain laws, and to the distribution of these with respect to the variables.

* See Note E.

† For an explanation of the upper left-hand indices attached to the V's in this and in the preceding Table, we must refer the reader to Note D, amongst those appended to the memoir.—NOTE BY TRANSLATOR.

We may deduce the following important consequence from these explanations, viz. that since the cards only indicate the nature of the operations to be performed, and the columns of Variables with which they are to be executed, these cards will themselves possess all the generality of analysis, of which they are in fact merely a translation. We shall now further examine some of the difficulties which the machine must surmount, if its assimilation to analysis is to be complete. There are certain functions which necessarily change in nature when they pass through zero or infinity, or whose values cannot be admitted when they pass these limits. When such cases present themselves, the machine is able, by means of a bell, to give notice that the passage through zero or infinity is taking place, and it then stops until the attendant has again set it in action for whatever process it may next be desired that it shall perform. If this process has been foreseen, then the machine, instead of ringing, will so dispose itself as to present the new cards which have relation to the operation that is to succeed the passage through zero and infinity. These new cards may follow the first, but may only come into play contingently upon one or other of the two circumstances just mentioned taking place.

Let us consider a term of the form $a\, b^n$; since the cards are but a translation of the analytical formula, their number in this particular case must be the same, whatever be the value of n; that is to say, whatever be the number of multiplications required for elevating b to the nth power (we are supposing for the moment that n is a whole number). Now, since the exponent n indicates that b is to be multiplied n times by itself, and all these operations are of the same nature, it will be sufficient to employ one single operation-card, viz. that which orders the multiplication.

But when n is given for the particular case to be calculated, it will be further requisite that the machine limit the number of its multiplications according to the given values. The process may be thus arranged. The three numbers a, b and n will be written on as many distinct columns of the store; we shall designate them V_0, V_1, V_2; the result $a\, b^n$ will place itself on the column V_3. When the number n has been introduced into the machine, a card will order a certain registering-apparatus to mark $(n-1)$, and will at the same time execute the multiplication of b by b. When this is completed, it will be found that the registering-apparatus has effaced a unit, and that it only marks $(n-2)$; while the machine will now again order the number b written on the column V_1 to multiply itself with the product b^2 written on the column V_3, which will give b^3. Another unit is then effaced from the registering-apparatus, and the same processes are continually repeated until it only marks zero. Thus the number b^n will be found inscribed on V_3, when the machine, pursuing its course of operations, will order the product of b^n by a; and the required calculation will have been completed without there being any necessity that the number of operation-cards used should vary with the value of n. If n were negative, the cards, instead of ordering the multiplication of a by b^n, would order its division; this we can easily conceive, since every number, being inscribed with its respective sign, is consequently capable of reacting on the nature of the operations to be executed.

Finally, if n were fractional, of the form $\dfrac{p}{q}$, an additional column would be used for the inscription of q, and the machine would bring into action two sets of processes, one for raising b to the power p, the other for extracting the qth root of the number so obtained.

Again, it may be required, for example, to multiply an expression of the form $a\, x^m + b\, x^n$ by another $A\, x^p + B\, x^q$, and then to reduce the product to the least number of terms, if any of the indices are equal. The two factors being ordered with respect to x, the general result of the multiplication would be $\mathrm{A}\,a\,x^{m+p} + \mathrm{A}\,b\,x^{n+p} + \mathrm{B}\,a\,x^{m+q} + \mathrm{B}\,b\,x^{n+q}$. Up to this point the process presents no difficulties; but suppose that we have $m = p$ and $n = q$, and that we wish to reduce the two middle terms to a single one $(\mathrm{A}\,b + \mathrm{B}\,a)\,x^{m+q}$. For this purpose, the cards may order $m + q$ and $n + p$ to be transferred into the mill, and there subtracted one from the other; if the remainder is nothing, as would be the case on the present hypothesis, the mill will order other cards to bring to it the coefficients $\mathrm{A}\,b$ and $\mathrm{B}\,a$, that it may add them together and give them in this state as a coefficient for the single term $x^{n+p} = x^{m+q}$.

This example illustrates how the cards are able to reproduce all the operations which intellect performs in order to attain a determinate result, if these operations are themselves capable of being precisely defined.

Let us now examine the following expression :—

$$2 \cdot \frac{2^2 \cdot 4^2 \cdot 6^2 \cdot 8^2 \cdot 10^2 \ldots \ldots (2\,n)^2}{1^2 \cdot 3^2 \cdot 5^2 \cdot 7^2 \cdot 9^2 \ldots \ldots (2\,n-1)^2 \cdot (2\,n+1)^2},$$

which we know becomes equal to the ratio of the circumference to the diameter, when n is infinite. We may require the machine not only to perform the calculation of this fractional expression, but further to give indication as soon as the value becomes identical with that of the ratio of the circumference to the diameter when n is infinite, a case in which the computation would be impossible. Observe that we should thus require of the machine to interpret a result not of itself evident, and that this is not amongst its attributes, since it is no thinking being. Nevertheless, when the cos of $n = \dfrac{1}{0}$ has been foreseen, a card may immediately order the substitution of the value of π (π being the ratio of the circumference to the diameter), without going through the series of calculations indicated. This would merely require that the machine contain a special card, whose office it should be to place the number π in a direct and independent manner on the column indicated to it. And here we should introduce the mention of a third species of cards, which may be called *cards of numbers*. There are certain numbers, such as those expressing the ratio of the circumference to the diameter, the Numbers of Bernoulli, &c., which frequently present themselves in calculations. To avoid the necessity for computing them every time they have to be used, certain cards may be combined specially in order to give these numbers ready made into the mill, whence they afterwards go and place themselves on those columns of the store that are destined for them. Through this means the machine will be susceptible of those simplifications afforded by the use of numerical tables. It would be equally possible to introduce, by means of these cards, the logarithms of numbers; but perhaps it might not be in this case either the shortest or the most appropriate method; for the machine might be able to perform the same calculations by other more expeditious combinations, founded on the rapidity with which it executes the first four operations of arithmetic. To give an idea of this rapidity, we need only mention that Mr. Babbage believes he can, by his engine, form the product of two numbers, each containing twenty figures, in *three minutes*.

Perhaps the immense number of cards required for the solution of any rather complicated problem may appear to be an obstacle; but this does not seem to be the case. There is no limit to the number of cards that can be used. Certain stuffs require for their fabrication not less than *twenty thousand* cards, and we may unquestionably far exceed even this quantity *.

Resuming what we have explained concerning the Analytical Engine, we may conclude that it is

* See Note F.

based on two principles: the first, consisting in the fact that every arithmetical calculation ultimately depends on four principal operations—addition, subtraction, multiplication, and division; the second, in the possibility of reducing every analytical calculation to that of the coefficients for the several terms of a series. If this last principle be true, all the operations of analysis come within the domain of the engine. To take another point of view: the use of the cards offers a generality equal to that of algebraical formulæ, since such a formula simply indicates the nature and order of the operations requisite for arriving at a certain definite result, and similarly the cards merely command the engine to perform these same operations; but in order that the mechanisms may be able to act to any purpose, the numerical data of the problem must in every particular case be introduced. Thus the same series of cards will serve for all questions whose sameness of nature is such as to require nothing altered excepting the numerical data. In this light the cards are merely a translation of algebraical formulæ, or, to express it better, another form of analytical notation.

Since the engine has a mode of acting peculiar to itself, it will in every particular case be necessary to arrange the series of calculations conformably to the means which the machine possesses; for such or such a process which might be very easy for a calculator may be long and complicated for the engine, and *vice versâ*.

Considered under the most general point of view, the essential object of the machine being to calculate, according to the laws dictated to it, the values of numerical coefficients which it is then to distribute appropriately on the columns which represent the variables, it follows that the interpretation of formulæ and of results is beyond its province, unless indeed this very interpretation be itself susceptible of expression by means of the symbols which the machine employs. Thus, although it is not itself the being that reflects, it may yet be considered as the being which executes the conceptions of intelligence*. The cards receive the impress of these conceptions, and transmit to the various trains of mechanism composing the engine the orders necessary for their action. When once the engine shall have been constructed, the difficulty will be reduced to the making out of the cards; but as these are merely the translation of algebraical formulæ, it will, by means of some simple notations, be easy to consign the execution of them to a workman. Thus the whole intellectual labour will be limited to the preparation of the formulæ, which must be adapted for calculation by the engine.

Now, admitting that such an engine can be constructed, it may be inquired: what will be its utility? To recapitulate; it will afford the following advantages:—First, rigid accuracy. We know that numerical calculations are generally the stumbling-block to the solution of problems, since errors easily creep into them, and it is by no means always easy to detect these errors. Now the engine, by the very nature of its mode of acting, which requires no human intervention during the course of its operations, presents every species of security under the head of correctness: besides, it carries with it its own check; for at the end of every operation it prints off, not only the results, but likewise the numerical data of the question; so that it is easy to verify whether the question has been correctly proposed. Secondly, economy of time: to convince ourselves of this, we need only recollect that the multiplication of two numbers, consisting each of twenty figures, requires at the very utmost three minutes. Likewise, when a long series of identical computations is to be performed, such as those required for the formation of numerical tables, the machine can be brought into play

* See Note G.

so as to give several results at the same time, which will greatly abridge the whole amount of the processes. Thirdly, economy of intelligence: a simple arithmetical computation requires to be performed by a person possessing some capacity; and when we pass to more complicated calculations, and wish to use algebraical formulæ in particular cases, knowledge must be possessed which presupposes preliminary mathematical studies of some extent. Now the engine, from its capability of performing by itself all these purely material operations, spares intellectual labour, which may be more profitably employed. Thus the engine may be considered as a real manufactory of figures, which will lend its aid to those many useful sciences and arts that depend on numbers. Again, who can foresee the consequences of such an invention? In truth, how many precious observations remain practically barren for the progress of the sciences, because there are not powers sufficient for computing the results! And what discouragement does the perspective of a long and arid computation cast into the mind of a man of genius, who demands time exclusively for meditation, and who beholds it snatched from him by the material routine of operations! Yet it is by the laborious route of analysis that he must reach truth; but he cannot pursue this unless guided by numbers; for without numbers it is not given us to raise the veil which envelopes the mysteries of nature. Thus the idea of constructing an apparatus capable of aiding human weakness in such researches, is a conception which, being realized, would mark a glorious epoch in the history of the sciences. The plans have been arranged for all the various parts, and for all the wheel-work, which compose this immense apparatus, and their action studied; but these have not yet been fully combined together in the drawings* and mechanical notation †. The confidence which the genius of Mr. Babbage must inspire, affords legitimate ground for hope that this enterprise will be crowned with success; and while we render homage to the intelligence which directs it, let us breathe aspirations for the accomplishment of such an undertaking.

---

* This sentence has been slightly altered in the translation in order to express more exactly the present state of the engine.—NOTE BY TRANSLATOR.

† The notation here alluded to is a most interesting and important subject, and would have well deserved a separate and detailed Note upon it amongst those appended to the Memoir. It has, however, been impossible, within the space allotted, even to touch upon so wide a field.—NOTE BY TRANSLATOR.

NOTES BY THE TRANSLATOR.

————

NOTE A.—Page 9.

THE particular function whose integral the Difference Engine was constructed to tabulate, is

$$\Delta^7 u_z = 0.$$

The purpose which that engine has been specially intended and adapted to fulfil, is the computation of nautical and astronomical tables. The integral of

$$\Delta^7 u_z = 0$$

being

$$u_z = a + b\,x + c\,x^2 + d\,x^3 + e\,x^4 + f\,x^5 + g\,x^6,$$

the constants a, b, c, &c. are represented on the seven columns of discs, of which the engine consists. It can therefore tabulate *accurately* and to an *unlimited extent*, all series whose general term is comprised in the above formula; and it can also tabulate *approximatively* between *intervals of greater or less extent*, all other series which are capable of tabulation by the Method of Differences.

The Analytical Engine, on the contrary, is not merely adapted for *tabulating* the results of one particular function and of no other, but for *developing and tabulating* any function whatever. In fact the engine may be described as being the material expression of any indefinite function of any degree of generality and complexity, such as for instance,

$$F\,(x, y, z, \log x, \sin y, x^p, \&c.),$$

which is, it will be observed, a function of all other possible functions of any number of quantities.

In this, which we may call the *neutral* or *zero* state of the engine, it is ready to receive at any moment, by means of cards constituting a portion of its mechanism (and applied on the principle of those used in the Jacquard-loom), the impress of whatever *special* function we may desire to develope or to tabulate. These cards contain within themselves (in a manner explained in the Memoir itself, pages 11 and 12) the law of development of the particular function that may be under consideration, and they compel the mechanism to act accordingly in a certain corresponding order. One of the simplest cases would be, for example, to suppose that

$$F\,(x, y, z, \&c. \&c.)$$

is the particular function

$$\Delta^n u_z = 0$$

which the Difference Engine tabulates for values of n only up to 7. In this case the cards would order the mechanism to go through that succession of operations which would tabulate

$$u_z = a + b\,x + c\,x^2 + \ldots m\,x^{n-1},$$

where n might be any number whatever.

These cards, however, have nothing to do with the regulation of the particular *numerical* data. They merely determine the *operations***** to be effected, which operations may of course be performed on an infinite variety of particular numerical values, and do not bring out any definite numerical results unless the numerical data of the problem have been impressed on the requisite portions of the train of mechanism. In the above example, the first essential step towards an arithmetical result would be the substitution of specific numbers for n, and for the other primitive quantities which enter into the function.

Again, let us suppose that for F we put two complete equations of the fourth degree between x and y. We

———

* We do not mean to imply that the *only* use made of the Jacquard cards is that of regulating the algebraical *operations*; but we mean to explain that *those* cards and portions of mechanism which regulate these *operations* are wholly independent of those which are used for other purposes. M. Menabrea explains that there are *three* classes of cards used in the engine for three distinct sets of objects, viz. *Cards of the Operations, Cards of the Variables,* and certain *Cards of Numbers.* (See pages 12 and 18.)

must then express on the cards the law of elimination for such equations. The engine would follow out those laws, and would ultimately give the equation of one variable which results from such elimination. Various *modes* of elimination might be selected ; and of course the cards must be made out accordingly. The following is one mode that might be adopted. The engine is able to multiply together any two functions of the form

$$a + b\,x + c\,x^2 + \ldots\ldots p\,x^n.$$

This granted, the two equations may be arranged according to the powers of y, and the coefficients of the powers of y may be arranged according to powers of x. The elimination of y will result from the successive multiplications and subtractions of several such functions. In this, and in all other instances, as was explained above, the particular *numerical* data and the *numerical* results are determined by means and by portions of the mechanism which act quite independently of those that regulate the *operations*.

In studying the action of the Analytical Engine, we find that the peculiar and independent nature of the considerations which in all mathematical analysis belong to *operations*, as distinguished from *the objects operated upon* and from the *results* of the operations performed upon those objects, is very strikingly defined and separated.

It is well to draw attention to this point, not only because its full appreciation is essential to the attainment of any very just and adequate general comprehension of the powers and mode of action of the Analytical Engine, but also because it is one which is perhaps too little kept in view in the study of mathematical science in general. It is, however, impossible to confound it with other considerations, either when we trace the manner in which that engine attains its results, or when we prepare the data for its attainment of those results. It were much to be desired, that when mathematical processes pass through the human brain instead of through the medium of inanimate mechanism, it were equally a necessity of things that the reasonings connected with *operations* should hold the same just place as a clear and well-defined branch of the subject of analysis, a fundamental but yet independent ingredient in the science, which they must do in studying the engine. The confusion, the difficulties, the contradictions which, in consequence of a want of accurate distinctions in this particular, have up to even a recent period encumbered mathematics in all those branches involving the consideration of negative and impossible quantities, will at once occur to the reader who is at all versed in this science, and would alone suffice to justify dwelling somewhat on the point, in connexion with any subject so peculiarly fitted to give forcible illustration of it as the Analytical Engine. It may be desirable to explain, that by the word *operation*, we mean *any process which alters the mutual relation of two or more things*, be this relation of what kind it may. This is the most general definition, and would include all subjects in the universe. In abstract mathematics, of course operations alter those particular relations which are involved in the considerations of number and space, and the *results* of operations are those peculiar results which correspond to the nature of the subjects of operation. But the science of operations, as derived from mathematics more especially, is a science of itself, and has its own abstract truth and value ; just as logic has its own peculiar truth and value, independently of the subjects to which we may apply its reasonings and processes. Those who are accustomed to some of the more modern views of the above subject, will know that a few fundamental relations being true, certain other combinations of relations must of necessity follow ; combinations unlimited in variety and extent if the deductions from the primary relations be carried on far enough. They will also be aware that one main reason why the separate nature of the science of operations has been little felt, and in general little dwelt on, is the *shifting* meaning of many of the symbols used in mathematical notation. First, the symbols of *operation* are frequently *also* the symbols of the *results* of operations. We may say that these symbols are apt to have both a *retrospective* and a *prospective* signification. They may signify either relations that are the consequence of a series of processes already performed, or relations that are yet to be effected through certain processes. Secondly, figures, the symbols of *numerical magnitude*, are frequently *also* the symbols of *operations*, as when they are the indices of powers. Wherever terms have a shifting meaning, independent sets of considerations are liable to become complicated together, and reasonings and results are frequently falsified. Now in the Analytical Engine, the operations which come under the first of the above heads are ordered and combined by means of a notation and of a train of mechanism which belong exclusively to themselves ; and with respect to the second head, whenever numbers meaning *operations* and not *quantities* (such as the indices of powers) are inscribed on any column or set of columns, those columns immediately act in a wholly separate and independent manner, becoming connected with the *operating mechanism* exclusively, and re-acting upon this. They never come into combination with numbers upon any other columns meaning *quantities*; though, of course, if there are numbers meaning *operations* upon *n* columns, these may *combine amongst each other*, and will often be required to do so, just as numbers meaning *quantities* combine with each other in any variety. It might have been arranged that all numbers meaning *operations* should have appeared on some separate portion of the engine from that which presents numerical *quantities*; but the present mode is in some cases more simple, and offers in reality quite as much distinctness when understood.

The operating mechanism can even be thrown into action independently of any object to operate upon (although of course no *result* could then be developed). Again, it might act upon other things besides *number*, were objects found whose mutual fundamental relations could be expressed by those of the abstract science of operations, and which should be also susceptible of adaptations to the action of the operating notation and mechanism of the engine. Supposing, for instance, that the fundamental relations of pitched sounds in the science of harmony and of musical composition were susceptible of such expression and adaptations, the engine might compose elaborate and scientific pieces of music of any degree of complexity or extent.

The Analytical Engine is an *embodying of the science of operations*, constructed with peculiar reference to abstract number as the subject of those operations. The Difference Engine is the embodying of *one particular and very limited set of operations*, which (see the notation used in Note B) may be expressed thus $(+, +, +, +, +, +)$, or thus, $6 (+)$. Six repetitions of the one operation, $+$, is, in fact, the whole sum and object of that engine. It has seven columns, and a number on any column can add itself to a number on the next column to its *right-hand*. So that, beginning with the column furthest to the left, six additions can be effected, and the result appears on the seventh column, which is the last on the right-hand. The *operating* mechanism of this engine acts in as separate and independent a manner as that of the Analytical Engine; but being susceptible of only one unvarying and restricted combination, it has little force or interest in illustration of the distinct nature of the *science of operations*. The importance of regarding the Analytical Engine under this point of view will, we think, become more and more obvious as the reader proceeds with M. Menabrea's clear and masterly article. The calculus of operations is likewise in itself a topic of so much interest, and has of late years been so much more written on and thought on than formerly, that any bearing which that engine, from its mode of constitution, may possess upon the illustration of this branch of mathematical science should not be overlooked. Whether the inventor of this engine had any such views in his mind while working out the invention, or whether he may subsequently ever have regarded it under this phase, we do not know; but it is one that forcibly occurred to ourselves on becoming acquainted with the means through which analytical combinations are actually attained by the mechanism. We cannot forbear suggesting one practical result which it appears to us must be greatly facilitated by the independent manner in which the engine orders and combines its *operations*: we allude to the attainment of those combinations into which *imaginary quantities* enter. This is a branch of its processes into which we have not had the opportunity of inquiring, and our conjecture therefore as to the principle on which we conceive the accomplishment of such results may have been made to depend, is very probably not in accordance with the fact, and less subservient for the purpose than some other principles, or at least requiring the cooperation of others. It seems to us obvious, however, that where operations are so independent in their mode of acting, it must be easy, by means of a few simple provisions and additions in arranging the mechanism, to bring out a *double* set of *results*, viz.—1st, the *numerical magnitudes* which are the results of operations performed on *numerical data*. (These results are the *primary* object of the engine.) 2ndly, the *symbolical results* to be attached to those numerical results, which symbolical results are not less the necessary and logical consequences of operations performed upon *symbolical data*, than are numerical results when the data are numerical*.

If we compare together the powers and the principles of construction of the Difference and of the Analytical

---

* In fact, such an extension as we allude to would merely constitute a further and more perfected development of any system introduced for making the proper combinations of the signs *plus* and *minus*. How ably M. Menabrea has touched on this restricted case is pointed out in Note B.

Engines, we shall perceive that the capabilities of the latter are immeasurably more extensive than those of the former, and that they in fact hold to each other the same relationship as that of analysis to arithmetic. The Difference Engine can effect but one particular series of operations, viz. that required for tabulating the integral of the special function

$$\Delta^n u_z = 0;$$

and as it can only do this for values of n up to $7*$, it cannot be considered as being the most *general* expression even of *one particular* function, much less as being the expression of any and all possible functions of all degrees of generality. The Difference Engine can in reality (as has been already partly explained) do nothing but *add*; and any other processes, not excepting those of simple subtraction, multiplication and division, can be performed by it only just to that extent in which it is possible, by judicious mathematical arrangement and artifices, to reduce them to a *series of additions*. The method of differences is, in fact, a method of additions; and as it includes within its means a larger number of results attainable by *addition* simply, than any other mathematical principle, it was very appropriately selected as the basis on which to construct *an Adding Machine*, so as to give to the powers of such a machine the widest possible range. The Analytical Engine, on the contrary, can either add, subtract, multiply or divide with equal facility; and performs each of these four operations in a direct manner, without the aid of any of the other three. This one fact implies everything; and it is scarcely necessary to point out, for instance, that while the Difference Engine can merely *tabulate*, and is incapable of *developing*, the Analytical Engine can *either tabulate or develope*.

The former engine is in its nature strictly *arithmetical*, and the results it can arrive at lie within a very clearly defined and restricted range, while there is no finite line of demarcation which limits the powers of the Analytical Engine. These powers are co-extensive with our knowledge of the laws of analysis itself, and need be bounded only by our acquaintance with the latter. Indeed we may consider the engine as the *material and mechanical representative* of analysis, and that our actual working powers in this department of human study will be enabled more effectually than heretofore to keep pace with our theoretical knowledge of its principles and laws, through the complete control which the engine gives us over the *executive manipulation* of algebraical and numerical symbols.

Those who view mathematical science, not merely as a vast body of abstract and immutable truths, whose intrinsic beauty, symmetry and logical completeness, when regarded in their connexion together as a whole, entitle them to a prominent place in the interest of all profound and logical minds, but as possessing a yet deeper interest for the human race, when it is remembered that this science constitutes the language through which alone we can adequately express the great facts of the natural world, and those unceasing changes of mutual relationship which, visibly or invisibly, consciously or unconsciously to our immediate physical perceptions, are interminably going on in the agencies of the creation we live amidst : those who thus think on mathematical truth as the instrument through which the weak mind of man can most effectually read his Creator's works, will regard with especial interest all that can tend to facilitate the translation of its principles into explicit practical forms.

The distinctive characteristic of the Analytical Engine, and that which has rendered it possible to endow mechanism with such extensive faculties as bid fair to make this engine the executive right-hand of abstract algebra, is the introduction into it of the principle which Jacquard devised for regulating, by means of punched cards, the most complicated patterns in the fabrication of brocaded stuffs. It is in this that the distinction between the two engines lies. Nothing of the sort exists in the Difference Engine. We may say most aptly, that

---

* The machine might have been constructed so as to tabulate for a higher value of n than seven. Since, however, every unit added to the value of n increases the extent of the mechanism requisite, there would on this account be a limit beyond which it could not be practically carried. Seven is sufficiently high for the calculation of all ordinary tables.

The fact that, in the Analytical Engine, the same extent of mechanism suffices for the solution of $\Delta^n u_z = 0$, whether $n = 7$, $n = 100,000$, or $n =$ any number whatever, at once suggests how entirely distinct must be the *nature of the principles* through whose application matter has been enabled to become the working agent of abstract mental operations in each of these engines respectively; and it affords an equally obvious presumption, that in the case of the Analytical Engine, not only are those principles in themselves of a higher and more comprehensive description, but also such as must vastly extend the *practical* value of the engine whose basis they constitute.

the Analytical Engine *weaves algebraical patterns* just as the Jacquard-loom weaves flowers and leaves. Here, it seems to us, resides much more of originality than the Difference Engine can be fairly entitled to claim. We do not wish to deny to this latter all such claims. We believe that it is the only proposal or attempt ever made to construct a calculating machine *founded on the principle of successive orders of differences*, and capable of *printing off its own results*; and that this engine surpasses its predecessors, both in the extent of the calculations which it can perform, in the facility, certainty and accuracy with which it can effect them, and in the absence of all necessity for the intervention of human intelligence *during the performance of its calculations*. Its nature is, however, limited to the strictly arithmetical, and it is far from being the first or only scheme for constructing *arithmetical* calculating machines with more or less of success.

The bounds of *arithmetic* were however outstepped the moment the idea of applying the cards had occurred; and the Analytical Engine does not occupy common ground with mere " calculating machines." It holds a position wholly its own; and the considerations it suggests are most interesting in their nature. In enabling mechanism to combine together *general* symbols in successions of unlimited variety and extent, a uniting link is established between the operations of matter and the abstract mental processes of the *most abstract* branch of mathematical science. A new, a vast, and a powerful language is developed for the future use of analysis, in which to wield its truths so that these may become of more speedy and accurate practical application for the purposes of mankind than the means hitherto in our possession have rendered possible. Thus not only the mental and the material, but the theoretical and the practical in the mathematical world, are brought into more intimate and effective connexion with each other. We are not aware of its being on record that anything partaking in the nature of what is so well designated the *Analytical* Engine has been hitherto proposed, or even thought of, as a practical possibility, any more than the idea of a thinking or of a reasoning machine.

We will touch on another point which constitutes an important distinction in the modes of operating of the Difference and Analytical Engines. In order to enable the former to do its business, it is necessary to put into its columns the series of numbers constituting the first terms of the several orders of differences for whatever is the particular table under consideration. The machine then works *upon* these as its data. But these data must themselves have been already computed through a series of calculations by a human head. Therefore that engine can only produce results depending on data which have been arrived at by the explicit and actual working out of processes that are in their nature different from any that come within the sphere of its own powers. In other words, an *analysing* process must have been gone through by a human mind in order to obtain the data upon which the engine then *synthetically* builds its results. The Difference Engine is in its character exclusively *synthetical*, while the Analytical Engine is equally capable of analysis or of synthesis.

It is true that the Difference Engine can calculate to a much greater extent with these few preliminary data, than the data themselves required for their own determination. The table of squares, for instance, can be calculated to any extent whatever, when the numbers *one* and *two* are furnished; and a very few differences computed at any part of a table of logarithms would enable the engine to calculate many hundreds or even thousands of logarithms. Still the circumstance of its requiring, as a previous condition, that any function whatever shall have been numerically worked out, makes it very inferior in its nature and advantages to an engine which, like the Analytical Engine, requires merely that we should know the *succession and distribution of the operations* to be performed; without there being any occasion *, in order to obtain data on which it can work, for our ever having gone through either the same particular operations which it is itself to effect, or any others. Numerical data must of course be given it, but they are mere arbitrary ones; not data that could only be arrived at through a systematic and necessary series of previous numerical calculations, which is quite a different thing.

To this it may be replied, that an analysing process must equally have been performed in order to furnish the Analytical Engine with the necessary *operative* data; and that herein may also lie a possible source of error. Granted that the actual mechanism is unerring in its processes, the *cards* may give it wrong orders. This is unquestionably the case; but there is much less chance of error, and likewise far less expenditure of time

* This subject is further noticed in Note F.

and labour, where operations only, and the distribution of these operations, have to be made out, than where explicit numerical results are to be attained. In the case of the Analytical Engine we have undoubtedly to lay out a certain capital of analytical labour in one particular line; but this is in order that the engine may bring us in a much larger return in another line. It should be remembered also that the cards, when once made out for any formula, have all the generality of algebra, and include an infinite number of particular cases.

We have dwelt considerably on the distinctive peculiarities of each of these engines, because we think it essential to place their respective attributes in strong relief before the apprehension of the public; and to define with clearness and accuracy the wholly different nature of the principles on which each is based, so as to make it self-evident to the reader (the mathematical reader at least) in what manner and degree the powers of the Analytical Engine transcend those of an engine, which, like the Difference Engine, can only work out such results as may be derived from *one restricted and particular series of processes*, such as those included in $\Delta^n u_z = 0$. We think this of importance, because we know that there exists considerable vagueness and inaccuracy in the mind of persons in general on the subject. There is a misty notion amongst most of those who have attended at all to it, that *two* "calculating machines" have been successively invented by the same person within the last few years; while others again have never heard but of the one original "calculating machine," and are not aware of there being any extension upon this. For either of these two classes of persons the above considerations are appropriate. While the latter require a knowledge of the fact that there *are two* such inventions, the former are not less in want of accurate and well-defined information on the subject. No very clear or correct ideas prevail as to the characteristics of each engine, or their respective advantages or disadvantages; and in meeting with those incidental allusions, of a more or less direct kind, which occur in so many publications of the day, to these machines, it must frequently be matter of doubt *which* "calculating machine" is referred to, or whether *both* are included in the general allusion.

We are desirous likewise of removing two misapprehensions which we know obtain, to some extent, respecting these engines. In the first place it is very generally supposed that the Difference Engine, after it had been completed up to a certain point, *suggested* the idea of the Analytical Engine; and that the second is in fact the improved offspring of the first, and *grew out* of the existence of its predecessor, through some natural or else accidental combination of ideas suggested by this one. Such a supposition is in this instance contrary to the facts; although it seems to be almost an obvious inference, wherever two inventions, similar in their nature and objects, succeed each other closely in order of *time*, and strikingly in order of *value*; more especially when the same individual is the author of both. Nevertheless the ideas which led to the Analytical Engine occurred in a manner wholly independent of any that were connected with the Difference Engine. These ideas are indeed in their own intrinsic nature independent of the latter engine, and might equally have occurred had it never existed nor been even thought of at all.

The second of the misapprehensions above alluded to relates to the well-known suspension, during some years past, of all progress in the construction of the Difference Engine. Respecting the circumstances which have interfered with the actual completion of either invention, we offer no opinion; and in fact are not possessed of the data for doing so, had we the inclination. But we know that some persons suppose these obstacles (be they what they may) to have arisen *in consequence* of the subsequent invention of the Analytical Engine while the former was in progress. We have ourselves heard it even *lamented* that an idea should ever have occurred at all, which had turned out to be merely the means of arresting what was already in a course of successful execution, without substituting the superior invention in its stead. This notion we can contradict in the most unqualified manner. The progress of the Difference Engine had long been suspended, before there were even the least crude glimmerings of any invention superior to it. Such glimmerings, therefore, and their subsequent development, were in no way the original *cause* of that suspension; although, where difficulties of some kind or other evidently already existed, it was not perhaps calculated to remove or lessen them that an invention should have been meanwhile thought of, which, while including all that the first was capable of, possesses powers so extended as to eclipse it altogether.

We leave it for the decision of each individual (*after he has possessed himself* of competent information as to the characteristics of each engine) to determine how far it ought to be matter of regret that such an accession has been made to the powers of human science, even if it *has* (which we greatly doubt) increased to a certain limited extent some already existing difficulties that had arisen in the way of completing a valuable but lesser work. We leave it for each to satisfy himself as to the wisdom of desiring the obliteration (were that now possible) of all records of the more perfect invention, in order that the comparatively limited one might be finished. The Difference Engine would doubtless fulfil all those practical objects which it was originally destined for. It would certainly calculate all the tables that are more directly necessary for the physical purposes of life, such as nautical and other computations. Those who incline to very strictly utilitarian views may perhaps feel that the peculiar powers of the Analytical Engine bear upon questions of abstract and speculative science, rather than upon those involving every-day and ordinary human interests. These persons being likely to possess but little sympathy, or possibly acquaintance, with any branches of science which they do not find to be *useful* (according to *their* definition of that word), may conceive that the undertaking of that engine, now that the other one is already in progress, would be a barren and unproductive laying out of yet more money and labour; in fact, a work of supererogation. Even in the utilitarian aspect, however, we do not doubt that very valuable practical results would be developed by the extended faculties of the Analytical Engine; some of which results we think we could now hint at, had we the space; and others, which it may not yet be possible to foresee, but which would be brought forth by the daily increasing requirements of science, and by a more intimate practical acquaintance with the powers of the engine, were it in actual existence.

On general grounds, both of an *à priori* description as well as those founded on the scientific history and experience of mankind, we see strong presumptions that such would be the case. Nevertheless all will probably concur in feeling that the completion of the Difference Engine would be far preferable to the non-completion of any calculating engine at all. With whomsoever or wheresoever may rest the present causes of difficulty that apparently exist towards either the completion of the old engine, or the commencement of the new one, we trust they will not ultimately result in this generation's being acquainted with these inventions through the medium of pen, ink and paper merely; and still more do we hope, that for the honour of our country's reputation in the future pages of history, these causes will not lead to the completion of the undertaking by some *other* nation or government. This could not but be matter of just regret; and equally so, whether the obstacles may have originated in private interests and feelings, in considerations of a more public description, or in causes combining the nature of both such solutions.

We refer the reader to the 'Edinburgh Review' of July 1834, for a very able account of the Difference Engine. The writer of the article we allude to has selected as his prominent matter for exposition, a wholly different view of the subject from that which M. Menabrea has chosen. The former chiefly treats it under its mechanical aspect, entering but slightly into the mathematical principles of which that engine is the representative, but giving, in considerable length, many details of the mechanism and contrivances by means of which it tabulates the various orders of differences. M. Menabrea, on the contrary, exclusively develops the analytical view; taking it for granted that mechanism is able to perform certain processes, but without attempting to explain *how*; and devoting his whole attention to explanations and illustrations of the manner in which analytical laws can be so arranged and combined as to bring every branch of that vast subject within the grasp of the assumed powers of mechanism. It is obvious that, in the invention of a calculating engine, these two branches of the subject are equally essential fields of investigation, and that on their mutual adjustment, one to the other, must depend all success. They must be made to meet each other, so that the weak points in the powers of either department may be compensated by the strong points in those of the other. They are indissolubly connected, though so different in their intrinsic nature, that perhaps the same mind might not be likely to prove equally profound or successful in both. We know those who doubt whether the powers of mechanism will in practice prove adequate in all respects to the demands made upon them in the working of such complicated trains of machinery as those of the above engines, and who apprehend that unforeseen practical difficulties and disturbances will arise in the way of accuracy and of facility of operation. The Difference Engine, however, appears to us to be in a great measure an answer to these doubts. It is complete as far as it goes, and it does work with all the anticipated success. The Analytical Engine, far from being more complicated, will in many respects be of simpler construction; and it is a remarkable circumstance attending it, that with very *simplified* means it is so much more powerful.

The article in the 'Edinburgh Review' was written some time previous to the occurrence of any ideas such as afterwards led to the invention of the Analytical Engine; and in the nature of the Difference Engine there is much less that would invite a writer to take exclusively, or even prominently, the mathematical view of it, than in that of the Analytical Engine; although mechanism has undoubtedly gone much further to meet mathematics, in the case of this engine, than of the former one. Some publication embracing the *mechanical* view of the Analytical Engine is a desideratum which we trust will be supplied before long.

Those who may have the patience to study a moderate quantity of rather dry details will find ample compensation, after perusing the article of 1834, in the clearness with which a succinct view will have been attained of the various practical steps through which mechanism can accomplish certain processes; and they will also find themselves still further capable of appreciating M. Menabrea's more comprehensive and generalized memoir. The very difference in the style and object of these two articles makes them peculiarly valuable to each other; at least for the purposes of those who really desire something more than a merely superficial and popular comprehension of the subject of calculating engines. A. A. L.

Note B.—Page 11.

That portion of the Analytical Engine here alluded to is called the storehouse. It contains an indefinite number of the columns of discs described by M. Menabrea. The reader may picture to himself a pile of rather large draughtsmen heaped perpendicularly one above another to a considerable height, each counter having the digits from 0 to 9 inscribed on its *edge* at equal intervals; and if he then conceives that the counters do not actually lie one upon another so as to be in contact, but are fixed at small intervals of vertical distance on a common axis which passes perpendicularly through their centres, and around which each disc can *revolve horizontally* so that any required digit amongst those inscribed on its margin can be brought into view, he will have a good idea of one of these columns. The *lowest* of the discs on any column belongs to the units, the next above to the tens, the next above this to the hundreds, and so on. Thus, if we wished to inscribe 1345 on a column of the engine, it would stand thus:—

$$1$$
$$3$$
$$4$$
$$5$$

In the Difference Engine there are seven of these columns placed side by side in a row, and the working mechanism extends behind them: the general form of the whole mass of machinery is that of a quadrangular prism (more or less approaching to the cube); the results always appearing on that perpendicular face of the engine which contains the columns of discs, opposite to which face a spectator may place himself. In the Analytical Engine there would be many more of these columns, probably at least two hundred. The precise form and arrangement which the whole mass of its mechanism will assume is not yet finally determined.

We may conveniently represent the columns of discs on paper in a diagram like the following:—

V_1 V_2 V_3 V_4 &c.
O O O O &c.
0 0 0 0
0 0 0 0 &c.
0 0 0 0
0 0 0 0
□ □ □ □ &c.

The V's are for the purpose of convenient reference to any column, either in writing or speaking, and are consequently numbered. The reason why the letter V is chosen for this purpose in preference to any other letter, is because these columns are designated (as the reader will find in proceeding with the Memoir) the *Variables*, and sometimes the *Variable columns*, or the *columns of Variables*. The origin of this appellation is, that the values on the columns are destined to change, that is to *vary*, in every conceivable manner.

But it is necessary to guard against the natural misapprehension that the columns are only intended to receive the values of the *variables* in an analytical formula, and not of the *constants*. The columns are called Variables on a ground wholly unconnected with the *analytical* distinction between constants and variables. In order to prevent the possibility of confusion, we have, both in the translation and in the notes, written Variable with a capital letter when we use the word to signify a *column of the engine,* and variable with a small letter when we mean the *variable of a formula.* Similarly, *Variable-cards* signify any cards that belong to a column of the engine.

To return to the explanation of the diagram: each circle at the top is intended to contain the algebraic sign + or —, either of which can be substituted* for the other, according as the number represented on the column below is positive or negative. In a similar manner any other purely *symbolical* results of algebraical processes might be made to appear in these circles. In Note A. the practicability of developing *symbolical* with no less ease than *numerical* results has been touched on. The zeros beneath the *symbolic* circles represent each of them a disc, supposed to have the digit 0 presented in front. Only four tiers of zeros have been figured in the diagram, but these may be considered as representing thirty or forty, or any number of tiers of discs that may be required. Since each disc can present any digit, and each circle any sign, the discs of every column may be so adjusted† as to express any positive or negative number whatever within the limits of the machine ; which limits depend on the *perpendicular* extent of the mechanism, that is, on the number of discs to a column.

Each of the squares below the zeros is intended for the inscription of any *general* symbol or combination of symbols we please; it being understood that the number represented on the column immediately above is the numerical value of that symbol, or combination of symbols. Let us, for instance, represent the three quantities a, n, x, and let us further suppose that $a=5$, $n=7$, $x=98$. We should have—

V_1	V_2	V_3	V_4 &c.
+ ‡	+	+	+
0	0	0	0
0	0	0	0 &c.
0	0	9	0
5	7	8	0 &c.
[a]	[n]	[x]	[ ]

We may now combine these symbols in a variety of ways, so as to form any required function or functions of them, and we may then inscribe each such function below brackets, every bracket uniting together those quantities (and those only) which enter into the function inscribed below it. We must also, when we have decided on the particular function whose numerical value we desire to calculate, assign another column to the right-hand for receiving the *results,* and must inscribe the function in the square below this column. In the above instance we might have any one of the following functions :—

$$a\,x^n, \quad x^{a\,n}, \quad a\,.\,n\,.\,x, \quad \frac{a}{n}\,x, \quad a+n+x, \&c. \&c.$$

Let us select the first. It would stand as follows, previous to calculation :—

V_1	V_2	V_3	V_4 &c.
+	+	+	+
0	0	0	0 &c.
0	0	0	0
0	0	9	0
5	7	8	0 &c.
[a]	[n]	[x]	[$a x^n$] &c.

$$a\,x_n$$

---

* A fuller account of the manner in which the *signs* are regulated is given in Mons. Menabrea's Memoir, pages 14, 15. He himself expresses doubts (in a note of his own at the bottom of the latter page) as to his having been likely to hit on the precise methods really adopted ; his explanation being merely a conjectural one. That it *does* accord precisely with the fact is a remarkable circumstance, and affords a convincing proof how completely Mons. Menabrea has been imbued with the true spirit of the invention. Indeed the whole of the above Memoir is a striking production, when we consider that Mons. Menabrea had had but very slight means for obtaining any adequate ideas respecting the Analytical Engine. It requires however a considerable acquaintance with the abstruse and complicated nature of such a subject, in order fully to appreciate the penetration of the writer who could take so just and comprehensive a view of it upon such limited opportunity.

† This adjustment is done by hand merely.

‡ It is convenient to omit the circles whenever the signs + or — can be actually represented.

The data being given, we must now put into the engine the cards proper for directing the operations in the case of the particular function chosen. These operations would in this instance be,—

First, six multiplications in order to get x^n ($= 98^7$ for the above particular data).

Secondly, one multiplication in order then to get $a . x^n$ ($= 5.98^7$).

In all, seven multiplications to complete the whole process. We may thus represent them :—

$$(\times, \times, \times, \times, \times, \times, \times), \text{ or } 7 (\times).$$

The multiplications would, however, at successive stages in the solution of the problem, operate on pairs of numbers, derived from *different* columns. In other words, the *same operation* would be performed on different *subjects of operation*. And here again is an illustration of the remarks made in the preceding Note on the independent manner in which the engine directs its *operations*. In determining the value of $a\, x^n$, the *operations* are *homogeneous*, but are distributed amongst different *subjects of operation*, at successive stages of the computation. It is by means of certain punched cards, belonging to the Variables themselves, that the action of the operations is so *distributed* as to suit each particular function. The *Operation-cards* merely determine the succession of operations in a general manner. They in fact throw all that portion of the mechanism included in the *mill* into a series of different *states*, which we may call the *adding state*, or the *multiplying state*, &c. respectively. In each of these states the mechanism is ready to act in the way peculiar to that state, on any pair of numbers which may be permitted to come within its sphere of action. Only *one* of these operating states of the mill can exist at a time; and the nature of the mechanism is also such that only *one pair of numbers* can be received and acted on at a time. Now, in order to secure that the mill shall receive a constant supply of the proper pairs of numbers in succession, and that it shall also rightly locate the result of an operation performed upon any pair, each Variable has cards of its own belonging to it. It has, first, a class of cards whose business it is to *allow* the number on the Variable to pass into the mill, there to be operated upon. These cards may be called the *Supplying-cards*. *They* furnish the mill with its proper food. Each Variable has, secondly, another class of cards, whose office it is to allow the Variable to *receive* a number *from* the mill. These cards may be called the *Receiving-cards*. *They* regulate the location of results, whether temporary or ultimate results. The Variable-cards in general (including both the preceding classes) might, it appears to us, be even more appropriately designated the Distributive-cards, since it is through their means that the action of the operations, and the results of this action, are rightly *distributed*.

There are *two varieties* of the *Supplying* Variable-cards, respectively adapted for fulfilling two distinct subsidiary purposes: but as these modifications do not bear upon the present subject, we shall notice them in another place.

In the above case of $a\, x^n$, the Operation-cards merely order seven multiplications, that is, they order the mill to be in the *multiplying state* seven successive times (without any reference to the particular columns whose numbers are to be acted upon). The proper Distributive Variable-cards step in at each successive multiplication, and cause the distributions requisite for the particular case.

For $x^{a\,n}$	the operations would be		$34 (\times)$
... $a . n . x$	... ...		$(\times, \times)$, or $2 (\times)$
... $\dfrac{a}{n} . x$	... ...		$(\div, \times)$
... $a + n + x$	... ...		$(+, +)$, or $2 (+)$

The engine might be made to calculate all these in succession. Having completed $a\, x^n$, the function $x^{a\,n}$ might be written under the brackets instead of $a\, x^n$, and a new calculation commenced (the appropriate Operation and Variable-cards for the new function of course coming into play). The results would then appear on V_5. So on for any number of different functions of the quantities a, n, x. Each *result* might either permanently remain on its column during the succeeding calculations, so that when all the functions had been computed, their values would simultaneously exist on V_4, V_5, V_6, &c.; or each result might (after being printed off, or used in any specified manner) be effaced, to make way for its successor. The square under V_4 ought, for the latter arrangement, to have the functions $a\, x^n$, $x^{a\,n}$, $a\, n\, x$, &c. successively inscribed in it.

Let us now suppose that we have *two* expressions whose values have been computed by the engine independently of each other (each having its own group of columns for data and results). Let them be $a\,x^n$, and $b\,p\,y$. They would then stand as follows on the columns :—

V_1	V_2	V_3	V_4	V_5	V_6	V_7	V_8	V_9
+	+	+	+	+	+	+	+	+
0	0	0	0	0	0	0	0	0
0	0	0	0	0	0	0	0	0
0	0	0	0	0	0	0	0	0
0	0	0	0	0	0	0	0	0
a	n	x	$a\,x^n$	b	p	y	$b\,p\,y$	$\dfrac{a\,x^n}{b\,p\,y}$

We may now desire to combine together these two *results*, in any manner we please; in which case it would only be necessary to have an additional card or cards, which should order the requisite operations to be performed with the numbers on the two result-columns, V_4 and V_8, and the *result of these further operations* to appear on a new column, V_9. Say that we wish to divide $a\,x^n$ by $b\,p\,y$. The numerical value of this division would then appear on the column V_9, beneath which we have inscribed $\dfrac{a\,x^n}{b\,p\,y}$. The whole series of operations from the beginning would be as follows (n being $= 7$):

$$\left\{ 7(\times),\, 2(\times),\, \div \right\},\ \text{or}\ \left\{ 9(\times),\, \div \right\}.$$

This example is introduced merely to show that we may, if we please, retain separately and permanently any *intermediate* results (like $a\,x^n$, $b\,p\,y$) which occur in the course of processes having an ulterior and more complicated result as their chief and final object $\left(\text{like } \dfrac{a\,x^n}{b\,p\,y}\right)$.

Any group of columns may be considered as representing a *general* function, until a *special* one has been implicitly impressed upon them through the introduction into the engine of the Operation and Variable-cards made out for a *particular* function. Thus, in the preceding example, V_1, V_2, V_3, V_5, V_6, V_7 represent the *general* function $\phi\,(a,\, n,\, b,\, p,\, x,\, y)$ until the function $\dfrac{a\,x^n}{b\,p\,y}$ has been determined on, and *implicitly* expressed by the placing of the right cards in the engine. The actual working of the mechanism, as regulated by these cards, then *explicitly* developes the value of the function. The inscription of a function under the brackets, and in the square under the result-column, in no way influences the processes or the results, and is merely a memorandum for the observer, to remind him of what is going on. It is the Operation and the Variable-cards only which in reality determine the function. Indeed it should be distinctly kept in mind, that the inscriptions within *any* of the squares are quite independent of the mechanism or workings of the engine, and are nothing but arbitrary memorandums placed there at pleasure to assist the spectator.

The further we analyse the manner in which such an engine performs its processes and attains its results, the more we perceive how distinctly it places in a true and just light the mutual relations and connexion of the various steps of mathematical analysis; how clearly it separates those things which are in reality distinct and independent, and unites those which are mutually dependent. A. A. L.

Note C.—Page 11.

Those who may desire to study the principles of the Jacquard-loom in the most effectual manner, viz. that of practical observation, have only to step into the Adelaide Gallery or the Polytechnic Institution. In each of these valuable repositories of scientific *illustration*, a weaver is constantly working at a Jacquard-loom, and is ready to give any information that may be desired as to the construction and modes of acting of his apparatus. The volume on the manufacture of silk, in Lardner's Cyclopædia, contains a chapter on the Jacquard-loom, which may also be consulted with advantage.

The mode of application of the cards, as hitherto used in the art of weaving, was not found, however, to be sufficiently powerful for all the simplifications which it was desirable to attain in such varied and complicated processes as those required in order to fulfil the purposes of an Analytical Engine. A method was devised of what was technically designated *backing* the cards in certain groups according to certain laws. The object of this extension is to secure the possibility of bringing any particular card or set of cards into use *any number of times successively* in the solution of one problem.   Whether this power shall be taken advantage of or not, in each particular instance, will depend on the nature of the operations which the problem under consideration may require.   The process is alluded to by M. Menabrea in page 16, and it is a very important simplification. It has been proposed to use it for the reciprocal benefit of that art, which, while it has itself no apparent connexion with the domains of abstract science, has yet proved so valuable to the latter, in suggesting the principles which, in their new and singular field of application, seem likely to place *algebraical* combinations not less completely within the province of mechanism, than are all those varied intricacies of which *intersecting threads* are susceptible.   By the introduction of the system of *backing* into the Jacquard-loom itself, patterns which should possess symmetry, and follow regular laws of any extent, might be woven by means of comparatively few cards.

Those who understand the mechanism of this loom will perceive that the above improvement is easily effected in practice, by causing the prism over which the train of pattern-cards is suspended to revolve *backwards* instead of *forwards*, at pleasure, under the requisite circumstances; until, by so doing, any particular card, or set of cards, that has done duty once, and passed on in the ordinary regular succession, is brought back to the position it occupied just before it was used the preceding time.   The prism then resumes its *forward* rotation, and thus brings the card or set of cards in question into play a second time.   This process may obviously be repeated any number of times.   A. A. L.

Note D.—Page 14.

We have represented the solution of these two equations, with every detail, in a diagram* similar to those used in Note B.; but additional explanations are requisite, partly in order to make this more complicated case perfectly clear, and partly for the comprehension of certain indications and notations not used in the preceding diagrams.   Those who may wish to understand Note G. completely, are recommended to pay particular attention to the contents of the present Note, or they will not otherwise comprehend the similar notation and indications when applied to a much more complicated case.

In all calculations, the columns of Variables used may be divided into three classes :—

1st. Those on which the data are inscribed:

2ndly. Those intended to receive the final results :

3rdly. Those intended to receive such intermediate and temporary combinations of the primitive data as are not to be permanently retained, but are merely needed for *working with*, in order to attain the ultimate results. Combinations of this kind might properly be called *secondary data*.   They are in fact so many *successive stages* towards the final result.   The columns which receive them are rightly named *Working-Variables*, for their office is in its nature purely *subsidiary* to other purposes.   They develope an intermediate and transient class of results, which unite the original data with the final results.

The Result-Variables sometimes partake of the nature of Working-Variables.   It frequently happens that a Variable destined to receive a final result is the recipient of one or more intermediate values successively, in the course of the processes.   Similarly, the Variables for data often become Working-Variables, or Result-Variables, or even both in succession.   It so happens, however, that in the case of the present equations the three sets of offices remain throughout perfectly separate and independent.

It will be observed, that in the squares below the *Working*-Variables nothing is inscribed.   Any one of these Variables is in many cases destined to pass through various values successively during the performance of a calculation (although in these particular equations no instance of this occurs).   Consequently no *one fixed* symbol, or combination of symbols, should be considered as properly belonging to a merely *Working*-Variable; and as a general rule their squares are left blank.   Of course in this, as in all other cases where we mention a *general* rule, it is understood that many particular exceptions may be expedient.

In order that all the indications contained in the diagram may be completely understood, we shall now

* See the diagram on the opposite page.

Diagram belonging to Note D.

Number of Operations	Nature of Operations	Variables for Data ${}^1V_0$	${}^1V_1$	${}^1V_2$	${}^1V_3$	${}^1V_4$	${}^1V_5$	Working Variables ${}^0V_6$	${}^0V_7$	${}^0V_8$	${}^0V_9$	${}^0V_{10}$	${}^0V_{11}$	${}^0V_{12}$	${}^0V_{13}$	${}^0V_{14}$	Variables for Results ${}^0V_{15}$	${}^0V_{16}$
		$+$ o o o o	$+$ o o o o	$+$ o o o o	$+$ o o o o	$+$ o o o o	$+$ o o o o	$+$ o o o o o	$+$ o o o o o	$+$ o o o o o	$+$ o o o o o	$+$ o o o o o	$+$ o o o o o	$+$ o o o o	$+$ o o o o	$+$ o o o o	$+$ o o o o	$+$ o o o o
		m	n	d	m'	n'	d'											
1	×	m	n	d	m'	n'	d'	mn'										
2	×		n	d	m'	n'	d'		$m'n$									
3	×			d	m'	n'	d'	0		dn'								
4	×		0				d'		0		$d'n$							
5	×	0					d'			0		$d'm$						
6	×			0	0		0				0		dm'					
7	−							0	0			0		$(mn'-m'n)$				
8	−									0	0		0		$(dn'-d'n)$			
9	−											0	0		0	$(d'm-dm')$		
10	÷													$(mn'-m'n)$	0		$\dfrac{dn'-d'n}{mn'-m'n}=x$	
11	÷													0		0		$\dfrac{d'm-dm'}{mn'-m'n}=y$

$$\frac{dn' - d'n}{mn' - m'n} = x$$

$$\frac{d'm - dm'}{mn' - m'n} = y$$

explain two or three points, not hitherto touched on. When the value on any Variable is called into use, one of two consequences may be made to result. Either the value may *return* to the Variable after it has been used, in which case it is ready for a second use if needed; or the Variable may be made zero. (We are of course not considering a third case, of not unfrequent occurrence, in which the same Variable is destined to receive the *result* of the very operation which it has just supplied with a number.) Now the ordinary rule is, that the value *returns* to the Variable; unless it has been foreseen that no use for that value can recur, in which case zero is substituted. At the *end* of a calculation, therefore, every column ought as a general rule to be zero, excepting those for results. Thus it will be seen by the diagram, that when m, the value on V_0, is used for the second time by Operation 5, V_0 becomes 0, since m is not again needed; that similarly, when $(m n' - m' n)$, on V_{12}, is used for the third time by Operation 11, V_{12} becomes zero, since $(m n' - m' n)$ is not again needed· In order to provide for the one or the other of the courses above indicated, there are *two* varieties of the *Supplying* Variable-cards. One of these varieties has provisions which cause the number given off from any Variable to *return* to that Variable after doing its duty in the mill. The other variety has provisions which cause *zero* to be substituted on the Variable, for the number given off. These two varieties are distinguished, when needful, by the respective appellations of the *Retaining* Supply-cards and the *Zero* Supply-cards. We see that the *primary* office (see Note B.) of both these varieties of cards is the same; they only differ in their *secondary* office.

Every Variable thus has belonging to it *one* class of *Receiving* Variable-cards and *two* classes of *Supplying* Variable-cards. It is plain however that only the *one* or the *other* of these two latter classes can be used by any one Variable for *one* operation; never *both* simultaneously; their respective functions being mutually incompatible.

It should be understood that the Variable-cards are not placed in *immediate contiguity* with the columns. Each card is connected by means of wires with the column it is intended to act upon.

Our diagram ought in reality to be placed side by side with M. Menabrea's corresponding table, so as to be compared with it, line for line belonging to each operation. But it was unfortunately inconvenient to print them in this desirable form. The diagram is, in the main, merely another manner of indicating the various relations denoted in M. Menabrea's table. Each mode has some advantages and some disadvantages. Combined, they form a complete and accurate method of registering every step and sequence in all calculations performed by the engine.

No notice has yet been taken of the *upper* indices which are added to the left of each V in the diagram; an addition which we have also taken the liberty of making to the V's in M. Menabrea's tables of pages 14, 16, since it does not *alter* anything therein represented by him, but merely *adds* something to the previous indications of those tables. The *lower* indices are obviously indices of *locality* only, and are wholly independent of the operations performed or of the results obtained, their value continuing unchanged during the performance of calculations. The *upper* indices, however, are of a different nature. Their office is to indicate any *alteration* in the value which a Variable represents; and they are of course liable to changes during the processes of a calculation. Whenever a Variable has only zeros upon it, it is called ^0V; the moment a value appears on it (whether that value be placed there arbitrarily, or appears in the natural course of a calculation), it becomes ^1V. If this value gives place to another value, the Variable becomes ^2V, and so forth. Whenever a *value* again gives place to *zero*, the Variable again becomes ^0V, even if it have been ^nV the moment before. If a *value* then again be substituted, the Variable becomes ^{n+1}V (as it would have done if it had not passed through the intermediate ^0V); &c. &c. Just before any calculation is commenced, and after the data have been given, and everything adjusted and prepared for setting the mechanism in action, the upper indices of the Variables for data are all unity, and those for the Working and Result-variables are all zero. In this state the diagram represents them*.

---

* We recommend the reader to trace the successive substitutions backwards from (1.) to (4.), in Mons. Menabrea's Table. This he will easily do by means of the upper and lower indices, and it is interesting to observe how each V successively ramifies (so to speak) into two other V's in some other column of the Table, until at length the V's of the original data are arrived at.

There are several advantages in having a set of indices of this nature; but these advantages are perhaps hardly of a kind to be immediately perceived, unless by a mind somewhat accustomed to trace the successive steps by means of which the engine accomplishes its purposes. We have only space to mention in a general way, that the whole notation of the tables is made more consistent by these indices, for they are able to mark a *difference* in certain cases, where there would otherwise be an apparent *identity* confusing in its tendency. In such a case as $V_n = V_p + V_n$ there is more clearness and more consistency with the usual laws of algebraical notation, in being able to write $^{m+1}V_n = {}^qV_p + {}^mV_n$. It is also obvious that the indices furnish a powerful means of tracing back the derivation of any result; and of registering various circumstances concerning that *series of successive substitutions*, of which every *result* is in fact merely the final consequence; circumstances that may in certain cases involve relations which it is important to observe, either for purely analytical reasons, or for practically adapting the workings of the engine to their occurrence. The series of substitutions which lead to the equations of the diagram are as follow :—

$$
\begin{array}{cccc}
(1.) & (2.) & (3.) & (4.) \\
\end{array}
$$
$$
{}^1V^*_{16} = \frac{{}^1V_{14}}{{}^1V_{12}} = \frac{{}^1V_{10} - {}^1V_{11}}{{}^1V_6 - {}^1V_7} = \frac{{}^1V_0 \cdot {}^1V_5 - {}^1V_2 \cdot {}^1V_3}{{}^1V_0 \cdot {}^1V_4 - {}^1V_3 \cdot {}^1V_1} = \frac{d'\,m - d\,m'}{m\,n' - m'\,n}
$$

$$
\begin{array}{cccc}
(1.) & (2.) & (3.) & (4.) \\
\end{array}
$$
$$
{}^1V_{15} = \frac{{}^1V_{13}}{{}^1V_{12}} = \frac{{}^1V_8 - {}^1V_9}{{}^1V_6 - {}^1V_7} = \frac{{}^1V_2 \cdot {}^1V_4 - {}^1V_5 \cdot {}^1V_1}{{}^1V_0 \cdot {}^1V_4 - {}^1V_3 \cdot {}^1V_1} = \frac{d\,n' - d'\,n}{m\,n' - m'\,n}
$$

There are *three* successive substitutions for each of these equations. The formulæ (2.), (3.) and (4.) are *implicitly* contained in (1.), which latter we may consider as being in fact the *condensed* expression of any of the former. It will be observed that every succeeding substitution must contain *twice* as many V's as its predecessor. So that if a problem require n substitutions, the successive series of numbers for the V's in the whole of them will be $2, 4, 8, 16 \ldots 2^n$.

The substitutions in the preceding equations happen to be of little value towards illustrating the power and uses of the upper indices; for, owing to the nature of these particular equations, the indices are all unity throughout. We wish we had space to enter more fully into the relations which these indices would in many cases enable us to trace.

M. Menabrea incloses the three centre columns of his table under the general title *Variable-cards*. The V's however in reality all represent the actual *Variable-columns* of the engine, and not the cards that belong to them. Still the title is a very just one, since it is through the special action of certain Variable-cards (when *combined* with the more generalized agency of the Operation-cards) that every one of the particular relations he has indicated under that title is brought about.

Suppose we wish to ascertain how often any *one* quantity, or combination of quantities, is brought into use during a calculation. We easily ascertain *this*, from the inspection of any vertical column or columns of the diagram in which that quantity may appear. Thus, in the present case, we see that all the data, and all the intermediate results likewise, are used twice, excepting $(m\,n' - m'\,n)$, which is used three times.

The *order* in which it is possible to perform the operations for the present example, enables us to effect all the eleven operations of which it consists with only *three Operation cards*; because the problem is of such a nature that it admits of each *class* of operations being performed in a group together; all the multiplications one after another, all the subtractions one after another, &c. The operations are $\left\{ 6(\times),\, 3(-),\, 2(\div) \right\}$.

Since the very definition of an operation implies that there must be *two* numbers to act upon, there are of course *two Supplying* Variable-cards necessarily brought into action for every operation, in order to furnish the two proper numbers. (See Note B.) Also, since every operation must produce a *result*, which must be placed *somewhere*, each operation entails the action of a *Receiving* Variable-card, to indicate the proper locality for the result. Therefore, at least three times as many Variable-cards as there are *operations* (not *Operation-cards*, for these, as we have just seen, are by no means always as numerous as the *operations*) are brought into use in every calculation. Indeed, under certain contingencies, a still larger proportion is requisite; such, for example, would probably be the case when the same result has to appear on more than one Variable simultaneously

* See note, page 34.

(which is not unfrequently a provision necessary for subsequent purposes in a calculation), and in some other cases which we shall not here specify. We see therefore that a great disproportion exists between the amount of *Variable* and of *Operation*-cards requisite for the working of even the simplest calculation.

All calculations do not admit, like this one, of the operations of the same nature being performed in groups together. Probably very few do so without exceptions occurring in one or other stage of the progress; and some would not admit it at all. The *order* in which the operations shall be performed in every particular case is a very interesting and curious question, on which our space does not permit us fully to enter. In almost every computation a great *variety* of arrangements for the succession of the processes is possible, and various considerations must influence the selection amongst them for the purposes of a Calculating Engine. One essential object is to choose that arrangement which shall tend to reduce to a minimum the *time* necessary for completing the calculation.

It must be evident how multifarious and how mutually complicated are the considerations which the workings of such an engine involve. There are frequently several distinct *sets of effects* going on simultaneously; all in a manner independent of each other, and yet to a greater or less degree exercising a mutual influence. To adjust each to every other, and indeed even to perceive and trace them out with perfect correctness and success, entails difficulties whose nature partakes to a certain extent of those involved in every question where *conditions* are very numerous and inter-complicated; such as for instance the estimation of the mutual relations amongst *statistical* phænomena, and of those involved in many other classes of facts.   A. A. L.

Note E.—Page 16.

This example has evidently been chosen on account of its brevity and simplicity, with a view merely to explain the *manner* in which the engine would proceed in the case of an *analytical calculation containing variables*, rather than to illustrate the *extent of its powers* to solve cases of a difficult and complex nature. The equations of page 12 are in fact a more complicated problem than the present one.

We have not subjoined any diagram of its development for this new example, as we did for the former one, because this is unnecessary after the full application already made of those diagrams to the illustration of M. Menabrea's excellent tables.

It may be remarked that a slight discrepancy exists between the formulæ

$$(a + b\,x^1)$$
$$(A + B\cos^1 x)$$

given in the Memoir as the *data* for calculation, and the *results* of the calculation as developed in the last division of the table which accompanies it. To agree perfectly with this latter, the data should have been given as

$$(a\,x^0 + b\,x^1)$$
$$(A\cos^0 x + B\cos^1 x).$$

The following is a more complicated example of the manner in which the engine would compute a trigonometrical function containing variables. To multiply

$$A + A_1\cos\theta + A_2\cos 2\theta + A_3\cos 3\theta + \ldots$$

by $$B + B_1\cos\theta.$$

Let the resulting products be represented under the general form

$$C_0 + C_1\cos\theta + C_2\cos 2\theta + C_3\cos 3\theta + \quad \ldots \ldots \ldots \quad (1.)$$

This trigonometrical series is not only in itself very appropriate for illustrating the processes of the engine, but is likewise of much practical interest from its frequent use in astronomical computations. Before proceeding further with it, we shall point out that there are three very distinct classes of ways in which it may be desired to deduce numerical values from any analytical formula.

First. We may wish to find the collective numerical value of the *whole formula*, without any reference to the quantities of which that formula is a function, or to the particular mode of their combination and distribution, of which the formula is the result and representative. Values of this kind are of a strictly arithmetical nature in the most limited sense of the term, and retain no trace whatever of the processes through which they have been deduced. In fact, any one such numerical value may have been attained from an *infinite variety* of data, or of problems. The values for x and y in the two equations (see Note D.) come under this class of numerical results.

Secondly. We may propose to compute the collective numerical value of *each term* of a formula, or of a series, and to keep these results separate. The engine must in such a case appropriate as many columns to *results* as there are terms to compute.

Thirdly. It may be desired to compute the numerical value of various *subdivisions of each term*, and to keep all these results separate. It may be required, for instance, to compute each coefficient separately from its variable, in which particular case the engine must appropriate *two* result-columns to *every term that contains both a variable and coefficient.*

There are many ways in which it may be desired in special cases to distribute and keep separate the numerical values of different parts of an algebraical formula: and the power of effecting such distributions to any extent is essential to the *algebraical* character of the Analytical Engine. Many persons who are not conversant with mathematical studies, imagine that because the business of the engine is to give its results in *numerical notation*, the *nature of its processes* must consequently be *arithmetical* and *numerical*, rather than *algebraical* and *analytical*. This is an error. The engine can arrange and combine its numerical quantities exactly as if they were *letters* or any other *general* symbols; and in fact it might bring out its results in algebraical *notation*, were provisions made accordingly. It might develope three sets of results simultaneously, viz. *symbolic* results (as already alluded to in Notes A. and B.); *numerical* results (its chief and primary object); and *algebraical* results in *literal* notation. This latter however has not been deemed a necessary or desirable addition to its powers, partly because the necessary arrangements for effecting it would increase the complexity and extent of the mechanism to a degree that would not be commensurate with the advantages, where the main object of the invention is to translate into *numerical* language general formulæ of analysis already known to us, or whose laws of formation are known to us. But it would be a mistake to suppose that because its *results* are given in the *notation* of a more restricted science, its *processes* are therefore restricted to those of that science. The object of the engine is in fact to give the *utmost practical efficiency* to the resources of *numerical interpretations* of the higher science of analysis, while it uses the processes and combinations of this latter.

To return to the trigonometrical series. We shall only consider the first four terms of the factor $(A + A_1 \cos \theta + \&c.)$, since this will be sufficient to show the method. We propose to obtain separately the numerical value of *each coefficient* C_0, C_1, &c. of (1.). The direct multiplication of the two factors gives

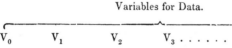

$$\left.\begin{array}{l} B\,A + B\,A_1 \cos\theta + B\,A_2 \quad \cos 2\theta + B\,A_3 \quad \cos 3\theta + \ldots\ldots\ldots \\ + B_1\,A \cos\theta + B_1\,A_1 \cos\theta \,.\, \cos\theta + B_1\,A_2 \cos 2\theta\,.\,\cos\theta + B_1\,A_3 \cos 3\theta\,.\,\cos\theta \end{array}\right\} \quad \cdots \cdots \quad (2.)$$

a result which would stand thus on the engine :—

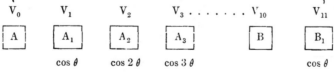

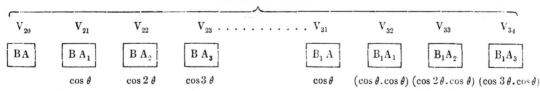

The variable belonging to each coefficient is written below it, as we have done in the diagram, by way of memorandum. The only further reduction which is at first apparently possible in the preceding result, would be the addition of V_{21} to V_{31} (in which case $B_1 A$ should be effaced from V_{31}). The whole operations from the beginning would then be—

First Series of Operations.	Second Series of Operations.	Third Series, which contains only one (final) operation.
$^1V_{10} \times {}^1V_0 = {}^1V_{20}$	$^1V_{11} \times {}^1V_0 = {}^1V_{31}$	$^1V_{21} + {}^1V_{31} = {}^2V_{21}$, and
$^1V_{10} \times {}^1V_1 = {}^1V_{21}$	$^1V_{11} \times {}^1V_1 = {}^1V_{32}$	V_{31} becomes $= 0$.
$^1V_{10} \times {}^1V_2 = {}^1V_{22}$	$^1V_{11} \times {}^1V_2 = {}^1V_{33}$	
$^1V_{10} \times {}^1V_3 = {}^1V_{23}$	$^1V_{11} \times {}^1V_3 = {}^1V_{34}$	

We do not enter into the same detail of *every* step of the processes as in the examples of Notes D. and G., thinking it unnecessary and tedious to do so. The reader will remember the meaning and use of the upper and lower indices, &c., as before explained.

To proceed: we know that

$$\cos n\theta \cdot \cos\theta = \tfrac{1}{2}\cos\overline{n+1}\,\theta + \tfrac{1}{2}\overline{n-1}\cdot\theta \quad \cdot \quad \cdot \quad \cdot \quad \cdot \quad \cdot \quad \cdot \quad \cdot \quad (3.)$$

Consequently, a slight examination of the second line of (2.) will show that by making the proper substitutions, (2.) will become

$\begin{array}{l} \text{B A} \\ \\ +\tfrac{1}{2}\,\text{B}_1\,\text{A}_1 \end{array}$	$\begin{array}{l} +\text{B A}_1 \quad \cdot \cos\theta \\ +\text{B}_1\,\text{A} \quad \cdot \cos\theta \\ \\ +\tfrac{1}{2}\,\text{B}_1\,\text{A}_2 \cdot \cos\theta \end{array}$	$\begin{array}{l} +\text{B A}_2 \quad \cdot \cos 2\theta \\ \\ +\tfrac{1}{2}\,\text{B}_1\,\text{A}_1 \cdot \cos 2\theta \\ \\ +\tfrac{1}{2}\,\text{B}_1\,\text{A}_3 \cdot \cos 2\theta \end{array}$	$\begin{array}{l} +\text{B A}_3 \quad \cdot \cos 3\theta \\ \\ \\ +\tfrac{1}{2}\,\text{B}_1\,\text{A}_2 \cdot \cos 3\theta \end{array}$	$+\tfrac{1}{2}\,\text{B}_1\,\text{A}_3 \cdot \cos 4\theta$
C_0	C_1	C_2	C_3	C_4

These coefficients should respectively appear on

V_{20}	V_{21}	V_{22}	V_{23}	$V_{24}.$

We shall perceive, if we inspect the particular arrangement of the results in (2.) on the Result-columns as represented in the diagram, that, in order to effect this transformation, each successive coefficient upon V_{32}, V_{33}, &c. (beginning with V_{32}), must through means of proper cards be divided by *two* [*]; and that one of the halves thus obtained must be added to the coefficient on the Variable which precedes it by ten columns, and the other half to the coefficient on the Variable which precedes it by twelve columns; V_{32}, V_{34}, &c. themselves becoming zeros during the process.

This series of operations may be thus expressed:—

Fourth Series.

$$\begin{cases} {}^1V_{32} \div 2 + {}^1V_{22} = {}^2V_{22} = \text{B A}_2 + \tfrac{1}{2}\,\text{B}_1\,\text{A}_1 \\ {}^1V_{32} \div 2 + {}^1V_{20} = {}^2V_{20} = \text{B A} + \tfrac{1}{2}\,\text{B}_1\,\text{A}_1 \cdot \cdots \cdots = C_0 \end{cases}$$
$$\begin{cases} {}^1V_{33} \div 2 + {}^1V_{23} = {}^2V_{23} = \text{B A}_3 + \tfrac{1}{2}\,\text{B}_1\,\text{A}_2 \cdot \cdots \cdots = C_3\dagger \\ {}^1V_{33} \div 2 + {}^2V_{21} = {}^3V_{21} = \text{B A}_1 + \text{B}_1\,\text{A} + \tfrac{1}{2}\,\text{B}_1\,\text{A}_2 = C_1 \end{cases}$$
$$\begin{cases} {}^1V_{34} \div 2 + {}^0V_{24} = {}^1V_{24} = \tfrac{1}{2}\,\text{B}_1\,\text{A}_3 \cdots \cdots \cdots = C_4 \\ {}^1V_{34} \div 2 + {}^2V_{22} = {}^3V_{22} = \text{B A}_2 + \tfrac{1}{2}\,\text{B}_1\,\text{A}_1 + \tfrac{1}{2}\,\text{B}_1\,\text{A}_3 = C_2. \end{cases}$$

The calculation of the coefficients C_0, C_1, &c. of (1.) would now be completed, and they would stand ranged in order on V_{20}, V_{21}, &c. It will be remarked, that from the moment the fourth series of operations is ordered, the Variables V_{31}, V_{32}, &c. cease to be *Result*-Variables, and become mere *Working*-Variables.

The substitution made by the engine of the processes in the second side of (3.) for those in the first side is an excellent illustration of the manner in which we may arbitrarily order it to substitute any function, number, or process, at pleasure, for any other function, number or process, on the occurrence of a specified contingency.

We will now suppose that we desire to go a step further, and to obtain the numerical value of each *complete*

---

[*] This division would be managed by ordering the number 2 to appear on any separate new column which should be conveniently situated for the purpose, and then directing this column (which is in the strictest sense a *Working*-Variable) to divide itself successively with V_{32}, V_{33}, &c.

† It should be observed, that were the rest of the factor $(\text{A} + \text{A}\cos\theta + \&c.)$ taken into account, instead of *four* terms only, C_3 would have the additional term $\tfrac{1}{2}\,\text{B}_1\,\text{A}_4$; and C_4 the two additional terms, B A_4, $\tfrac{1}{2}\,\text{B}_1\,\text{A}_5$. This would indeed have been the case had even *six* terms been multiplied.

term of the product (1.); that is, of each *coefficient and variable united*, which for the $(n + 1)$th term would be $C_n . \cos n\theta$.

We must for this purpose place the variables themselves on another set of columns, V_{41}, V_{42}, &c., and then order their successive multiplication by V_{21}, V_{22}, &c., each for each. There would thus be a final series of operations as follows:—

Fifth and Final Series of Operations.

$$^2V_{20} \times {}^0V_{40} = {}^1V_{40}$$
$$^3V_{21} \times {}^0V_{41} = {}^1V_{41}$$
$$^3V_{22} \times {}^0V_{42} = {}^1V_{42}$$
$$^2V_{23} \times {}^0V_{43} = {}^1V_{43}$$
$$^1V_{24} \times {}^0V_{44} = {}^1V_{44}$$

(N.B. that V_{40} being intended to receive the coefficient on V_{20} which has *no* variable, will only have $\cos 0\,\theta$ ($= 1$) inscribed on it, preparatory to commencing the fifth series of operations.)

From the moment that the fifth and final series of operations is ordered, the Variables V_{20}, V_{21}, &c. then in their turn cease to be *Result*-Variables and become mere *Working*-Variables; V_{40}, V_{41}, &c. being now the recipients of the ultimate results.

We should observe, that if the variables $\cos\theta$, $\cos 2\theta$, $\cos 3\theta$, &c. are furnished, they would be placed directly upon V_{41}, V_{42}, &c., like any other data. If not, a separate computation might be entered upon in a separate part of the engine, in order to calculate them, and place them on V_{41}, &c.

We have now explained how the engine might compute (1.) in the most direct manner, supposing we knew nothing about the *general* term of the resulting series. But the engine would in reality set to work very differently, whenever (as in this case) we *do* know the law for the general term.

The first two terms of (1.) are

$$(\mathrm{B\,A} + \tfrac{1}{2}\mathrm{B_1\,A_1}) + (\overline{\mathrm{B\,A_1} + \mathrm{B_1\,A} + \tfrac{1}{2}\mathrm{B_1\,A_2}} . \cos\theta) \quad . \quad . \quad . \quad . \quad . \quad . \quad . \quad . \quad (4.)$$

and the general term for all after these is

$$(\mathrm{B\,A}_n + \tfrac{1}{2}\mathrm{B_1} . \overline{\mathrm{A}_{n-1} + \mathrm{A}_{n+2}}) \cos n\theta \quad . \quad . \quad . \quad . \quad . \quad . \quad . \quad . \quad . \quad . \quad . \quad (5.)$$

which is the coefficient of the $(n + 1)$th term. The engine would calculate the first two terms by means of a separate set of suitable Operation-cards, and would then need another set for the third term; which last set of Operation-cards would calculate all the succeeding terms *ad infinitum*, merely requiring certain new Variable-cards for each term to direct the operations to act on the proper columns. The following would be the successive sets of operations for computing the coefficients of $n + 2$ terms:—

$$(\times, \times, \div, +), (\times, \times, \times, \div, +, +), n(\times, +, \times, \div, +).$$

Or we might represent them as follows, according to the numerical order of the operations:—

$$(1, 2 \dots 4), (5, 6 \dots 10), n(11, 12 \dots 15).$$

The brackets, it should be understood, point out the relation in which the operations may be *grouped*, while the comma marks *succession*. The symbol $+$ might be used for this latter purpose, but this would be liable to produce confusion, as $+$ is also necessarily used to represent one class of the actual operations which are the subject of that succession. In accordance with this meaning attached to the comma, care must be taken when any one group of operations recurs more than once, as is represented above by $n(11 \dots 15)$, not to insert a comma after the number or letter prefixed to that group. $n, (11 \dots 15)$ would stand for *an operation n, followed by the group of operations* $(11 \dots 15)$; instead of denoting *the number of groups which are to follow each other*.

Wherever a *general term* exists, there will be a *recurring group* of operations, as in the above example. Both for brevity and for distinctness, a *recurring group* is called a *cycle*. A *cycle* of operations, then, must be understood to signify any *set of operations* which is repeated *more than once*. It is equally a *cycle*, whether it be repeated *twice* only, or an indefinite number of times; for it is the fact of a *repetition occurring at all* that constitutes it such. In many cases of analysis there is a *recurring group* of one or more *cycles*; that is, a *cycle of a cycle*, or a *cycle of cycles*. For instance: suppose we wish to divide a series by a series,

(1.)
$$\frac{a + b\,x + c\,x^2 + \,\cdots}{a' + b'\,x + c'\,x^2 + \,\cdots},$$

it being required that the result shall be developed, like the dividend and the divisor, in successive powers of x. A little consideration of (1.), and of the steps through which algebraical division is effected, will show that (if the denominator be supposed to consist of p terms) the first partial quotient will be completed by the following operations:—

(2.)
$$\left\{(\div),\, p\,(\times,\, -)\right\} \quad \text{or} \quad \left\{(1),\, p\,(2,\, 3)\right\},$$

that the second partial quotient will be completed by an exactly similar set of operations, which acts on the remainder obtained by the first set, instead of on the original dividend. The whole of the processes therefore that have been gone through, by the time the *second* partial quotient has been obtained, will be,—

(3.)
$$2\left\{(\div),\, p\,(\times,\, -)\right\} \quad \text{or} \quad 2\left\{(1),\, p\,(2,\, 3)\right\},$$

which is a cycle that includes a cycle, or a cycle of the second order. The operations for the *complete* division, supposing we propose to obtain n terms of the series constituting the quotient, will be,—

(4.)
$$n\left\{(\div),\, p\,(\times,\, -)\right\} \quad \text{or} \quad n\left\{(1),\, p\,(2,\, 3)\right\}.$$

It is of course to be remembered that the process of algebraical division in reality continues *ad infinitum*, except in the few exceptional cases which admit of an exact quotient being obtained. The number n in the formula (4.) is always that of the number of terms we propose to ourselves to obtain; and the nth partial quotient is the coefficient of the $(n-1)$th power of x.

There are some cases which entail *cycles of cycles of cycles*, to an indefinite extent. Such cases are usually very complicated, and they are of extreme interest when considered with reference to the engine. The algebraical development in a series of the nth function of any given function is of this nature. Let it be proosed to obtain the nth function of

(5.)
$$\phi\,(a,\, b,\, c\,\cdots\cdots\, x),\ x \text{ being the variable.}$$

We should premise, that we suppose the reader to understand what is meant by an nth function. We suppose him likewise to comprehend distinctly the difference between developing *an nth function algebraically*, and merely *calculating an nth function arithmetically*. If he does not, the following will be by no means very intelligible; but we have not space to give any preliminary explanations. To proceed: the law, according to which the successive functions of (5.) are to be developed, must of course first be fixed on. This law may be of very various kinds. We may propose to obtain our results in successive *powers* of x, in which case the general form would be

$$C + C_1\,x + C_2\,x^2 + \&c.;$$

or in successive powers of n itself, the index of the function we are ultimately to obtain, in which case the general form would be

$$C + C_1\,n + C_2\,n^2 + \&c.,$$

and x would only enter in the coefficients. Again, other functions of x or of n instead of *powers* might be selected. It might be in addition proposed, that the coefficients themselves should be arranged according to given functions of a certain quantity. Another mode would be to make equations arbitrarily amongst the coefficients only, in which case the several functions, according to either of which it might be possible to develope the nth function of (5.), would have to be determined from the combined consideration of these equations and of (5.) itself.

The *algebraical* nature of the engine (so strongly insisted on in a previous part of this Note) would enable it to follow out any of these various modes indifferently; just as we recently showed that it can distribute and separate the numerical results of any one prescribed series of processes, in a perfectly arbitrary manner. Were it otherwise, the engine could merely *compute the arithmetical nth function*, a result which, like any other purely arithmetical results, would be simply a collective number, bearing no traces of the data or the processes which had led to it.

Secondly, the *law* of development for the nth function being selected, the next step would obviously be to develope (5.) itself, according to this law. This result would be the first function, and would be obtained by a determinate series of processes. These in most cases would include amongst them one or more *cycles* of operations.

The third step (which would consist of the various processes necessary for effecting the actual substitution of the series constituting the *first function*, for the *variable* itself) might proceed in either of two ways. It might make the substitution either wherever x occurs in the original (5.), or it might similarly make it wherever x occurs in the first function itself which is the equivalent of (5.). In some cases the former mode might be best, and in others the latter.

Whichever is adopted, it must be understood that the result is to appear arranged in a series following the law originally prescribed for the development of the nth function. This result constitutes the second function; with which we are to proceed exactly as we did with the first function, in order to obtain the third function, and so on, $n - 1$ times, to obtain the nth function. We easily perceive that since every successive function is arranged in a series *following the same law*, there would (after the *first* function is obtained) be a *cycle of a cycle of a cycle*, &c. of operations*, one, two, three, up to $n - 1$ times, in order to get the nth function. We say, *after the first function is obtained*, because (for reasons on which we cannot here enter) the *first* function might in many cases be developed through a set of processes peculiar to itself, and not recurring for the remaining functions.

We have given but a very slight sketch of the principal *general* steps which would be requisite for obtaining an nth function of such a formula as (5.). The question is so exceedingly complicated, that perhaps few persons can be expected to follow, to their own satisfaction, so brief and general a statement as we are here restricted to on this subject. Still it is a very important case as regards the engine, and suggests ideas peculiar to itself, which we should regret to pass wholly without allusion. Nothing could be more interesting than to follow out, in every detail, the solution by the engine of such a case as the above; but the time, space and labour this would necessitate, could only suit a very extensive work.

To return to the subject of *cycles* of operations: some of the notation of the integral calculus lends itself very aptly to express them: (2.) might be thus written:—

(6.) $$(\div), \ \Sigma(+1)^p \ (\times, -) \ \text{or} \ (1), \ \Sigma(+1)^p \ (2, 3),$$

where p stands for the variable; $(+1)^p$ for the function of the variable, that is, for $\phi\, p$; and the limits are from 1 to p, or from 0 to $p - 1$, each increment being equal to unity. Similarly, (4.) would be,—

(7.) $$\Sigma(+1)^n \left\{ (\div), \ \Sigma(+1)^p \ (\times, -) \right\}$$

the limits of n being from 1 to n, or from 0 to $n - 1$,

(8.) $$\text{or} \ \Sigma(+1)^n \left\{ (1), \ \Sigma(+1)^p \ (2, 3) \right\}.$$

Perhaps it may be thought that this notation is merely a circuitous way of expressing what was more simply and as effectually expressed before; and, in the above example, there may be some truth in this. But there is another description of cycles which *can* only effectually be expressed, in a condensed form, by the preceding notation. We shall call them *varying cycles*. They are of frequent occurrence, and include successive cycles of operations of the following nature:—

(9.) $$p\,(1, 2, ..m), \ \overline{p-1}\,(1, 2...m), \ \overline{p-2}\,(1, 2..m)...\overline{p-n}\,(1, 2..m),$$

where each cycle contains the same group of operations, but in which the number of repetitions of the group varies according to a fixed rate, with every cycle. (9.) can be well expressed as follows:—

(10.) $$\Sigma p\,(1, 2, ..\ m), \ \text{the limits of } p \text{ being from } p - n \text{ to } p.$$

---

* A cycle that includes n other cycles, successively *contained one within another*, is called a cycle of the $n + 1$th order. A cycle may simply *include* many other cycles, and yet only be of the second order. If a series follows a certain law for a certain number of terms, and then another law for another number of terms, there will be a cycle of operations for every new law; but these cycles will not be *contained one within another*,—they merely *follow each other*. Therefore their number may be infinite without influencing the *order* of a cycle that includes a repetition of such a series.

Independent of the intrinsic advantages which we thus perceive to result in certain cases from this use of the notation of the integral calculus, there are likewise considerations which make it interesting, from the connections and relations involved in this new application. It has been observed in some of the former Notes, that the processes used in analysis form a logical system of much higher generality than the applications to number merely. Thus, when we read over any algebraical formula, considering it exclusively with reference to the processes of the engine, and putting aside for the moment its abstract signification as to the relations of quantity, the symbols +, ×, &c. in reality represent (as their immediate and proximate effect, when the formula is applied to the engine) that a certain prism which is a part of the mechanism (see Note C.) turns a new face, and thus presents a new card to act on the bundles of levers of the engine; the new card being perforated with holes, which are arranged according to the peculiarities of the operation of addition, or of multiplication, &c. Again, the *numbers* in the preceding formula (8.), each of them really represents one of these very pieces of card that are hung over the prism.

Now in the use made in the formulæ (7.), (8.) and (10.), of the notation of the integral calculus, we have glimpses of a similar new application of the language of the *higher* mathematics. Σ, in reality, here indicates that when a certain number of cards have acted in succession, the prism over which they revolve must *rotate backwards*, so as to bring those cards into their former position; and the limits 1 to n, 1 to p, &c., regulate how often this backward rotation is to be repeated.   A. A. L.

Note F.—Page 18.

There is in existence a beautiful woven portrait of Jacquard, in the fabrication of which 24,000 cards were required.

The power of *repeating* the cards, alluded to by M. Menabrea in page 16, and more fully explained in Note C., reduces to an immense extent the number of cards required. It is obvious that this mechanical improvement is especially applicable wherever *cycles* occur in the mathematical operations, and that, in preparing data for calculations by the engine, it is desirable to arrange the order and combination of the processes with a view to obtain them as much as possible *symmetrically* and in cycles, in order that the mechanical advantages of the *backing* system may be applied to the utmost. It is here interesting to observe the manner in which the value of an *analytical* resource is *met* and *enhanced* by an ingenious *mechanical* contrivance. We see in it an instance of one of those mutual *adjustments* between the purely mathematical and the mechanical departments, mentioned in Note A. as being a main and essential condition of success in the invention of a calculating engine. The nature of the resources afforded by such adjustments would be of two principal kinds. In some cases, a difficulty (perhaps in itself insurmountable) in the one department would be overcome by facilities in the other; and sometimes (as in the present case) a strong point in the one would be rendered still stronger and more available by combination with a corresponding strong point in the other.

As a mere example of the degree to which the combined systems of cycles and of backing can diminish the *number* of cards requisite, we shall choose a case which places it in strong evidence, and which has likewise the advantage of being a perfectly different *kind* of problem from those that are mentioned in any of the other Notes. Suppose it be required to eliminate nine variables from ten simple equations of the form—

$$a\,x_0 + b\,x_1 + c\,x_2 + d\,x_3 + \ldots \ldots = p \qquad (1.)$$
$$a^1 x_0 + b^1 x_1 + c^1 x_2 + d^1 x_3 + \ldots \ldots = p^1 \qquad (2.)$$
$$\&c. \qquad \&c. \qquad \&c. \qquad \&c.$$

We should explain, before proceeding, that it is not our object to consider this problem with reference to the actual arrangement of the data on the Variables of the engine, but simply as an abstract question of the *nature* and *number* of the *operations* required to be performed during its complete solution.

The first step would be the elimination of the first unknown quantity x_0 between the first two equations. This would be obtained by the form—

$$(a^1 a - a\,a^1)\,x_0 + (a^1 b - a\,b^1)\,x_1 + (a^1 c - a\,c^1)\,x_2 +$$
$$+ (a^1 d - a\,d^1)\,x_3 + \ldots \ldots \ldots \ldots \ldots \ldots = a^1 p - a\,p^1.$$

for which the operations $10\,(\times,\times,-)$ would be needed. The second step would be the elimination of x_0 between the second and third equations, for which the operations would be precisely the same. We should then have had altogether the following operations :—

$$10\,(\times,\times,-),10\,(\times,\times,-),=20\,(\times,\times,-).$$

Continuing in the same manner, the total number of operations for the complete elimination of x_0 between all the successive pairs of equations would be—

$$9.10\,(\times,\times,-)=90\,(\times,\times,-).$$

We should then be left with nine simple equations of nine variables from which to eliminate the next variable x_1, for which the total of the processes would be—

$$8.9\,(\times,\times,-)=72\,(\times,\times,-).$$

We should then be left with eight simple equations of eight variables from which to eliminate x_2, for which the processes would be—

$$7.8\,(\times,\times,-)=56\,(\times,\times,-),$$

and so on. The total operations for the elimination of all the variables would thus be—

$$9.10+8.9+7.8+6.7.+5.6+4.5+3.4+2.3+1.2=330.$$

So that *three* Operation-cards would perform the office of 330 such cards.

If we take n simple equations containing $n-1$ variables, n being a number unlimited in magnitude, the case becomes still more obvious, as the same three cards might then take the place of thousands or millions of cards.

We shall now draw further attention to the fact, already noticed, of its being by no means necessary that a formula proposed for solution should ever have been actually worked out, as a condition for enabling the engine to solve it. Provided we know the *series of operations* to be gone through, that is sufficient. In the foregoing instance this will be obvious enough on a slight consideration. And it is a circumstance which deserves particular notice, since herein may reside a latent value of such an engine almost incalculable in its possible ultimate results. We already know that there are functions whose numerical value it is of importance for the purposes both of abstract and of practical science to ascertain, but whose determination requires processes so lengthy and so complicated, that, although it is possible to arrive at them through great expenditure of time, labour and money, it is yet on these accounts practically almost unattainable; and we can conceive there being some results which it may be *absolutely impossible* in practice to attain with any accuracy, and whose precise determination it may prove highly important for some of the future wants of science, in its manifold, complicated and rapidly-developing fields of inquiry, to arrive at.

Without, however, stepping into the region of conjecture, we will mention a particular problem which occurs to us at this moment as being an apt illustration of the use to which such an engine may be turned for determining that which human brains find it difficult or impossible to work out unerringly. In the solution of the famous problem of the Three Bodies, there are, out of about 295 coefficients of lunar perturbations given by M. Clausen (Astro[e]. Nachrichten, No. 406) as the result of the calculations by Burg, of two by Damoiseau, and of one by Burckhardt, fourteen coefficients that differ in the nature of their algebraic sign ; and out of the remainder there are only 101 (or about one-third) that agree precisely both in signs and in amount. These discordances, which are generally small in individual magnitude, may arise either from an erroneous determination of the abstract coefficients in the development of the problem, or from discrepancies in the data deduced from observation, or from both causes combined. The former is the most ordinary source of error in astronomical computations, and this the engine would entirely obviate.

We might even invent laws for series or formulæ in an arbitrary manner, and set the engine to work upon them, and thus deduce numerical results which we might not otherwise have thought of obtaining ; but this would hardly perhaps in any instance be productive of any great practical utility, or calculated to rank higher than as a kind of philosophical amusement.

A. A. L.

NOTE G.—Page 19.

It is desirable to guard against the possibility of exaggerated ideas that might arise as to the powers of the Analytical Engine. In considering any new subject, there is frequently a tendency, first, to *overrate* what we find to be already interesting or remarkable; and, secondly, by a sort of natural reaction, to *undervalue* the true state of the case, when we do discover that our notions have surpassed those that were really tenable.

The Analytical Engine has no pretensions whatever to *originate* anything. It can do whatever we *know how to order it* to perform. It can *follow* analysis; but it has no power of *anticipating* any analytical relations or truths. Its province is to assist us in making *available* what we are already acquainted with. This it is calculated to effect primarily and chiefly of course, through its executive faculties; but it is likely to exert an *indirect* and reciprocal influence on science itself in another manner. For, in so distributing and combining the truths and the formulæ of analysis, that they may become most easily and rapidly amenable to the mechanical combinations of the engine, the relations and the nature of many subjects in that science are necessarily thrown into new lights, and more profoundly investigated. This is a decidedly indirect, and a somewhat *speculative*, consequence of such an invention. It is however pretty evident, on general principles, that in devising for mathematical truths a new form in which to record and throw themselves out for actual use, views are likely to be induced, which should again react on the more theoretical phase of the subject. There are in all extensions of human power, or additions to human knowledge, various *collateral* influences, besides the main and primary object attained.

To return to the executive faculties of this engine: the question must arise in every mind, are they *really* even able to *follow* analysis in its whole extent? No reply, entirely satisfactory to all minds, can be given to this query, excepting the actual existence of the engine, and actual experience of its practical results. We will however sum up for each reader's consideration the chief elements with which the engine works:—

1. It performs the four operations of simple arithmetic upon any numbers whatever.

2. By means of certain artifices and arrangements (upon which we cannot enter within the restricted space which such a publication as the present may admit of), there is no limit either to the *magnitude* of the *numbers* used, or to the *number* of *quantities* (either variables or constants) that may be employed.

3. It can combine these numbers and these quantities either algebraically or arithmetically, in relations unlimited as to variety, extent, or complexity.

4. It uses algebraic *signs* according to their proper laws, and developes the logical consequences of these laws.

5. It can arbitrarily substitute any formula for any other; effacing the first from the columns on which it is represented, and making the second appear in its stead.

6. It can provide for singular values. Its power of doing this is referred to in M. Menabrea's memoir, page 17, where he mentions the passage of values through zero and infinity. The practicability of causing it arbitrarily to change its processes at any moment, on the occurrence of any specified contingency (of which its substitution of $(\frac{1}{2}\cos.\overline{n+1}\,\theta + \frac{1}{2}\cos.\overline{n-1}\,\theta)$ for $(\cos.n\theta.\cos.\theta)$, explained in Note E., is in some degree an illustration), at once secures this point.

The subject of integration and of differentiation demands some notice. The engine can effect these processes in either of two ways:—

First. We may order it, by means of the Operation and of the Variable-cards, to go through the various steps by which the required *limit* can be worked out for whatever function is under consideration.

Secondly. It may (if we know the form of the limit for the function in question) effect the integration or differentiation by direct* substitution. We remarked in Note B., that any *set* of columns on which numbers are

---

* The engine cannot of course compute limits for perfectly *simple* and *uncompounded* functions, except in this manner. It is obvious that it has no power of representing or of manipulating with any but *finite* increments or decrements; and consequently that wherever the computation of limits (or of any other functions) depends upon the *direct* introduction of quantities which either increase or decrease *indefinitely*, we are absolutely beyond the sphere of its powers. Its nature and arrangements are remarkably adapted for taking into account all *finite* increments or decrements (however small or large), and for developing the true and logical modifications of form or value dependent upon differences of this nature.

inscribed, represents merely a *general* function of the several quantities, until the special function have been impressed by means of the Operation and Variable-cards. Consequently, if instead of requiring the value of the function, we require that of its integral, or of its differential coefficient, we have merely to order whatever particular combination of the ingredient quantities may constitute that integral or that coefficient. In $a\,x^n$, for instance, instead of the quantities

$$\begin{array}{cccc} \mathrm{V_0} & \mathrm{V_1} & \mathrm{V_2} & \mathrm{V_3} \\ \boxed{a} & \boxed{n} & \boxed{x} & \boxed{a\,x^n} \end{array}$$

$$\underbrace{\phantom{\mathrm{V_0}\quad\mathrm{V_1}\quad\mathrm{V_2}}}_{a\,x^n}$$

being ordered to appear on $\mathrm{V_3}$ in the combination $a\,x^n$, they would be ordered to appear in that of

$$a\,n\,x^{n-1}.$$

They would then stand thus:—

$$\begin{array}{cccc} \mathrm{V_0} & \mathrm{V_1} & \mathrm{V_2} & \mathrm{V_3} \\ \boxed{a} & \boxed{n} & \boxed{x} & \boxed{a\,n\,x^{n-1}} \end{array}$$

$$\underbrace{\phantom{\mathrm{V_0}\quad\mathrm{V_1}\quad\mathrm{V_2}}}_{a\,n\,x^{n-1}}$$

Similarly, we might have $\dfrac{a}{n+1}\,x^{n+1}$, the integral of $a\,x^n$.

An interesting example for following out the processes of the engine would be such a form as

$$\int \frac{x^n\,d\,x}{\sqrt{a^2 - x^2}},$$

or any other cases of integration by successive reductions, where an integral which contains an operation repeated n times can be made to depend upon another which contains the same $n-1$ or $n-2$ times, and so on until by continued reduction we arrive at a certain *ultimate* form, whose value has then to be determined.

The methods in Arbogast's *Calcul des Dérivations* are peculiarly fitted for the notation and the processes of the engine. Likewise the whole of the Combinatorial Analysis, which consists first in a purely numerical calculation of indices, and secondly in the distribution and combination of the quantities according to laws prescribed by these indices.

We will terminate these Notes by following up in detail the steps through which the engine could compute the Numbers of Bernoulli, this being (in the form in which we shall deduce it) a rather complicated example of its powers. The simplest manner of computing these numbers would be from the direct expansion of

$$\frac{x}{\epsilon^x - 1} = \frac{1}{1 + \dfrac{x}{2} + \dfrac{x^2}{2.3} + \dfrac{x^3}{2.3.4} + \&\mathrm{c.}} \quad \cdots\cdots\cdots\cdots\cdots\cdots \quad (1.)$$

which is in fact a particular case of the development of

$$\frac{a + b\,x + c\,x^2 + \&\mathrm{c.}}{a' + b'\,x + c'\,x^2 + \&\mathrm{c.}}$$

mentioned in Note E. Or again, we might compute them from the well-known form

$$\mathrm{B}_{2n-1} = 2 \cdot \frac{1.2.3....2n}{(2\,\pi)^{2n}} \cdot \left\{ 1 + \frac{1}{2^{2n}} + \frac{1}{3^{2n}} + ... \right\} \quad \cdots\cdots\cdots\cdots\cdots \quad (2.)$$

The engine may indeed be considered as including the whole Calculus of Finite Differences; many of whose theorems would be especially and beautifully fitted for development by its processes, and would offer peculiarly interesting considerations. We may mention, as an example, the calculation of the Numbers of Bernoulli by means of the *Differences of Zero*.

or from the form

$$B_{2n-1} = \frac{\pm 2n}{(2^{2n}-1)2^{n-1}} \begin{cases} \frac{1}{2}n^{2n-1} \\ -(n-1)^{2n-1}\left\{1+\frac{1}{2}\cdot\frac{2n}{1}\right\} \\ +(n-2)^{2n-1}\left\{1+\frac{2n}{1}+\frac{1}{2}\cdot\frac{2n.(2n-1)}{1.2}\right\} \\ -(n-3)^{2n-1}\left\{\begin{array}{l}1+\frac{2n}{1}+\frac{2n.2n-1}{1.2}+ \\ +\frac{1}{2}\cdot\frac{2n.(2n-1).(2n-2)}{1.2.3}\end{array}\right\} \\ +\cdots \quad\cdots \quad\cdots \quad\cdots \end{cases} \quad \ldots\ldots\ldots\ldots (3.)$$

or from many others. As however our object is not simplicity or facility of computation, but the illustration of the powers of the engine, we prefer selecting the formula below, marked (8.). This is derived in the following manner :—

If in the equation

$$\frac{x}{e^x-1} = 1 - \frac{x}{2} + B_1\frac{x^2}{2} + B_3\frac{x^4}{2.3.4} + B_5\frac{x^6}{2.3.4.5.6} + \ldots\ldots\ldots\ldots\ldots (4.)$$

(in which B_1, B_3 ...., &c. are the Numbers of Bernoulli), we expand the denominator of the first side in powers of x, and then divide both numerator and denominator by x, we shall derive

$$1 = \left(1 - \frac{x}{2} + B_1\frac{x^2}{2} + B_3\frac{x^4}{2.3.4} + \ldots\right)\left(1 + \frac{x}{2} + \frac{x^2}{2.3} + \frac{x^3}{2.3.4}\ldots\right) \ldots\ldots\ldots\ldots (5.)$$

If this latter multiplication be actually performed, we shall have a series of the general form

$$1 + D_1x + D_2x^2 + D_3x^3 + \ldots\ldots\ldots\ldots\ldots\ldots\ldots\ldots\ldots\ldots\ldots\ldots (6.)$$

in which we see, first, that all the coefficients of the powers of x are severally equal to zero; and secondly, that the general form for D_{2n}, the coefficient of the $2n+1$th *term* (that is of x^{2n} any *even* power of x), is the following :—

$$\begin{aligned}\frac{1}{2.3\ldots 2n+1} - \frac{1}{2}\cdot\frac{1}{2.3..2n} + \frac{B_1}{2}\cdot\frac{1}{2.3..2n-1} + \frac{B_3}{2.3.4}\cdot\frac{1}{2.3..2n-3} + \\ +\frac{B_5}{2.3.4.5.6}\cdot\frac{1}{2.3\ldots 2n-5} + \ldots + \frac{B_{2n-1}}{2.3\ldots 2n}\cdot 1 = 0\end{aligned} \quad \ldots\ldots\ldots\ldots (7.)$$

Multiplying every term by $(2.3\ldots 2n)$, we have

$$\begin{aligned}0 = -\frac{1}{2}\cdot\frac{2n-1}{2n+1} + B_1\left(\frac{2n}{-}\right) + B_3\left(\frac{2n.2n-1.2n-2}{2.3.4}\right) + \\ + B_5\left(\frac{2n.2n-1\ldots\ldots 2n-4}{2.3.4.5.6}\right) + \ldots + B_{2n-1}\end{aligned} \quad \ldots\ldots\ldots\ldots (8.)$$

which it may be convenient to write under the general form :—

$$0 = A_0 + A_1B_1 + A_3B_3 + A_5B_5 + \ldots + B_{2n-1} \ldots\ldots\ldots\ldots\ldots\ldots\ldots (9.)$$

A_1, A_3, &c. being those functions of n which respectively belong to B_1, B_3, &c.

We might have derived a form nearly similar to (8.), from D_{2n-1} the coefficient of any *odd* power of x in (6.) ; but the general form is a little different for the coefficients of the *odd* powers, and not quite so convenient.

On examining (7.) and (8.), we perceive that, when these formulæ are isolated from (6.), whence they are derived, and considered in themselves separately and independently, n may be any whole number whatever; although when (7.) occurs *as one of the* D's in (6.), it is obvious that n is then not arbitrary, but is always a certain function of the *distance of that* D *from the beginning*. If that distance be $= d$, then

$$2n + 1 = d, \text{ and } n = \frac{d-1}{2} \text{ (for any } even \text{ power of } x)$$

$$2n = d, \text{ and } n = \frac{d}{2}\text{(for any } odd \text{ power of } x).$$

It is with the *independent* formula (8.) that we have to do. Therefore it must be remembered that the conditions for the value of n are now modified, and that n is a perfectly *arbitrary* whole number. This circumstance, combined with the fact (which we may easily perceive) that whatever n is, every term of (8.) after the $(n + 1)$th is $= 0$, and that the $(n + 1)$th term itself is always $= B_{2n-1} \cdot \dfrac{1}{1} = B_{2n-1}$, enables us to find the value (either numerical or algebraical) of any nth Number of Bernoulli B_{2n-1}, *in terms of all the preceding ones*, if we but know the values of $B_1, B_3 \ldots B_{2n-3}$. We append to this Note a Diagram and Table, containing the details of the computation for B_7, (B_1, B_3, B_5 being supposed given).

On attentively considering (8.), we shall likewise perceive that we may derive from it the numerical value of *every* Number of Bernoulli in succession, from the very beginning, *ad infinitum*, by the following series of computations :—

1st Series.—Let $n = 1$, and calculate (8.) for this value of n. The result is B_1.

2nd Series.—Let $n = 2$. Calculate (8.) for this value of n, substituting the value of B_1 just obtained. The result is B_3.

3rd Series.—Let $n = 3$. Calculate (8.) for this value of n, substituting the values of B_1, B_3 before obtained. The result is B_5. And so on, to any extent.

The diagram* represents the columns of the engine when just prepared for computing B_{2n-1} (in the case of $n = 4$); while the table beneath them presents a complete simultaneous view of all the successive changes which these columns then severally pass through in order to perform the computation. (The reader is referred to Note D. for explanations respecting the nature and notation of such tables.)

Six numerical *data* are in this case necessary for making the requisite combinations. These data are $1, 2$, $n (=4), B_1, B_3, B_5$. Were $n = 5$, the additional datum B_7 would be needed. Were $n = 6$, the datum B_9 would be needed ; and so on. Thus the actual *number of data* needed will always be $n + 2$, for $n = n$; and out of these $n + 2$ data, $(\overline{n + 2} - 3)$ of them are successive Numbers of Bernoulli. The reason why the Bernoulli Numbers used as data are nevertheless placed on *Result*-columns in the diagram, is because they may properly be supposed to have been previously computed in succession by the *engine* itself; under which circumstances each B will appear as a *result*, previous to being used as a *datum* for computing the succeeding B. Here then is an instance (of the kind alluded to in Note D.) of the same Variables filling more than one office in turn. It is true that if we consider our computation of B_7 as a perfectly *isolated* calculation, we may conclude B_1, B_3, B_5 to have been arbitrarily placed on the columns; and it would then perhaps be more consistent to put them on V_4, V_5, V_6 as data and not results. But we are not taking this view. On the contrary, we suppose the engine to be *in the course of* computing the Numbers to an indefinite extent, from the very beginning ; and that we merely single out, by way of example, *one amongst* the successive but distinct series' of computations it is thus performing. Where the B's are fractional, it must be understood that they are computed and appear in the notation of *decimal* fractions. Indeed this is a circumstance that should be noticed with reference to all calculations. In any of the examples already given in the translation and in the Notes, some of the *data*, or of the temporary or permanent results, might be fractional, quite as probably as whole numbers. But the arrangements are so made, that the nature of the processes would be the same as for whole numbers.

In the above table and diagram we are not considering the *signs* of any of the B's, merely their numerical magnitude. The engine would bring out the sign for each of them correctly of course, but we cannot enter on *every* additional detail of this kind as we might wish to do. The circles for the signs are therefore intentionally left blank in the diagram.

Operation-cards 1, 2, 3, 4, 5, 6 prepare $-\dfrac{1}{2} \cdot \dfrac{2n-1}{2n+1}$. Thus, Card 1 multiplies *two* into n, and the three *Receiving* Variable-cards belonging respectively to V_4, V_5, V_6, allow the result $2n$ to be placed on each of these latter columns (this being a case in which a triple receipt of the result is needed for subsequent purposes) ; we see that the upper indices of the two Variables used, during Operation 1, remain unaltered.

We shall not go through the details of every operation singly, since the table and diagram sufficiently indicate them ; we shall merely notice some few peculiar cases.

---

* See the diagram at the end of these Notes.

By Operation 6, a *positive* quantity is turned into a *negative* quantity, by simply subtracting the quantity from a column which has only zero upon it. (The sign at the top of V_8 would become — during this process.)

Operation 7 will be unintelligible, unless it be remembered that if we were calculating for $n = 1$ instead of $n = 4$, Operation 6 would have completed the computation of B_1 itself; in which case the engine, instead of continuing its processes, would have to put B_1 on V_{21}; and then either to stop altogether, or to begin Operations 1, 2....7 all over again for value of $n (= 2)$, in order to enter on the computation of B_3; (having however taken care, previous to this recommencement, to make the number on V_3 equal to *two*, by the addition of unity to the former $n = 1$ on that column). Now Operation 7 must either bring out a result equal to zero (if $n = 1$); or a result *greater* than *zero*, as in the present case; and the engine follows the one or the other of the two courses just explained, contingently on the one or the other result of Operation 7. In order fully to perceive the necessity of this *experimental* operation, it is important to keep in mind what was pointed out, that we are not treating a perfectly isolated and independent computation, but one out of a series of antecedent and prospective computations.

Cards 8, 9, 10 produce $-\dfrac{1}{2} \cdot \dfrac{2n-1}{2n+1} + B_1 \dfrac{2n}{2}$. In Operation 9 we see an example of an upper index which again becomes a value after having passed from preceding values to zero. V_{11} has successively been $^0V_{11}$, $^1V_{11}$, $^2V_{11}$, $^0V_{11}$, $^3V_{11}$; and, from the nature of the office which V_{11} performs in the calculation, its index will continue to go through further changes of the same description, which, if examined, will be found to be regular and periodic.

Card 12 has to perform the same office as Card 7 did in the preceding section; since, if n had been $= 2$, the 11th operation would have completed the computation of B_3.

Cards 13 to 20 make A_3. Since A_{2n-1} always consists of $2n-1$ factors, A_3 has three factors; and it will be seen that Cards 13, 14, 15, 16 make the second of these factors, and then multiply it with the first; and that 17, 18, 19, 20 make the third factor, and then multiply this with the product of the two former factors.

Card 23 has the office of Cards 11 and 7 to perform, since if n were $= 3$, the 21st and 22nd operations would complete the computation of B_5. As our case is B_7, the computation will continue one more stage; and we must now direct attention to the fact, that in order to compute A_7 it is merely necessary precisely to repeat the group of Operations 13 to 20; and then, in order to complete the computation of B_7, to repeat Operations 21, 22.

It will be perceived that every unit added to n in B_{2n-1}, entails an additional repetition of operations (13...23) for the computation of B_{2n-1}. Not only are all the *operations* precisely the same however for every such repetition, but they require to be respectively supplied with numbers from the very *same pairs of columns*: with only the one exception of Operation 21, which will of course need B_5 (from V_{23}) instead of B_3 (from V_{22}). This identity in the *columns* which supply the requisite numbers must not be confounded with identity in the *values* those columns have upon them and give out to the mill. Most of those values undergo alterations during a performance of the operations (13...23), and consequently the columns present a new set of values for the *next* performance of (13...23) to work on.

At the termination of the *repetition* of operations (13...23) in computing B_7, the alterations in the values on the Variables are, that

$$V_6 = 2n - 4 \text{ instead of } 2n - 2.$$
$$V_7 = 6 \ldots\ldots\ldots\ldots 4.$$
$$V_{10} = 0 \ldots\ldots\ldots\ldots 1.$$
$$V_{13} = A_0 + A_1 B_1 + A_3 B_3 + A_5 B_5 \text{ instead of } A_0 + A_1 B_1 + A_3 B_3.$$

In this state the only remaining processes are, first, to transfer the value which is on V_{13} to V_{24}; and secondly, to reduce V_6, V_7, V_{13} to zero, and to add* *one* to V_3, in order that the engine may be ready to com-

---

* It is interesting to observe, that so complicated a case as this calculation of the Bernoullian Numbers nevertheless presents a remarkable simplicity in one respect; viz. that during the processes for the computation of *millions* of these Numbers, no other arbitrary modification would be requisite in the arrangements, excepting the above simple and uniform provision for causing one of the data periodically to receive the finite increment unity.

mence computing B_9. Operations 24 and 25 accomplish these purposes. It may be thought anomalous that Operation 25 is represented as leaving the upper index of V_3 still $=$ unity; but it must be remembered that these indices always begin anew for a separate calculation, and that Operation 25 places upon V_3 the *first* value *for the new calculation*.

It should be remarked, that when the group (13...23) is *repeated*, changes occur in some of the *upper* indices during the course of the repetition: for example, $^3V_6$ would become $^4V_6$ and $^5V_6$.

We thus see that when $n = 1$, nine Operation-cards are used; that when $n = 2$, fourteen Operation-cards are used; and that when $n > 2$, twenty-five Operation-cards are used; but that no *more* are needed, however great n may be; and not only this, but that these same twenty-five cards suffice for the successive computation of all the Numbers from B_1 to B_{2n-1} inclusive. With respect to the number of *Variable*-cards, it will be remembered, from the explanations in previous Notes, that an average of three such cards to each *operation* (not however to each Operation-*card*) is the estimate. According to this, the computation of B_1 will require twenty-seven Variable-cards; B_3 forty-two such cards; B_5 seventy-five; and for every succeeding B after B_5, there would be thirty-three additional Variable-cards (since each repetition of the group (13...23) adds eleven to the number of operations required for computing the previous B). But we must now explain, that whenever there is a *cycle of operations*, and if these merely require to be supplied with numbers from the *same pairs of columns*, and likewise each operation to place its *result* on the *same* column for every repetition of the whole group, the process then admits of a *cycle of Variable-cards* for effecting its purposes. There is obviously much more symmetry and simplicity in the arrangements, when cases do admit of repeating the Variable as well as the Operation-cards. Our present example is of this nature. The only exception to a *perfect identity* in *all* the processes and columns used, for every repetition of Operations (13...23), is, that Operation 21 always requires one of its factors from a new column, and Operation 24 always puts its result on a new column. But as these variations follow the same law at each repetition (Operation 21 always requiring its factor from a column *one* in advance of that which it used the previous time, and Operation 24 always putting its result on the column *one* in advance of that which received the previous result), they are easily provided for in arranging the recurring group (or cycle) of Variable-cards.

We may here remark, that the average estimate of three Variable-cards coming into use to each operation, is not to be taken as an absolutely and literally correct amount for all cases and circumstances. Many special circumstances, either in the nature of a problem, or in the arrangements of the engine under certain contingencies, influence and modify this average to a greater or less extent; but it is a very safe and correct *general* rule to go upon. In the preceding case it will give us seventy-five Variable-cards as the total number which will be necessary for computing any B after B_3. This is very nearly the precise amount really used, but we cannot here enter into the minutiæ of the few particular circumstances which occur in this example (as indeed at some one stage or other of probably most computations) to modify slightly this number.

It will be obvious that the very *same* seventy-five Variable-cards may be repeated for the computation of every succeeding Number, just on the same principle as admits of the repetition of the thirty-three Variable-cards of Operations (13...23) in the computation of any *one* Number. Thus there will be a *cycle of a cycle* of Variable-cards.

If we now apply the notation for cycles, as explained in Note E., we may express the operations for computing the Numbers of Bernoulli in the following manner:—

$(1...7), (24, 25) \ldots\ldots\ldots\ldots\ldots$ gives $B_1 \quad = $ 1st number; (n being $=1$).
$(1...7), (8...12), (24, 25) \ldots\ldots\ldots\ldots B_3 \quad = $ 2nd $\ldots\ldots$; ($n \ldots = 2$).
$(1...7), (8...12), (13...23), (24, 25) \ldots\ldots B_5 \quad = $ 3rd $\ldots\ldots$; ($n \ldots = 3$).
$(1...7), (8...12), 2(13...23), (24, 25) \ldots\ldots B_7 \quad = $ 4th $\ldots\ldots$; ($n \ldots = 4$).
$\ldots\ldots\ldots\ldots\ldots\ldots\ldots\ldots\ldots\ldots\ldots\ldots\ldots\ldots\ldots\ldots\ldots\ldots$
$\ldots\ldots\ldots\ldots\ldots\ldots\ldots\ldots\ldots\ldots\ldots\ldots\ldots\ldots\ldots\ldots\ldots\ldots$
$(1...7), (8...12), \Sigma(+1)^{n-2}(13...23), (24,25)...B_{2n-1} = n$th $\ldots\ldots$; ($n \ldots = n$).

Again,

$$(1...7), (24, 25), \sum_{\text{limits 1 to } n} (+1)^n \left\{ (1...7), (8...12), \sum_{\text{limits 0 to } (n+2)} (n+2) (13...23), (24, 25) \right\}$$

represents the total operations for computing every number in succession, from B_1 to B_{2n-1} inclusive.

In this formula we see a *varying cycle* of the *first* order, and an ordinary cycle of the *second* order. The latter cycle in this case includes in it the varying cycle.

On inspecting the ten Working-Variables of the diagram, it will be perceived, that although the *value* on any one of them (excepting V_4 and V_5) goes through a series of changes, the *office* which each performs is in this calculation *fixed* and *invariable*. Thus V_6 always prepares the *numerators* of the factors of any A ; V_7 the *denominators*. V_8 always receives the $(2n-3)$th factor of A_{2n-1}, and V_9 the $(2n-1)$th. V_{10} always decides which of two courses the succeeding processes are to follow, by feeling for the value of n through means of a subtraction; and so on; but we shall not enumerate further. It is desirable in all calculations so to arrange the processes, that the *offices* performed by the Variables may be as uniform and fixed as possible.

Supposing that it was desired not only to tabulate B_1, B_3, &c., but A_0, A_1, &c. ; we have only then to appoint another series of Variables, V_{41}, V_{42}, &c., for receiving these latter results as they are successively produced upon V_{11}. Or again, we may, instead of this, or in addition to this second series of results, wish to tabulate the value of each successive *total* term of the series (8.), viz. A_0, $A_1 B_1$, $A_3 B_3$, &c. We have then merely to multiply each B with each corresponding A, as produced, and to place these successive products on Result-columns appointed for the purpose.

The formula (8.) is interesting in another point of view. It is one particular case of the general Integral of the following Equation of Mixed Differences :—

$$\frac{d^2}{dx^2}\left(z_{n+1} x^{2n+2} \right) = (2n+1)(2n+2) z^n x^{2n}$$

for certain special suppositions respecting z, x and n.

The *general* integral itself is of the form,

$$z_n = f(n) \cdot x + f_1(n) + f_2(n) \cdot x^{-1} + f_3(n) \cdot x^{-3} + \ldots$$

and it is worthy of remark, that the engine might (in a manner more or less similar to the preceding) calculate the value of this formula upon most *other* hypotheses for the functions in the integral with as much, or (in many cases) with more ease than it can formula (8.).   A. A. L.

Number of Operation.	Nature of Operation.	Variables acted upon.	Variables receiving results.	Indication of change in the value on any Variable.	Statement of Results.	Data.						
						${}^1V_1$ O 0 0 1 [1]	${}^1V_2$ O 0 0 2 [2]	${}^1V_3$ O 0 0 4 [n]	${}^0V_4$ O 0 0 0	${}^0V_5$ O 0 0 0	${}^0V_6$ O 0 0 0	${}^0V_7$ O 0 0 0
1	×	${}^1V_2 \times {}^1V_3$	${}^1V_4, {}^1V_5, {}^1V_6$	$\begin{cases}{}^1V_2={}^1V_2\\{}^1V_3={}^1V_3\end{cases}$	$=2n$	...	2	n	$2n$	$2n$	$2n$	
2	−	${}^1V_4-{}^1V_1$	${}^2V_4$	$\begin{cases}{}^1V_4={}^2V_4\\{}^1V_1={}^1V_1\end{cases}$	$=2n-1$	1	...	...	$2n-1$			
3	+	${}^1V_5+{}^1V_1$	${}^2V_5$	$\begin{cases}{}^1V_5={}^2V_5\\{}^1V_1={}^1V_1\end{cases}$	$=2n+1$	1	...	...	...	$2n+1$		
4	÷	${}^2V_5\div{}^2V_4$	${}^1V_{11}$	$\begin{cases}{}^2V_5={}^0V_5\\{}^2V_4={}^0V_4\end{cases}$	$=\dfrac{2n-1}{2n+1}$	...	...	...	0	0	...	...
5	÷	${}^1V_{11}\div{}^1V_2$	${}^2V_{11}$	$\begin{cases}{}^1V_{11}={}^2V_{11}\\{}^1V_2={}^1V_2\end{cases}$	$=\dfrac{1}{2}\cdot\dfrac{2n-1}{2n+1}$	...	2	...	...	...	...	...
6	−	${}^0V_{13}-{}^2V_{11}$	${}^1V_{13}$	$\begin{cases}{}^2V_{11}={}^0V_{11}\\{}^0V_{13}={}^1V_{13}\end{cases}$	$=-\dfrac{1}{2}\cdot\dfrac{2n-1}{2n+1}=A_0$	...	...	...	...	...	...	...
7	−	${}^1V_3-{}^1V_1$	${}^1V_{10}$	$\begin{cases}{}^1V_3={}^1V_3\\{}^1V_1={}^1V_1\end{cases}$	$=n-1(=3)$	1	...	n	...	...	...	...
8	+	${}^1V_2+{}^0V_7$	${}^1V_7$	$\begin{cases}{}^1V_2={}^1V_2\\{}^0V_7={}^1V_7\end{cases}$	$=2+0=2$	...	2	...	...	...	...	2
9	÷	${}^1V_6\div{}^1V_7$	${}^3V_{11}$	$\begin{cases}{}^1V_6={}^1V_6\\{}^0V_{11}={}^3V_{11}\end{cases}$	$=\dfrac{2n}{2}=A_1$	...	...	...	...	...	$2n$	2
10	×	${}^1V_{21}\times {}^3V_{11}$	${}^1V_{12}$	$\begin{cases}{}^1V_{21}={}^1V_{21}\\{}^3V_{11}={}^3V_{11}\end{cases}$	$=B_1\cdot\dfrac{2n}{2}=B_1A_1$	...	...	...	...	...	...	...
11	+	${}^1V_{12}+{}^1V_{13}$	${}^2V_{13}$	$\begin{cases}{}^1V_{12}={}^0V_{12}\\{}^1V_{13}={}^2V_{13}\end{cases}$	$=-\dfrac{1}{2}\cdot\dfrac{2n-1}{2n+1}+B_1\cdot\dfrac{2n}{2}$	...	...	...	...	...	...	...
12	−	${}^1V_{10}-{}^1V_1$	${}^2V_{10}$	$\begin{cases}{}^1V_{10}={}^2V_{10}\\{}^1V_1={}^1V_1\end{cases}$	$=n-2(=2)$	1	...	...	...	...	...	...
13	−	${}^1V_6-{}^1V_1$	${}^2V_6$	$\begin{cases}{}^1V_6={}^2V_6\\{}^1V_1={}^1V_1\end{cases}$	$=2n-1$	1	...	...	...	...	$2n-1$	
14	+	${}^1V_1+{}^1V_7$	${}^2V_7$	$\begin{cases}{}^1V_1={}^1V_1\\{}^1V_7={}^2V_7\end{cases}$	$=2+1=3$	1	...	...	...	...	...	3
15	÷	${}^2V_6\div {}^2V_7$	${}^1V_8$	$\begin{cases}{}^2V_6={}^2V_6\\{}^2V_7={}^2V_7\end{cases}$	$=\dfrac{2n-1}{3}$	...	...	...	...	...	$2n-1$	3
16	×	${}^1V_8\times {}^3V_{11}$	${}^4V_{11}$	$\begin{cases}{}^1V_8={}^0V_8\\{}^3V_{11}={}^4V_{11}\end{cases}$	$=\dfrac{2n}{2}\cdot\dfrac{2n-1}{3}$	...	...	...	...	...	...	...
17	−	${}^2V_6-{}^1V_1$	${}^3V_6$	$\begin{cases}{}^2V_6={}^3V_6\\{}^1V_1={}^1V_1\end{cases}$	$=2n-2$	1	...	...	...	...	$2n-2$	
18	+	${}^1V_1+{}^2V_7$	${}^3V_7$	$\begin{cases}{}^2V_7={}^3V_7\\{}^1V_1={}^1V_1\end{cases}$	$=3+1=4$	1	...	...	...	...	...	4
19	÷	${}^3V_6\div {}^3V_7$	${}^1V_9$	$\begin{cases}{}^3V_6={}^3V_6\\{}^3V_7={}^3V_7\end{cases}$	$=\dfrac{2n-2}{4}$	...	...	...	...	...	$2n-2$	4
20	×	${}^1V_9\times {}^4V_{11}$	${}^5V_{11}$	$\begin{cases}{}^1V_9={}^0V_9\\{}^4V_{11}={}^5V_{11}\end{cases}$	$=\dfrac{2n}{2}\cdot\dfrac{2n-1}{3}\cdot\dfrac{2n-2}{4}=A_3$	...	...	...	...	...	...	...
21	×	${}^1V_{22}\times {}^5V_{11}$	${}^0V_{12}$	$\begin{cases}{}^1V_{22}={}^1V_{22}\\{}^0V_{12}={}^2V_{12}\end{cases}$	$=B_3\cdot\dfrac{2n}{2}\cdot\dfrac{2n-1}{3}\cdot\dfrac{2n-2}{3}=B_3A_3$	...	...	...	...	...	...	...
22	+	${}^2V_{12}+{}^2V_{13}$	${}^3V_{13}$	$\begin{cases}{}^2V_{12}={}^0V_{12}\\{}^2V_{13}={}^3V_{13}\end{cases}$	$=A_0+B_1A_1+B_3A_3$	...	...	...	...	...	...	...
23	−	${}^2V_{10}-{}^1V_1$	${}^3V_{10}$	$\begin{cases}{}^2V_{10}={}^3V_{10}\\{}^1V_1={}^1V_1\end{cases}$	$=n-3(=1)$	1	...	...	...	...	...	...
					Here follows a repetition of Oper							
24	+	${}^4V_{13}+{}^0V_{24}$	${}^1V_{24}$	$\begin{cases}{}^4V_{13}={}^0V_{13}\\{}^0V_{24}={}^0V_{24}\end{cases}$	$=B_7$	...	...	...	...	...	...	...
25	+	${}^1V_1+{}^1V_3$	${}^1V_3$	$\begin{cases}{}^1V_1={}^1V_1\\{}^1V_3={}^1V_3\\{}^5V_6={}^0V_6\quad\text{by a Variable-card.}\\{}^5V_7={}^0V_7\quad\text{by a Variable-card.}\end{cases}$	$=n+1=4+1=5$	1	...	$n+1$	...	...	0	0

			Working Variables.				Result Variables.		
$^0V_8$ ◯ 0 0 0 ⬜	$^0V_9$ ◯ 0 0 0 ⬜	$^0V_{10}$ ◯ 0 0 0 ⬜	$^0V_{11}$ ◯ 0 0 0	$^0V_{12}$ ◯ 0 0 0	$^0V_{13}$ .... ◯ 0 0 0	$^1V_{21}$ ◯ 0 0 0 B_1 in a decimal fraction. B_1	$^1V_{22}$ ◯ 0 0 0 B_3 in a decimal fraction. B_3	$^1V_{23}$ ◯ 0 0 0 B_5 in a decimal fraction. B_5	$^0V_{24}...$ ◯ 0 0 0 B_7
...	...	...	$\dfrac{2n-1}{2n+1}$						
...	...	...	$\dfrac{1}{2}\cdot\dfrac{2n-1}{2n+1}$						
...	...	...	0	....	$-\dfrac{1}{2}\cdot\dfrac{2n-1}{2n+1}=A_0$				
...	...	$n-1$							
...	...	...	$\dfrac{2n}{2}=A_1$						
...	...	...	$\dfrac{2n}{2}=A_1$	$B_1.\dfrac{2n}{2}=B_1A_1$	....	B_1			
...	...	...	....	0	$\left\{-\dfrac{1}{2}\cdot\dfrac{2n-1}{2n+1}+B_1.\dfrac{2n}{2}\right\}$				
...	...	$n-2$							
$\dfrac{2n-1}{3}$ 0	...	...	$\dfrac{2n}{2}\cdot\dfrac{2n-1}{3}$						
...	$\dfrac{2n-2}{4}$ 0	...	$\left\{\dfrac{2n}{2}\cdot\dfrac{2n-1}{3}\cdot\dfrac{2n-2}{3} = A_3\right\}$						
...	...	...	0	B_3A_3	....	......	B_3		
...	...	...	....	0	$\left\{A_3+B_1A_1+B_3A_3\right\}$				
...	...	$n-3$							

erations thirteen to twenty-three.

...	...	...	....	....	....	......	......	......	B

BABBAGE'S CALCULATING ENGINE.

Excerpt from the "Edinburgh Review," July, 1834, No. CXX.*

Art. I.—1. *Letter to Sir Humphry Davy, Bart., P.R.S., on the application of Machinery to Calculate and Print Mathematical Tables.* By Charles Babbage, Esq., F.R.S., 4to. Printed by order of the House of Commons. No. 370. 22 May, 1823.

2. *On the Application of Machinery to the Calculation of Astronomical and Mathematical Tables.* By Charles Babbage, Esq. Memoirs Astron. Soc. Vol. I. Part 2. London: 1822.

3. *Address to the Astronomical Society, by Henry Thomas Colebrooke, Esq., F.R.S. President, on presenting the first gold medal of the Society to Charles Babbage, Esq., for the invention of the Calculating Engine.* Memoirs Astron. Soc. Vol. I. Part 2. London: 1822.

4. *On the determination of the General Term of a new Class of Infinite Series.* By Charles Babbage, Esq. Transactions Camb. Phil. Soc. Cambridge: 1824.

5. *On Errors common to many Tables of Logarithms.* By Charles Babbage, Esq. Memoirs Astron. Soc. London: 1827.

6. *On a Method of Expressing by Signs the Action of Machinery.* By Charles Babbage, Esq. Phil. Trans. London: 1826.

7. *Report by the Committee appointed by the Council of the Royal Society to consider the subject referred to in a communication received by them from the Treasury, respecting Mr. Babbage's Calculating Engine, and to report thereupon.* London: 1829.

There is no position in society more enviable than that of the few who unite a moderate independence with high intellectual qualities. Liberated from the necessity of seeking their support by a profession, they are unfettered by its restraints, and are enabled to direct the powers of their minds, and to concentrate their intellectual energies on those objects exclusively to which they feel that their powers may be applied with the greatest advantage to the community, and with the most lasting reputation to themselves. On the other hand, their middle station and limited income rescue them from those allurements to frivolity and dissipation, to which rank and wealth ever expose their possessors. Placed in such favourable circumstances, Mr. Babbage selected science as the field of his ambition; and his mathematical researches have conferred on him a high reputation, wherever the exact sciences are studied and appreciated. The suffrages of the mathematical world have been ratified in his own country, where he has been elected to the Lucasian Professorship in his own University— a chair, which, though of inconsiderable emolument, is one on which Newton has conferred everlasting celebrity. But it has been the fortune of this mathematician to surround himself with fame of another and more popular kind, and which rarely falls to the lot of those who devote their lives to the cultivation of the abstract sciences. This distinction he owes to the announcement, some years since, of his celebrated project of a Calculating Engine. A proposition to reduce arithmetic to the dominion of mechanism—to substitute an automaton for a compositor—to throw the powers of thought into wheel-work could not fail to awaken the attention of the world. To bring the practicability of such a project within the compass of popular belief was not easy: to do so by bringing it within the compass of popular comprehension was not possible. It transcended the imagination of the public in general to conceive its possibility; and the sentiments of wonder with which it was received, were only prevented from

* The author of this review was Dr. Dionysius Lardner.

merging into those of incredulity, by the faith reposed in the high attainments of its projector. This extraordinary undertaking was, however, viewed in a very different light by the small section of the community, who, being sufficiently versed in mathematics, were acquainted with the principle upon which it was founded. By reference to that principle, they perceived at a glance the practicability of the project; and being enabled by the nature of their attainments and pursuits to appreciate the immeasurable importance of its results, they regarded the invention with a proportionately profound interest. The production of numerical tables, unlimited in quantity and variety, restricted to no particular species, and limited by no particular law; extending not merely to the boundaries of existing knowledge, but spreading their powers over the undefined regions of future discovery—were results, the magnitude and the value of which the community in general could neither comprehend nor appreciate. In such a case, the judgment of the world could only rest upon the authority of the philosophical part of it; and the fiat of the scientific community swayed for once political councils. The British Government, advised by the Royal Society, and a committee formed of the most eminent mechanicians and practical engineers, determined on constructing the projected mechanism at the expense of the nation, to be held as national property.

Notwithstanding the interest with which this invention has been regarded in every part of the world, it has never yet been embodied in a written, much less in a published form. We trust, therefore, that some credit will be conceded to us for having been the first to make the public acquainted with the object, principle, and structure of a piece of machinery, which, though at present unknown (except as to a few of its probable results), must, when completed, produce important effects, not only on the progress of science, but on that of civilization.

The calculating machinery thus undertaken for the public gratuitously (so far as Mr. Babbage is concerned), has now attained a very advanced stage towards completion; and a portion of it has been put together, and performs various calculations;—affording a practical demonstration that the anticipations of those, under whose advice Government has acted, have been well founded.

There are, nevertheless, many persons who, admitting the great ingenuity of the contrivance, have, notwithstanding, been accustomed to regard it more in the light of a philosophical curiosity, than an instrument for purposes practically useful. This mistake (than which it is not possible to imagine a greater) has arisen mainly from the ignorance which prevails of the extensive utility of those numerical tables which it is the purpose of the engine in question to produce. There are also some persons who, not considering the time requisite to bring any invention of this magnitude to perfection in all its details, incline to consider the delays which have taken place in its progress as presumptions against its practicability. These persons should, however, before they arrive at such a conclusion, reflect upon the time which was necessary to bring to perfection engines infinitely inferior in complexity and mechanical difficulty. Let them remember that—not to mention the *invention* of that machine—the *improvements* alone introduced into the steam-engine by the celebrated Watt, occupied a period of not less than twenty years of the life of that distinguished person, and involved an expenditure of capital amounting to £50,000.* The calculating machinery is a contrivance new even in its details. Its inventor did not take it up already imperfectly formed, after having received the contributions of human ingenuity exercised upon it for a century or more. It has not, like almost all other great mechanical inventions, been gradually advanced to its present state through a series of failures, through difficulties encountered and overcome by a succession of projectors. It is not an object on which the light of various minds has thus been shed. It is, on the contrary, the production of solitary and individual thought—begun, advanced through each successive stage of improvement, and brought to perfection by one mind. Yet this creation of genius, from its first rude conception to its present state, has cost little more than half the time, and not one-third of the expense, consumed in bringing the steam-engine (previously far advanced in the course of improvement) to that state of comparative perfection in which it was left by Watt. Short as the period of time has been which the inventor has devoted to this enterprise, it has, nevertheless, been demonstrated, to the satisfaction of many scientific men of the first eminence, that the design in all its details, reduced, as it is, to a system of mechanical drawings, is complete; and requires only to be constructed in conformity with those plans, to realize all that its inventor has promised.

With a view to remove and correct erroneous impressions, and at the same time to convert the vague sense of wonder at what seems incomprehensible, with which this project is contemplated by the public in general, into a more rational and edifying sentiment, it is our purpose in the present article,

First, To show the immense importance of any method by which numerical tables, absolutely

* Watt commenced his investigations respecting the steam-engine in 1763, between which time and the year 1782 inclusive, he took out several patents for improvements in details. Bolton and Watt had expended the above sum on their improvements before they began to receive any return.

accurate in every individual copy, may be produced with facility and cheapness. This we shall establish by coveying to the reader some notion of the number and variety of tables published in every country of the world to which civilization has extended, a large portion of which have been produced at the public expense ; by showing also, that they are nevertheless rendered inefficient, to a greater or less extent, by the prevalence of errors in them ; that these errors pervade not merely tables produced by individual labour and enterprise, but that they vitiate even those on which national resources have been prodigally expended, and to which the highest mathematical ability, which the most enlightened nations of the world could command, has been unsparingly and systematically directed.

Secondly, To attempt to convey to the reader a general notion of the mathematical principle on which the calculating machinery is founded, and of the manner in which this principle is brought into practical operation, both in the process of calculating and printing. It would be incompatible with the nature of this review, and indeed impossible without the aid of numerous plans, sections, and elevations, to convey clear and precise notions of the details of the means by which the process of reasoning is performed by inanimate matter, and the arbitrary and capricious evolutions of the fingers of typographical compositors are reduced to a system of wheel-work. We are, nevertheless, not without hopes of conveying, even to readers unskilled in mathematics, some satisfactory notions of a general nature on this subject.

Thirdly, To explain the actual state of the machinery at the present time ; what progress has been made towards its completion ; and what are the probable causes of those delays in its progress, which must be a subject of regret to all friends of science. We shall indicate what appears to us the best and most practicable course to prevent the unnecessary recurrence of such obstructions for the future, and to bring this noble project to a speedy and successful issue.

Viewing the infinite extent and variety of the tables which have been calculated and printed, from the earliest periods of human civilization to the present time, we feel embarrassed with the difficulties of the task which we have imposed on ourselves ;—that of attempting to convey to readers unaccustomed to such speculations, anything approaching to an adequate idea of them. These tables are connected with the various sciences, with almost every department of the useful arts, with commerce in all its relations ; but, above all, with Astronomy and Navigation. So important have they been considered, that in many instances large sums have been appropriated by the most enlightened nations in the production of them ; and yet so numerous and insurmountable have been the difficulties attending the attainment of this end, that after all, even navigators, putting aside every other department of art and science, have, until very recently, been scantily and imperfectly supplied with the tables indispensably necessary to determine their position at sea,

The first class of tables which naturally present themselves, are those of Multiplication. A great variety of extensive multiplication tables have been published from an early period in different countries ; and especially tables of *Powers*, in which a number is multiplied by itself successively. In Dodson's *Calculator* we find a table of multiplication extending as far as 10 times 1000.* In 1775, a still more extensive table was published to 10 times 10,000. The Board of Longitude subsequently employed the late Dr. Hutton to calculate and print various numerical tables, and among others, a multiplication table extending as far as 100 times 1000 ; tables of the squares of numbers, as far as 25,400 ; tables of cubes, and of the first ten powers of numbers, as far as 100.† In 1814, Professor Barlow, of Woolwich, published, in an octavo volume, the squares, cubes, square roots, cube roots, and reciprocals of all numbers from 1 to 10,000 ; a table of the first ten powers of all numbers from 1 to 100, and of the fourth and fifth powers of all numbers from 100 to 1000.

Tables of Multiplication to a still greater extent have been published in France. In 1785, was published an octavo volume of tables of the squares, cubes, square roots, and cube roots of all numbers from 1 to 10,000 ; and similar tables were again published in 1801. In 1817, multiplication tables were published in Paris by Voisin ; and similar tables, in two quarto volumes, in 1824, by the French Board of Longitude, extending as far as a thousand times a thousand. A table of squares was published in 1810, in Hanover ; in 1812, at Leipzig ; in 1825, at Berlin ; and in 1827, at Ghent. A table of cubes was published in 1827, at Eisenach ; in the same year a similar table at Ghent ; and one of the squares of all numbers as far as 10,000, was published in that year, in quarto, at Bonn. The Prussian Government has caused a multiplication table to be calculated and printed, extending as far as 1000 times 1000. Such are a few of the tables of this class which have been published in different countries.

---

* Dodson's " Calculator." 4to. London : 1747.
† Hutton's " Tables of Products and Powers." Folio. London : 1781.

This class of tables may be considered as purely arithmetical, since the results which they express involve no other relations than the arithmetical dependence of abstract numbers upon each other. When numbers, however, are taken in a concrete sense, and are applied to express peculiar modes of quantity,—such as angular, linear, superficial, and solid magnitudes,—a new set of numerical relations arise, and a large number of computations are required.

To express angular magnitude, and the various relations of linear magnitude with which it is connected, involves the consideration of a vast variety of Geometrical and Trigonometrical tables; such as tables of the natural sines, co-sines, tangents, secants, co-tangents, &c. &c.; tables of arcs and angles in terms of the radius; tables for the immediate solution of various cases of triangles, &c. Volumes without number of such tables have been from to time computed and published. It is not sufficient, however, for the purposes of computation to tabulate these immediate trigonometrical functions. Their squares* and higher powers, their square roots, and other roots, occur so frequently, that it has been found expedient to compute tables for them, as well as for the same functions of abstract numbers.

The measurement of linear, superficial, and solid magnitudes, in the various forms and modifications in which they are required in the arts, demands another extensive catalogue of numerical tables. The surveyor, the architect, the builder, the carpenter, the miner, the gauger, the naval architect, the engineer, civil and military, all require the aid of peculiar numerical tables, and such have been published in all countries.

The increased expedition and accuracy which was introduced into the art of computation by the invention of Logarithms, greatly enlarged the number of tables previously necessary. To apply the logarithmic method, it was not merely necessary to place in the hands of the computist extensive tables of the logarithms of the natural numbers, but likewise to supply him with tables in which he might find already calculated the logarithms of those arithmetical, trigonometrical, and geometrical functions of numbers, which he has most frequent occasion to use. It would be a circuitous process, when the logarithm of a sine or co-sine of an angle is required, to refer, first to the table of sines, or co-sines, and thence to the table of the logarithms of natural numbers. It was therefore found expedient to compute distinct tables of the logarithms of the sines, co-sines, tangents, &c., as well as of various other functions frequently required, such as sums, differences, &c.

Great as is the extent of the tables we have just enumerated, they bear a very insignificant proportion to those which remain to be mentioned. The above are, for the most part, general in their nature, not belonging particularly to any science or art. There is a much greater variety of tables, whose importance is no way inferior, which are, however, of a more special nature: Such are, for example, tables of interest, discount, and exchange, tables of annuities, and other tables necessary in life insurances; tables of rates of various kinds necessary in general commerce. But the science in which, above all others, the most extensive and accurate tables are indispensable, is Astronomy; with the improvement and perfection of which is inseparably connected that of the kindred art of Navigation. We scarcely dare hope to convey to the general reader anything approaching to an adequate notion of the multiplicity and complexity of the tables necessary for the purposes of the astronomer and navigator. We feel, nevertheless, that the truly national importance which must attach to any perfect and easy means of producing those tables cannot be at all estimated, unless we state some of the previous calculations necessary in order to enable the mariner to determine with the requisite certainty and precision the place of his ship.

In a word, then, all the purely arithmetical, trigonometrical, and logarithmic tables already mentioned, are necessary, either immediately or remotely, for this purpose. But in addition to these, a great number of tables, exclusively astronomical, are likewise indispensable. The predictions of the astronomer, with respect to the positions and motions of the bodies of the firmament, are the means, and the only means, which enable the mariner to prosecute his art. By these he is enabled to discover the distance of his ship from the Line, and the extent of his departure from the meridian of Greenwich, or from any other meridian to which the astronomical predictions refer. The more numerous, minute and accurate these predictions can be made, the greater will be the facilities which can be furnished to the mariner. But the computation of those tables, in which the future position of celestial objects are registered, depend themselves upon an infinite variety of other tables which never reach the hands of the mariner. It cannot be said that there is any table whatever, necessary for the astronomer, which is unnecessary for the navigator.

The purposes of the marine of a country whose interests are so inseparably connected as ours are with the improvement of the art of navigation, would be very inadequately fulfilled, if our navi-

---

* The squares of the sines of angles are extensively used in the calculations connected with the theory of the tides. Not aware that tables of these squares existed, Bouvard, who calculated the tides for Laplace, underwent the labour of calculating the square of each individual sine in every case in which it occurred.

gators were merely supplied with the means of determining by *Nautical Astronomy* the position of a ship at sea. It has been well observed by the Committee of the Astronomical Society, to whom the recent improvement of the Nautical Almanac was confided, that it is not by those means merely by which the seaman is enabled to determine the position of his vessel at sea, that the full intent and purpose of what is usually called *Nautical Astronomy* are answered. This object is merely a part of that comprehensive and important subject; and might be attained by a very cheap publication, and without the aid of expensive instruments. A not less important and much more difficult part of nautical science has for its object to determine the precise position of various interesting and important points on the surface of the earth,—such as remarkable headlands, ports, and islands; together with the general trending of the coast between well-known harbours. It is not necessary to point out here how important such knowledge is to the mariner. This knowledge, which may be called *Nautical Geography*, cannot be obtained by the methods of observation used on board ship, but requires much more delicate and accurate instruments, firmly placed upon the solid ground, besides all the astronomical aid which can be afforded by the best tables, arranged in the most convenient form for immediate use. This was Dr. Maskelyne's view of the subject, and his opinion has been confirmed by the repeated wants and demands of those distinguished navigators who have been employed in several recent scientific expeditions.*

Among the tables *directly* necessary for navigation, are those which predict the position of the centre of the sun from hour to hour. These tables include the sun's right ascension and declination, daily, at noon, with the hourly change in these quantities. They also include the equation of time, together with its hourly variation.

Tables of the moon's place for every hour, are likewise necessary, together with the change of declination for every ten minutes. The lunar method of determining the longitude depends upon tables containing the predicted distances of the moon from the sun, the principal planets, and from certain conspicuous fixed stars; which distances being observed by the mariner, he is enabled thence to discover the *time* at the meridian from which the longitude is measured; and, by comparing that time with the time known or discoverable in his actual situation, he infers his longitude. But not only does the prediction of the position of the moon, with respect to these celestial objects, require a vast number of numerical tables, but likewise the observations necessary to be made by the mariner, in order to determine the lunar distances, also require several tables. To predict the exact position of any fixed star, requires not less than ten numerical tables peculiar to that star; and if the mariner be furnished (as is actually the case) with tables of the predicted distances of the moon from one hundred such stars, such predictions must require not less than a thousand numerical tables. Regarding the range of the moon through the firmament, however, it will be readily conceived that a hundred stars form but a scanty supply; especially when it is considered that an accurate method of determining the longitude, consists in observing the extinction of a star by the dark edge of the moon. Within the limits of the lunar orbit there are not less than one thousand stars, which are so situated as to be in the moon's path, and therefore to exhibit, at some period or other, those desirable occultations. These stars are also of such magnitudes, that their occultations may be distinctly observed from the deck, even when subject to all the unsteadiness produced by an agitated sea. To predict the occultations of such stars, would require not less than ten thousand tables. The stars from which lunar distances might be taken are still more numerous; and we may safely pronounce, that, great as has been the improvement effected recently in our Nautical Almanac, it does not yet furnish more than a small fraction of that aid to navigation (in the large sense of that term), which, with greater facility, expedition, and economy in the calculation and printing of tables, it might be made to supply.

Tables necessary to determine the places of the planets are not less necessary than those for the sun, moon and stars. Some notion of the number and complexity of these tables may be formed, when we state that the positions of the two principal planets (and these the most necessary for the navigator), Jupiter and Saturn, require each not less than one hundred and sixteen tables. Yet it is not only necessary to predict the position of these bodies, but it is likewise expedient to tabulate the motions of the four satellites of Jupiter, to predict the exact times at which they enter his shadow, and at which their shadows cross his disc, as well as the times at which they are interposed between him and the Earth, and he between them and the Earth.

Among the extensive classes of tables here enumerated, there are several which are in their nature permanent and unalterable, and would never require to be recomputed, if they once could be computed with perfect accuracy on accurate data; but the data on which such computations are conducted, can only be regarded as approximations to truth, within limits the extent of which must necessarily vary with our knowledge of astronomical science. It has accordingly happened, that one

---

* Report of the Committee of the Astronomical Society, prefixed to the " Nautical Almanac " for 1834.

set of tables after another has been superseded with each advance of astronomical science. Some striking examples of this may not be uninstructive. In 1765, the Board of Longitude paid to the celebrated Euler the sum of 300*l.*, for furnishing general formulæ for the computation of lunar tables. Professor Mayer was employed to calculate the tables upon these formulæ, and the sum of 3000*l.* was voted for them by the British Parliament, to his widow, after his decease. These tables had been used for ten years, from 1766 to 1776, in computing the Nautical Almanac, when they were superseded by new and improved tables, composed by Mr. Charles Mason, under the direction of Dr. Maskelyne, from calculations made by order of the Board of Longitude, on the observations of Dr. Bradley. A farther improvement was made by Mason in 1780; but a much more extensive improvement took place in the lunar calculations by the publication of the tables of the Moon, by M. Bürg, deduced from Laplace's theory, in 1806. Perfect, however, as Bürg's tables were considered, at the time of their publication, they were, within the short period of six years, superseded by a more accurate set of tables published by Burckhardt in 1812; and these also have since been followed by the tables of Damoiseau. Professor Schumacher has calculated by the latter tables his ephemeris of the Planetary Lunar Distances, and astronomers will hence be enabled to put to the strict test of observation the merits of the tables of Burckhardt and Damoiseau.*

The solar tables have undergone, from time to time, similar changes. The solar tables of Mayer were used in the computation of the Nautical Almanac, from its commencement in 1767, to 1804, inclusive. Within the six years immediately succeeding 1804, not less than three successive sets of solar tables appeared, each improving on the other; the first by Baron de Zach, the second by Delambre, under the direction of the French Board of Longitude, and the third by Carlini. The last, however, differ only in arrangement from those of Delambre.

Similar observations will be applicable to the tables of the principal planets. Bouvard published, in 1808, tables of Jupiter and Saturn; but from the improved state of astronomy, he found it necessary to recompute these tables in 1821.

Although it is now about thirty years since the discovery of the four new planets, Ceres, Pallas, Juno, and Vesta, it was not till recently that tables of their motions were published. They have lately appeared in Encke's Ephemeris.

We have thus attempted to convey some notion (though necessarily a very inadequate one) of the immense extent of numerical tables which it has been found necessary to calculate and print for the purposes of the arts and sciences. We have before us a catalogue of the tables contained in the library of one private individual, consisting of not less than one hundred and forty volumes. Among these there are no duplicate copies; and we observe that many of the most celebrated voluminous tabular works are not contained among them. They are confined exclusively to arithmetical and trigonometrical tables; and, consequently, the myriad of astronomical and nautical tables are totally excluded from them. Nevertheless, they contain an extent of printed surface covered with figures amounting to above sixteen thousand square feet. We have taken at random forty of these tables, and have found that the number of errors *acknowledged* in the respective errata, amounts to above *three thousand seven hundred.*

To be convinced of the necessity which has existed for accurate numerical tables, it will only be necessary to consider at what an immense expenditure of labour and of money even the imperfect ones which we possess have been produced.

To enable the reader to estimate the difficulties which attend the attainment even of a limited degree of accuracy, we shall now explain some of the expedients which have been from time to time resorted to for the attainment of numerical correctness in calculating and printing them.

Among the scientific enterprises which the ambition of the French nation aspired to during the Republic, was the construction of a magnificent system of numerical tables. Their most distinguished mathematicians were called upon to contribute to the attainment of this important object; and the superintendence of the undertaking was confided to the celebrated Prony, who co-operated with the government in the adoption of such means as might be expected to ensure the production of a system of logarithmic and trigonometric tables, constructed with such accuracy that they should form a monument of calculation the most vast and imposing that had ever been executed, or even conceived. To accomplish this gigantic task, the principle of the division of labour, found to be so powerful in manufactures, was resorted to with singular success. The persons employed in the work were divided into three sections: the first consisted of half a dozen of the most eminent analysts. Their duty was to investigate the most convenient mathematical formulæ, which should enable the computors to proceed with the greatest expedition and accuracy by the method of Differences, of which we shall speak more fully hereafter. These formulæ, when decided upon by this first section, were handed over to the second section, which consisted of eight or ten properly qualified mathematicians. It was the duty of

* A comparison of the results for 1834, will be found in the "Nautical Almanac" for 1835.

this second section to convert into numbers certain general or algebraical expressions which occurred in the formulæ, so as to prepare them for the hands of the computers. Thus prepared, these formulæ were handed over to the third section, who formed a body of nearly one hundred computers. The duty of this numerous section was to compute the numbers finally intended for the tables. Every possible precaution was, of course, taken to ensure the numerical accuracy of the results. Each number was calculated by two or more distinct and independent computers, and its truth and accuracy determined by the coincidence of the results thus obtained.

The body of tables thus calculated occupied in manuscript *seventeen* folio volumes.*

As an example of the precautions which have been considered necessary to guard against errors in the calculation of numerical tables, we shall further state those which were adopted by Mr. Babbage, previously to the publication of his tables of logarithms. In order to render the terminal figure of tables in which one or more decimal places are omitted as accurate as it can be, it has been the practice to compute one or more of the succeeding figures; and if the first omitted figure be greater than 4, then the terminal figure is always increased by 1, since the value of the tabulated number is by such means brought nearer to the truth.† The tables of Callet, which were among the most accurate published logarithms, and which extended to seven places of decimals, were first carefully compared with the tables of Vega, which extended to ten places, in order to discover whether Callet had made the above correction of the final figure in every case where it was necessary. This previous precaution being taken, and the corrections which appeared to be necessary being made in a copy of Callet's tables, the proofs of Mr. Babbage's tables were submitted to the following test: They were first compared, number by number, with the corrected copy of Callet's logarithms; secondly with Hutton's logarithms; and thirdly, with Vega's logarithms. The corrections thus suggested being marked in the proofs, corrected revises were received back. These revises were then again compared, number by number, first with Vega's logarithms; secondly, with the logarithms of Callet; and thirdly, as far as the first 20,000 numbers, with the corresponding ones in Brigg's logarithms. They were now returned to the printer, and were stereotyped; proofs were taken from the stereotyped plates, which were put through the following ordeal: They were first compared once more with the logarithms of Vega as far as 47,500; they were then compared with the whole of the logarithms of Gardner; and next with the whole of Taylor's logarithms; and as a last test, they were transferred to the hands of a different set of readers, and were once more compared with Taylor. That these precautions were by no means superfluous may be collected from the following circumstances mentioned by Mr. Babbage: In the sheets read immediately previous to stereotyping, thirty-two errors were detected; after stereotyping, eight more were found, and corrected in the plates.

By such elaborate and expensive precautions many of the errors of computation and printing may certainly be removed; but it is too much to expect that in general such measures can be adopted; and we accordingly find by far the greater number of tables disfigured by errors, the extent of which is rather to be conjectured than determined. When the nature of a numerical table is considered— page after page densely covered with figures, and with nothing else—the chances against the detection of any single error will be easily comprehended; and it may therefore be fairly presumed, that for one error which may happen to be detected, there must be a great number which escape detection. Notwithstanding this difficulty, it is truly surprising how great a number of numerical errors have been detected by individuals no otherwise concerned in the tables than in their use. Mr. Baily states that he has himself detected in the solar and lunar tables, from which our Nautical Almanac was for a long period computed, more than five hundred errors. In the multiplication table already mentioned, computed by Dr. Hutton for the Board of Longitude, a single page was examined and recomputed: it was found to contain about forty errors.

In order to make the calculations upon the numbers found in the Ephemeral Tables published in the Nautical Almanac, it is necessary that the mariner should be supplied with certain permanent tables. A volume of these, to the number of about thirty, was accordingly computed, and published at national expense, by order of the Board of Longitude, entitled " Tables requisite to be used with

---

* These tables were never published. The printing of them was commenced by Didot, and a small portion was actually stereotyped, but never published. Soon after the commencement of the undertaking, the sudden fall of the assignats rendered it impossible for Didot to fulfil his contract with the government. The work was accordingly abandoned, and has never since been resumed. We have before us a copy of 100 pages folio of the portion which was printed at the time the work was stopped, given to a friend on a late occasion by Didot himself. It was remarked in this, as in other similar cases, that the computers who committed fewest errors were those who understood nothing beyond the process of addition.

† Thus, suppose the number expressed at full length were 3·1415927. If the table extend to no more than four places of decimals, we should tabulate the number 3·1416 and not 3·1415. The former would be evidently nearer to the true number, 3·1415927.

" the Nautical Ephemeris for finding the latitude and longitude at sea." In the first edition of these requisite tables, there were detected, by one individual, above a thousand errors.

The tables published by the Board of Longitude for the correction of the observed distances of the moon from certain fixed stars, are followed by a table of acknowledged errata, extending to seven folio pages, and containing more than eleven hundred errors. Even this table of errata itself is not correct: a considerable number of errors have been detected in it, so that errata upon errata have become necessary.

One of the tests most frequently resorted to for the detection of errors in numerical tables, has been the comparison of tables of the same kind, published by different authors. It has been generally considered that those numbers in which they are found to agree must be correct; inasmuch as the chances are supposed to be very considerable against two or more independent computers falling into precisely the same errors. How far this coincidence may be safely assumed as a test of accuracy we shall presently see.

A few years ago, it was found desirable to compute some very accurate logarithmic tables for the use of the great national survey of Ireland, which was then, and still is in progress; and on that occasion a careful comparison of various logarithmic tables was made. Six remarkable errors were detected, which were found to be common to several apparently independent sets of tables. This singular coincidence led to an unusually extensive examination of the logarithmic tables published both in England and in other countries; by which it appeared that thirteen sets of tables, published in London between the years 1633 and 1822, all agreed in these six errors. Upon extending the inquiry to foreign tables, it appeared that two sets of tables published at Paris, one at Gouda, one at Avignon, one at Berlin, and one at Florence, were infected by exactly the same six errors. The only tables which were found free from them were those of Vega, and the more recent impressions of Callet. It happened that the Royal Society possessed a set of tables of logarithms printed in the Chinese character, and on Chinese paper, consisting of two volumes: these volumes contained no indication or acknowledgment of being copied from any other work. They were examined; and the result was the detection in them of the same six errors.*

It is quite apparent that this remarkable coincidence of error must have arisen from the various tables being copied successively one from another. The earliest work in which they appeared was Vlacq's Logarithms (folio, Gouda, 1628); and from it, doubtless, those which immediately succeeded it in point of time were copied; from which the errors were subsequently transcribed into all the other, including the Chinese logarithms.

The most certain and effectual check upon errors which arise in the process of computation, is to cause the same computations to be made by separate and independent computers; and this check is rendered still more decisive if they make their computations by different methods. It is, nevertheless, a remarkable fact, that several computers, working separately and independently, do frequently commit precisely the same error; so that falsehood in this case assumes that character of consistency, which is regarded as the exclusive attribute of truth. Instances of this are familiar to most persons who have had the management of the computation of tables. We have reason to know, that M. Prony experienced it on many occasions in the management of the great French tables, when he found three, and even a greater number of computers, working separately and independently, to return him the same numerical result, and *that result wrong*. Mr. Stratford, the conductor of the Nautical Almanac, to whose talents and zeal that work owes the execution of its recent improvements, has more than once observed a similar occurrence. But one of the most signal examples of this kind, of which we are aware, is related by Mr. Baily. The catalogue of stars published by the Astronomical Society was computed by two separate and independent persons, and was afterwards compared and examined with great care and attention by Mr. Stratford. On examining this catalogue, and recalculating a portion of it, Mr. Baily discovered an error in the case of the star, κ Cephei. Its right ascension was calculated *wrongly*, and yet *consistently*, by two computers working separately. Their numerical results agreed precisely in every figure; but Mr. Stratford, on examining the catalogue, failed to detect the error. Mr. Baily having reason, from some discordancy which he observed, to suspect an error, recomputed the place of the star with a view to discover it; and he himself, in the first instance, obtained precisely *the same erroneous numerical result*. It was only on going over the operation a second time that he *accidentally* discovered that all had inadvertently committed the same error. †

It appears, therefore, that the coincidence of different tables, even when it is certain that they could not have been copied one from another, but must have been computed independently, is not a decisive test of their correctness, neither is it possible to ensure accuracy by the device of separate and independent computation.

* Memoirs Ast. Soc., vol. iii. p. 65.　　　　　† Ibid., vol. iv. p. 290.

Besides the errors incidental to the process of computation, there are further liabilities in the process of *transcribing* the final results of each calculation into the fair copy of the table designed for the printer. The next source of error lies with the compositor, in transferring this copy into type. But the liabilities to error do not stop even here; for it frequently happens, that after the press has been fully corrected, errors will be produced in the process of printing. A remarkable instance of this occurs in one of the six errors detected in so many different tables already mentioned. In one of these cases, the last five figures of two successive numbers of a logarithmic table were the following:—

35875
10436.

Now, both of these are erroneous; the figure 8 in the first line should be 4, and the figure 4 in the second should be 8. It is evident that the types, as first composed, were correct; but in the course of printing the two types 4 and 8 being loose, adhered to the inking-balls, and were drawn out; the pressman in replacing them transposed them, putting the 8 *above* and the 4 *below*, instead of *vice versâ*. It would be a curious inquiry, were it possible to obtain all the copies of the original edition of Vlacq's Logarithms, published at Gouda in 1628, from which this error appears to have been copied in all the subsequent tables, to ascertain whether it extends through the entire edition. It would probably, nay almost certainly, be discovered that some of the copies of that edition are correct in this number, while others are incorrect; the former having been worked off before the transposition of the types.

It is a circumstance worthy of notice, that this error in Vlacq's tables has produced a corresponding error in a variety of other tables deduced from them, *in which nevertheless the erroneous figures in Vlacq are omitted.* In no less than sixteen sets of tables published at various times since the publication of Vlacq, in which the logarithms extend only to seven places of figures, the error just mentioned in the *eighth place* in Vlacq causes a corresponding error in the *seventh* place. When the last three figures are omitted in the first of the above numbers, the seventh figure should be 5, inasmuch as the first of the omitted figures is under 5: the erroneous insertion, however, of the figure 8 in Vlacq has caused the figure 6 to be substituted for 5 in the various tables just alluded to. For the same reason, the erroneous occurrence of 4 in the second number has caused the adoption of a 0 instead of a 1 in the seventh place in the other tables. The only tables in which this error does not occur are those of Vega, the more recent editions of Callet, and the still later Logarithms of Mr. Babbage.

The *Opus Palatinum*, a work published in 1596, containing an extensive collection of trigonometrical tables, affords a remarkable instance of a tabular error; which, as it is not generally known, it may not be uninteresting to mention here. After that work had been for several years in circulation in every part of Europe, it was discovered that the commencement of the table of co-tangents and co-secants was vitiated by an error of considerable magnitude. In the first co-tangent the last nine places of figures were incorrect; but from the manner in which the numbers of the table were computed, the error was gradually, though slowly, diminished, until at length it became extinguished in the eighty-sixth page. After the detection of this extensive error, Pitiscus undertook the recomputation of the eighty-six erroneous pages. His corrected calculation was printed; and the erroneous part of the remaining copies of the *Opus Palatinum* was cancelled. But as the corrected table of Pitiscus was not published until 1607,—thirteen years after the original work,—the erroneous part of the volume was cancelled in comparatively few copies, and consequently correct copies of the work are now exceedingly rare. Thus, in the collection of tables published by M. Schulze,* the whole of the erroneous part of the *Opus Palatinum* has been adopted; he having used the copy of that work which exists in the library of the Academy of Berlin, and which is one of those copies in which the incorrect part was not cancelled. The corrected copies of this work may be very easily distinguished at present from the erroneous ones; it happened that the former were printed with a very bad and worn-out type, and upon a paper of a quality inferior to that of the original work. On comparing the first eighty-six pages of the volume with the succeeding ones, they are, therefore, immediately distinguishable in the corrected copies. Besides this test, there is another, which it may not be uninteresting to point out:—At the bottom of page 7 in the corrected copies, there is an error in the position of the words *basis* and *hypothenusa*, their places being interchanged. In the original uncorrected work this error does not exist.

At the time when the calculation and publication of Taylor's Logarithms were undertaken, it so happened that a similar work was in progress in France; and it was not until the calculation of the

* "Reçueil des Tables Logarithmiques et Trigonometriques." Par J. C. Schulze. 2 vols. Berlin: 1778.

K

French work was completed, that its author was informed of the publication of the English work. This circumstance caused the French calculator to relinquish the publication of his tables. The manuscript subsequently passed into the library of Delambre, and, after his death, was purchased at the sale of his books, by Mr. Babbage, in whose possession it now is. Some years ago it was thought advisable to compare these manuscript tables with Taylor's Logarithms, with a view to ascertain the errors in each, but especially in Taylor. The two works were peculiarly well suited for the attainment of this end: as the circumstances under which they were produced rendered it quite certain that they were computed independently of each other. The comparison was conducted under the direction of the late Dr. Young, and the result was the detection of the following nineteen errors in Taylor's Logarithms. To enable those who used Taylor's Logarithms to make the necessary corrections in them, the corrections of the detected errors appeared as follows in the " Nautical Almanac " for 1832.

ERRATA, *detected in* TAYLOR'S *Logarithms.* London: 4to, 1792.

				° ′ ″		
1 . E	. . .	Co-tangent of	. . .	1.35.55	. . .	*for* 43671 *read* 42671
2 . M	. . .	Co-tangent of	. . .	4. 4.49	. . .	— 66976 — 66979
3 . . .	. . .	Sine of	. . .	4.23.38	. . .	— 43107 — 43007
4 . . .	. . .	Sine of	. . .	4.23.39	. . .	— 43381 — 43281
5 . S	. . .	Sine of	. . .	6.45.52	. . .	— 10001 — 11001
6 . Kk	. . .	Co-sine of	. . .	14.18. 3	. . .	— 3398 — 3298
7 . Ss	. . .	Tangent of	. . .	18. 1.56	. . .	— 5064 — 6064
8 . Aaa	. .	Co-tangent of	. . .	21.11.14	. . .	— 6062 — 5962
9 . Ggg	. .	Tangent of	. . .	23.48.19	. . .	— 6087 — 5987
10 . . .	. . .	Co-tangent of	. . .	23.48.19	. . .	— 3913 — 4013
11 . Iii	. .	Sine of	. . .	25. 5. 4	. . .	— 3173 — 3183
12 . . .	. . .	Sine of	. . .	25. 5. 5	. . .	— 3218 — 3228
13 . . .	. . .	Sine of	. . .	25. 5. 6	. . .	— 3263 — 3273
14 . . .	. . .	Sine of	. . .	25. 5. 7	. . .	— 3308 — 3318
15 . . .	. . .	Sine of	. . .	25. 5. 8	. . .	— 3353 — 3363
16 . . .	. . .	Sine of	. . .	25. 5. 9	. . .	— 3398 — 3408
17 . Qqq	. .	Tangent of	. . .	28.19.39	. . .	— 6302 — 6402
18 . $4H$	. .	Tangent of	. . .	35.55.51	. . .	— 1681 — 1581
19 . $4K$	. .	Co-sine of	. . .	37.29. 2	. . .	— 5503 — 5603

An error being detected in this list of ERRATA, we find, in the Nautical Almanac for the year 1833, the following ERRATUM of the ERRATA of Taylor's Logarithms:

' In the list of ERRATA detected in Taylor's Logarithms, for cos. 4° 18′ 3″ *read* cos. 14° 18′ 2″.'

Here, however, confusion is worse confounded; for a new error, not before existing, and of much greater magnitude, is introduced ! It will be necessary, in the Nautical Almanac for 1836 (that for 1835 is already published), to introduce the following

ERRATUM of the ERRATUM of the ERRATA of TAYLOR'S *Logarithms.* For cos. 4° 18′ 3″, *read* cos. 14° 18′ 3″.

If proof were wanted to establish incontrovertibly the utter impracticability of precluding numerical errors in works of this nature, we should find it in this succession of error upon error, produced, in spite of the universally acknowledged accuracy and assiduity of the persons at present employed in the construction and management of the Nautical Almanac. It is only by the *mechanical fabrication of tables* that such errors can be rendered impossible.

On examining this list with attention, we have been particularly struck with the circumstances in which these errors appear to have originated. It is a remarkable fact, that of the above nineteen errors, eighteen have arisen from mistakes in *carrying*. Errors 5, 7, 10, 11, 12, 13, 14, 15, 16, 17, 19, have arisen from a carriage being neglected; and errors 1, 3, 4, 6, 8, 9, and 18, from a carriage being made where none should take place. In four cases, namely, errors 8, 9, 10, and 16, this has caused *two* figures to be wrong. The only error of the nineteen which appears to have been a press error is the second; which has evidently arisen from the type 9 being accidentally inverted, and thus becoming a 6. This may have originated with the compositor, but more probably it took place in the press-work; the type 9 being accidentally drawn out of the form by the inking-ball, as mentioned in a former case, and on being restored to its place, inverted by the pressman.

There are two cases among the above errata, in which an error, committed in the calculation of one number, has evidently been the cause of other errors. In the third erratum, a wrong carriage was made, in computing the sine of 4° 23′ 38″. The next number of the table was vitiated by this error; for we find the next erratum to be in the sine of 4° 23′ 39″, in which the figure similarly placed is 1 in excess. A still more extensive effect of this kind appears in errata 11, 12, 13, 14, 15, 16. A carriage was neglected in computing the sine of 25° 5′ 4″, and this produced a corresponding error in the five following numbers of the table, which are those corrected in the five following errata.

This frequency of errors arising in the process of carrying, would afford a curious subject of metaphysical speculation respecting the operation of the faculty of memory. In the arithmetical process, the memory is employed in a twofold way;—in ascertaining each successive figure of the calculated result by the recollection of a table committed to memory at an early period of life; and by another act of memory, in which the number *carried* from column to column is retained. It is a curious fact, that this latter circumstance, occurring only the moment before, and being in its nature little complex, is so much more liable to be forgotten or mistaken than the results of rather complicated tables. It appears, that among the above errata, the errors 5, 7, 10, 11, 17, 19, have been produced by the computer forgetting a carriage; while the errors 1, 3, 6, 8, 9, 18, have been produced by his making a carriage improperly. Thus, so far as the above list of errata affords grounds for judging, it would seem, (contrary to what might be expected,) that the error by which improper carriages are made is as frequent as that by which necessary carriages are overlooked.

We trust that we have succeeded in proving, first, the great national and universal utility of numerical tables, by showing the vast number of them, which have been calculated and published; secondly, that more effectual means are necessary to obtain such tables suitable to the present state of the arts, sciences and commerce, by showing that the existing supply of tables, vast as it certainly is, is still scanty, and utterly inadequate to the demands of the community;—that it is rendered inefficient, not only in quantity, but in quality, by its want of numerical correctness; and that such numerical correctness is altogether unattainable until some more perfect method be discovered, not only of calculating the numerical results, but of tabulating these,—of reducing such tables to type, and of printing that type so as to intercept the possibility of error during the press-work. Such are the ends which are proposed to be attained by the calculating machinery invented by Mr. Babbage.

The benefits to be derived from this invention cannot be more strongly expressed than they have been by Mr. Colebrooke, President of the Astronomical Society, on the occasion of presenting the gold medal voted by that body to Mr. Babbage:—" In no department of science, or of the arts, " does this discovery promise to be so eminently useful as in that of astronomy, and its kindred " sciences, with the various arts dependent on them. In none are computations more operose " than those which astronomy in particular requires;—in none are preparatory facilities more " needful;—in none is error more detrimental. The practical astronomer is interrupted in his " pursuit, and diverted from his task of observation by the irksome labours of computation, " or his diligence in observing becomes ineffectual for want of yet greater industry of calculation. " Let the aid which tables previously computed afford, be furnished to the utmost extent which " mechanism has made attainable through Mr. Babbage's invention, and the most irksome " portion of the astronomer's task is alleviated, and a fresh impulse is given to astronomical " research."

The first step in the progress of this singular invention was the discovery of some common principle which pervaded numerical tables of every description; so that by the adoption of such a principle as the basis of the machinery, a corresponding degree of generality would be conferred upon its calculations. Among the properties of numerical functions, several of a general nature exist; and it was a matter of no ordinary difficulty, and requiring no common skill, to select one which might, in all respects, be preferable to the others. Whether or not that which was selected by Mr. Babbage affords the greatest practical advantages, would be extremely difficult to decide—perhaps impossible, unless some other projector could be found possessed of sufficient genius, and sustained by sufficient energy of mind and character, to attempt the invention of calculating machinery on other principles. The principle selected by Mr. Babbage as the basis of that part of the machinery which calculates, is the Method of Differences; and he has, in fact, literally thrown this mathematical principle into wheel-work. In order to form a notion of the nature of the machinery, it will be necessary, first, to convey to the reader some idea of the mathematical principle just alluded to.

A numerical table, of whatever kind, is a series of numbers which possess some common character, and which proceed increasing or decreasing according to some general law. Supposing such a series continually to increase, let us imagine each number in it to be subtracted from that which follows it, and the remainders thus successively obtained to be ranged beside the first, so as to form another table: these numbers are called the *first differences*. If we suppose these likewise to increase continually, we may obtain a third table from them by a like process, subtracting each number from the succeeding one: this series is called the *second differences*. By adopting a like method of proceeding, another series may be obtained, called the *third differences*; and so on. By continuing this process, we shall at length obtain a series of differences, of some order, more or less high, according to the nature of the original table, in which we shall find the same number constantly repeated, to whatever extent the original table may have been continued; so that if the next series of differences had been obtained in the same manner as the preceding ones, every term of it would be 0. In some cases this would continue to whatever extent the original table might be carried; but in all cases a series of differences would be obtained, which would continue constant for a very long succession of terms.

As the successive series of differences are derived from the original table, and from each other, by *subtraction*, the same succession of series may be reproduced in the other direction by *addition*. But let us suppose that the first number of the original table, and of each of the series of differences, including the last, be given: all the numbers of each of the series may thence be obtained by the mere process of addition. The second term of the original table will be obtained by adding to the first the first term of the first difference series; in like manner, the second term of the first difference series will be obtained by adding to the first term, the first term of the third difference series, and so on. The second terms of all the series being thus obtained, the third terms may be obtained by a like process of addition; and so the series may be continued. These observations will perhaps be rendered more clearly intelligible when illustrated by a numerical example. The following is the commencement of a series of the fourth powers of the natural numbers :—

No.						Table.
1	.	.	.	.	.	1
2	.	.	.	.	.	16
3	.	.	.	.	.	81
4	.	.	.	.	.	256
5	.	.	.	.	.	625
6	.	.	.	.	.	1296
7	.	.	.	.	.	2401
8	.	.	.	.	.	4096
9	.	.	.	.	.	6561
10	.	.	.	.	.	10,000
11	.	.	.	.	.	14,641
12	.	.	.	.	.	20,736
13	.	.	.	.	.	28,561

By subtracting each number from the succeeding one in this series, we obtain the following series of first differences :—

$$
\begin{array}{c}
15 \\
65 \\
175 \\
369 \\
671 \\
1105 \\
1695 \\
2465 \\
3439 \\
4641 \\
6095 \\
7825
\end{array}
$$

In like manner, subtracting each term of this series from the succeeding one, we obtain the following series of second differences :—

$$
\begin{array}{c}
50 \\
110 \\
194 \\
302 \\
434 \\
590 \\
770 \\
974 \\
1202 \\
1454 \\
1730
\end{array}
$$

Proceeding with this series in the same way, we obtain the following series of third differences :—

$$
\begin{array}{c}
60 \\
84 \\
108 \\
132 \\
156 \\
180 \\
204 \\
228 \\
252 \\
276
\end{array}
$$

Proceeding in the same way with these, we obtain the following for the series of fourth differences :—

$$
\begin{array}{c}
24 \\
24 \\
24 \\
24 \\
24 \\
24 \\
24 \\
24 \\
24
\end{array}
$$

It appears, therefore, that in this case the series of fourth differences consists of a constant repetition of the number 24. Now, a slight consideration of the succession of arithmetical operations by which we have obtained this result, will show, that by reversing the process, we could obtain the table of fourth powers by the mere process of addition. Beginning with the first numbers in each successive series of differences, and designating the table and the successive differences by the letters T, D^1 D^2 D^3 D^4, we have then the following to begin with :—

T	D^1	D^2	D^3	D^4
1	15	50	60	24

Adding each number to the number on its left, and repeating 24, we get the following as the second terms of the several series :—

T	D^1	D^2	D^3	D^4
16	65	110	84	24

And, in the same manner, the third and succeeding terms as follows :—

No.	T	D¹	D²	D³	D⁴
1	1	15	50	60	24
2	16	65	110	84	24
3	81	175	194	108	24
4	256	369	302	132	24
5	625	671	434	156	24
6	1296	1105	590	180	24 .
7	2401	1695	770	204	24
8	4096	2465	974	228	24
9	6561	3439	1202	252	24
10	10000	4641	1454	276	
11	14641	6095	1730		
12	20736	7825			
13	28561				

There are numerous tables in which, as already stated, to whatever order of differences we may proceed, we should not obtain a series of rigorously constant differences; but we should always obtain a certain number of differences which to a given number of decimal places would remain constant for a long succession of terms. It is plain that such a table might be calculated by addition in the same manner as those which have a difference rigorously and continuously constant; and if at every point where the last difference requires an increase, that increase be given to it, the same principle of addition may again be applied for a like succession of terms, and so on.

By this principle it appears, that all tables in which each series of differences continually increases, may be produced by the operation of addition alone; provided the first terms of the table, and of each series of differences, be given in the first instance. But it sometimes happens, that while the table continually increases, one or more series of differences may continually diminish. In this case, the series of differences are found by subtracting each term of the series, not from that which follows, but from that which precedes it; and consequently, in the re-production of the several series, when their first terms are given, it will be necessary in some cases to obtain them by *addition*, and in others by *subtraction*. It is possible, however, still to perform all the operations by addition alone; this is effected in performing the operation of subtraction, by substituting for the subtrahend its *arithmetical complement*, and adding that, omitting the unit of the highest order in the result. This process, and its principle, will be readily comprehended by an example. Let it be required to subtract 357 from 768.

The common process would be as follows :—

$$
\begin{array}{lr}
\text{From} \quad . \quad . \quad . \quad . & 768 \\
\text{Subtract} \quad . \quad . \quad . & 357 \\
\hline
\text{Remainder} \quad . \quad . \quad . & 411 \\
\hline
\end{array}
$$

The *arithmetical complement* of 357, or the number by which it falls short of 1000, is 643. Now, if this number be added to 768, and the first figure on the left be struck out of the sum, the process will be as follows :—

$$
\begin{array}{lr}
\text{To} \quad . \quad . \quad . \quad . & 768 \\
\text{Add} \quad . \quad . \quad . \quad . & 643 \\
\hline
\text{Sum} \quad . \quad . \quad . \quad . & 1411 \\
\hline
\text{Remainder sought} \quad . \quad . & 411 \\
\hline
\end{array}
$$

The principle on which this process is founded is easily explained. In the latter process we have first added 643, and then subtracted 1000. On the whole, therefore, we have subtracted 357, since the number actually subtracted exceeds the number previously added by that amount.

Since, therefore, subtraction may be effected in this manner by addition, it follows that the calculation of all series, so far as an order of differences can be found in them which continues constant, may be conducted by the process of addition alone.

It also appears, from what has been stated, that each addition consists only of two operations. However numerous the figures may be of which the several pairs of numbers to be thus added may consist, it is obvious that the operation of adding them can only consist of repetitions of the process of adding one digit to another; and of carrying one from the column of inferior units to the column of units next superior when necessary. If we would therefore reduce such a process to machinery, it would only be necessary to discover such a combination of moving parts as are capable of performing these two processes of *adding* and *carrying* on two single figures; for, this being once accomplished, the process of adding two numbers, consisting of any number of digits, will be effected by repeating the same mechanism as often as there are pairs of digits to be added. Such was the simple form to which Mr. Babbage reduced the problem of discovering the calculating machinery; and we shall now proceed to convey some notion of the manner in which he solved it.

For the sake of illustration, we shall suppose that the table to be calculated shall consist of numbers not exceeding six places of figures; and we shall also suppose that the difference of the fifth order is the constant difference. Imagine, then, six rows of wheels, each wheel carrying upon it a dial-plate like that of a common clock, but consisting of *ten* instead of *twelve* divisions; the several divisions being marked 1, 2, 3, 4, 5, 6, 7, 8, 9, 0. Let these dials be supposed to revolve whenever the wheels to which they are attached are put in motion, and to turn in such a direction that the series of increasing numbers shall pass under the index which appears over each dial:—thus, after 0 passes the index, 1 follows, then 2, 3, and so on, as the dial revolves. In Fig. 1 are represented six horizontal rows of such dials.

The method of differences, as already explained, requires, that in proceeding with the calculation, this apparatus should perform continually the addition of the number expressed upon each row of dials, to the number expressed upon the row immediately above it. Now, we shall first explain how this process of addition may be conceived to be performed by the motion of the dials; and in doing so, we shall consider separately the processes of addition and carriage, considering the addition first, and then the carriage.

Let us first suppose the line D^1 to be added to the line T. To accomplish this, let us imagine that while the dials on the line D^1 are quiescent, the dials on the line T are put in motion, in such a manner, that as many divisions on each dial shall pass under its index, as there are units in the number at the index immediately below it. It is evident that this condition supposes that if 0 be at any index on the line D^1, the dial immediately above it in the line T shall not move. Now the motion here supposed, would bring under the indices on the line T such a number as would be produced by adding the number D^1 to T, neglecting all the carriages; for a carriage should have taken place in every case in which the figure 9 of every dial in the line T had passed under the index during the adding motion. To accomplish this carriage, it would be necessary that the dial immediately on the left of any dial in which 9 passes under the index, should be advanced one division, independently of those divisions which it may have been advanced by the addition of the number immediately below it. This effect may be conceived to take place in either of two ways. It may be either produced at the moment when the division between 9 and 0 of any dial passes under the index; in which case the process of carrying would go on simultaneously with the process of adding; or the process of carrying may be postponed in every instance until the process of addition, without carrying, has been completed; and then by another distinct and independent motion of the machinery, a carriage may be made by advancing one division all those dials on the right of which a dial had, during the previous addition, passed from 9 to 0 under the index. The latter is the method adopted in the calculating machinery, in order to enable its inventor to construct the carrying machinery independent of the adding mechanism.

Having explained the motion of the dials by which the addition, excluding the carriages of the number on the row D^1, may be made to the number on the row T, the same explanation may be applied to the number on the row D^2 to the number on the row D^1; also, of the number D^3 to the number on the row D^4, and so on. Now it is possible to suppose the additions of all the rows, except the first, to be made to all the rows except the last, simultaneously; and after these additions have been made, to conceive all the requisite carriages to be also made by advancing the proper dials one division forward. This would suppose all the dials in the scheme to receive their adding motion together; and, this being accomplished, the requisite dials to receive their carrying motions together. The production of so great a number of simultaneous motions throughout any machinery, would be attended with great mechanical difficulties, if indeed it be practicable. In the calculating machinery it is not attempted. The additions are performed in two successive periods of time, and the carriages in two other periods of time, in the following manner. We shall suppose one complete revolution of the axis which moves the machinery, to make one complete set of additions and carriages; it will then make them in the following order:—

The first quarter of a turn of the axis will add the second, fourth, and sixth rows to the first, third, and fifth, omitting the carriages; this it will do by causing the dials on the first, third, and fifth rows, to turn through as many divisions as are expressed by the numbers at the indices below them, as already mentioned.

The second quarter of a turn will cause the carriages consequent on the previous addition, to be made by moving forward the proper dials one division.

(During these two quarters of a turn, the dials of the first, third, and fifth row alone have been moved; those of the second, fourth, and sixth, have been quiescent.)

The third quarter of a turn will produce the addition of the third and fifth rows to the second and fourth, omitting the carriages; which it will do by causing the dials of the second and fourth rows to turn through as many divisions as are expressed by the numbers at the indices immediately below them.

The fourth and last quarter of a turn will cause the carriages consequent on the previous addition, to be made by moving the proper dials forward one division.

This evidently completes one calculation, since all the rows except the first have been respectively added to all the rows except the last.

To illustrate this: let us suppose the table to be computed to be that of the fifth powers of the natural numbers, and the computation to have already proceeded so far as the fifth power of 6, which is 7776. This number appears, accordingly, in the highest row, being the place appropriated to the number of the table to be calculated. The several differences as far as the fifth, which is in this case constant, are exhibited on the successive rows of dials in such a manner as to be adapted to the process of addition by alternate rows, in the manner already explained. The process of addition will commence by the motion of the dials in the first, third, and fifth rows, in the following manner: The dial A, fig. 1, must turn through one division, which will bring the number 7 to the index; the dial B must turn through three divisions, which will bring 0 to the index; this will render a carriage necessary, but that carriage will not take place during the present motion of the dial. The dial C will remain unmoved, since 0 is at the index below it; the dial D must turn through nine divisions; and as, in doing so, the division between 9 and 0 must pass under the index, a carriage must subsequently take place upon the dial to the left; the remaining dials of the row T, fig. 1, will remain unmoved. In the row D^2 the dial A^2 will remain unmoved, since 0 is at the index below it; the dial B^2 will be moved through five divisions, and will render a subsequent carriage on the dial to the left necessary; the dial C^2 will be moved through five divisions; the dial D^2 will be moved through three divisions, and the remaining dials of this row will remain unmoved. The dials of the row D^4 will be moved according to the same rules; and the whole scheme will undergo a change exhibited in fig. 2; a mark (*) being introduced on those dials to which a carriage is rendered necessary by the addition which has just taken place.

The second quarter of a turn of the moving axis will move foward through one division all the dials which in fig. 2 are marked (*), and the scheme will be converted into the scheme expressed in fig. 3.

In the third quarter of a turn, the dial A^1, fig. 3, will remain unmoved, since 0 is at the index below it; the dial B^1 will be moved forward through three divisions; C^1 through nine divisions, and so on; and in like manner the dials of the row D^3 will be moved forward through the number of divisions expressed at the indices in the row D^4. This change will convert the arrangement into that expressed in fig. 4, the dials to which a carriage is due, being distinguished as before by (*).

The fourth quarter of a turn of the axis will move forward one division all the dials marked (*); and the arrangement will finally assume the form exhibited in fig. 5, in which the calculation is completed. The first row T in this expresses the fifth power of 7; and the second expresses the number which must be added to the first row, in order to produce the fifth power of 8; the numbers in each row being prepared for the change which they must undergo, in order to enable them to continue the computation according to the method of alternate addition here adopted.

Having thus explained what it is that the mechanism is required to do, we shall now attempt to convey at least a general notion of some of the mechanical contrivances by which the desired ends are attained. To simplify the explanation, let us first take one particular instance—the dials B and B^1, fig. 1, for example. Behind the dial B is a bolt, which, at the commencement of the process, is shot between the teeth of a wheel which drives the dial B: during the first quarter of a turn this bolt is made to revolve, and if it continued to be engaged in the teeth of the said wheel, it would cause the dial B to make a complete revolution; but it is necessary that the dial B should only move through three divisions, and, therefore, when three divisions of this dial have passed under its index, the aforesaid bolt must be withdrawn : this is accomplished by a small wedge, which is placed in a fixed position on the wheel behind the dial B¹, and that position is such that this wedge will press upon the bolt in such a manner, that at the moment when three divisions of the dial B have passed under the index, it shall withdraw the bolt from the teeth of the wheel which it drives. The bolt will continue to revolve during the remainder of the first quarter of a turn of the axis, but it will no longer drive the dial B, which will remain quiescent. Had the figure at the index of the dial B¹ been any other, the wedge which withdraws the bolt would have assumed a different position, and would have withdrawn the bolt at a different time, but at a time always corresponding with the number under the index of the dial B¹ : thus, if 5 had been under the index of the dial B¹, then the bolt would have been withdrawn from between the teeth of the wheel which it drives, when five divisions of the dial B had passed under the index, and so on. Behind each dial in the row D¹ there is a similar bolt and a similar withdrawing wedge, and the action upon the dial above is transmitted and suspended in precisely the same manner. Like observations will be applicable to all the dials in the scheme here referred to, in reference to their adding actions upon those above them.

There is, however, a particular case which here merits notice : it is the case in which 0 is under the index of the dial from which the addition is to be transmitted upwards. As in that case nothing is to be added, a mechanical provision should be made to prevent the bolt from engaging in the teeth of the wheel which acts upon the dial above : the wedge which causes the bolt to be withdrawn, is thrown into such a position as to render it impossible that the bolt should be shot, or that it should enter between the teeth of the wheel, which in other cases it drives. But inasmuch as the usual means of shooting the bolt would still act, a strain would necessarily take place in the parts of the mechanism, owing to the bolt not yielding to the usual impulse. A small shoulder is therefore provided, which puts aside, in this case, the piece by which the bolt is usually struck, and allows the striking implement to pass without encountering the head of the bolt or any other obstruction. This mechanism is brought into play in the scheme, fig. 1, in the cases of all those dials in which 0 is under the index.

Such is the general description of the nature of the mechanism by which the adding process, apart from the carriages, is effected. During the first quarter of a turn, the bolts which drive the dials in the first, third, and fifth rows, are caused to revolve, and to act upon these dials, so long as they are permitted by the position of the several wedges on the second, fourth, and sixth rows of dials, by which these bolts are respectively withdrawn ; and, during the third quarter of a turn, the bolts which drive the dials of the second and fourth rows are made to revolve and act upon these dials so long as the wedges on the dials of the third and fifth rows, which withdraw them, permit. It will hence be perceived, that, during the first and third quarters of a turn, the process of addition is continually passing upwards through the machinery ; alternately from the even to the odd rows, and from the odd to the even rows, counting downwards.

We shall now attempt to convey some notion of the mechanism by which the process of carrying is effected during the second and fourth quarters of a turn of the axis. As before, we shall first explain it in reference to a particular instance. During the first quarter of a turn the wheel B², fig. 1, is caused by the adding bolt to move through five divisions ; and the fifth of these divisions, which passes under the index, is that between 9 and 0. On the axis of the wheel C², immediately to the left of B², is fixed a wheel, called in mechanics a ratchet wheel, which is driven by a claw which constantly rests in its teeth. This claw is in such a position as to permit the wheel C² to move in obedience to the action of the adding bolt, but to resist its motion in the contrary direction. It is drawn back by a spiral spring, but its recoil is prevented by a hook which sustains it; which hook, however, is capable of being withdrawn, and when withdrawn, the aforesaid spiral spring would draw back the claw, and make it fall through one tooth of the ratchet wheel. Now, at the moment that the division between 9 and 0 on the dial B² passes under the index, a thumb placed on the axis of this dial touches a trigger which raises out of the notch the hook which sustains the claw just mentioned, and allows it to fall back by the recoil of the spring, and to drop into the next tooth of the ratchet wheel. This process, however, produces no immediate effect upon the position of the wheel C², and is merely preparatory to an action intended to take place during the second quarter of a turn of the moving axis. It is in effect a memorandum taken by the machine of a carriage to be made in the next quarter of a turn.

During the second quarter of a turn, a finger placed on the axis of the dial B² is made to revolve, and it encounters the heel of the above-mentioned claw. As it moves forward it drives the claw before it ; and this claw, resting in the teeth of the ratchet wheel fixed upon the axis of the dial C², drives forward that wheel, and with it the dial. But the length and position of the finger which drives the claw limits its action, so as to move the claw forward through such a space only as will cause the dial C² to advance through a single division; at which point it is again caught and

L

retained by the hook. This will be added to the number under its index, and the requisite carriage from B² to C² will be accomplished.

In connection with every dial is placed a similar ratchet wheel with a similar claw, drawn by a similar spring, sustained by a similar hook, and acted upon by a similar thumb and trigger; and therefore the necessary carriages, throughout the whole machinery, take place in the same manner and by similar means.

During the second quarter of a turn, such of the carrying claws as have been allowed to recoil in the first, third, and fifth rows, are drawn up by the fingers on the axes of the adjacent dials; and, during the fourth quarter of a turn, such of the carrying claws on the second and fourth rows as have been allowed to recoil during the third quarter of a turn, are in like manner drawn up by the carrying fingers on the axes of the adjacent dials. It appears that the carriages proceed alternately from right to left along the horizontal rows during the second and fourth quarters of a turn; in the one, they pass along the first, third, and fifth rows, and in the other, along the second and fourth.

There are two systems of waves of mechanical action continually flowing from the bottom to the top; and two streams of similar action constantly passing from the right to the left. The crests of the first system of adding waves fall upon the last difference, and upon every alternate one proceeding upwards; while the crests of the other system touch upon the intermediate differences. The first stream of carrying action passes from right to left along the highest row and every alternate row, while the second stream passes along the intermediate rows.

Such is a very rapid and general outline of this machinery. Its wonders, however, are still greater in its details than even in its broader features. Although we despair of doing it justice by any description which can be attempted here, yet we should not fulfil the duty we owe to our readers, if we did not call their attention at least to a few of the instances of consummate skill which are scattered with a prodigality characteristic of the highest order of inventive genius, throughout this astonishing mechanism.

In the general description which we have given of the mechanism for *carrying*, it will be observed, that the preparation for every carriage is stated to be made during the previous addition, by the disengagement of the carrying claw before mentioned, and by its consequent recoil, urged by the spiral spring with which it is connected; but it may, and does, frequently happen, that though the process of addition may not have rendered a carriage necessary, one carriage may itself produce the necessity for another. This is a contingency not provided against in the mechanism as we have described it: the case would occur in the scheme represented in fig. 1, if the figure under the index of C² were 4 instead of 3. The addition of the number 5 at the index of C³ would, in this case, in the first quarter of a turn, bring 9 to the index of C²; this would obviously render no carriage necessary, and of course no preparation would be made for one by the mechanism—that is to say, the carrying claw of the wheel D² would not be detached. Meanwhile a carriage upon C² has been rendered necessary by the addition made in the first quarter of a turn to B². This carriage takes places in the ordinary way, and would cause the dial C², in the second quarter of a turn, to advance from 9 to 0: this would make the necessary preparation for a carriage from C² to D². But unless some special arrangement were made for the purpose, that carriage would not take place during the second quarter of a turn. This peculiar contingency is provided against by an arrangement of singular mechanical beauty, and which, at the same time, answers another purpose—that of equalizing the resistance opposed to the moving power by the carrying mechanism. The fingers placed on the axes of the several dials in the row D², do not act at the same instant on the carrying claws adjacent to them; but they are so placed that their action may be distributed throughout the second quarter of a turn in regular succession. Thus the finger on the axis of the dial A² first encounters the claw upon B², and drives it through one tooth immediately forwards; the finger on the axis of B² encounters the claw upon C², and drives it through one tooth; the action of the finger on C² on the claw on D² next succeeds, and so on. Thus, while the finger on B² acts on C², and causes the division from 9 to 0 to pass under the index, the thumb on C² at the same instant acts on the trigger, and detaches the carrying claw on D², which is forthwith encountered by the carrying finger on C², and driven forward one tooth. The dial D² accordingly moves forward one division, and 5 is brought under the index. This arrangement is beautifully effected by placing the several fingers, which act upon the carrying claws, *spirally* on their axes, so that they come into action in regular succession.

We have stated that, at the commencement of each revolution of the moving axis, the bolts which drive the dials of the first, third, and fifth rows, are shot. The process of shooting these bolts must therefore have taken place during the last quarter of the preceding revolution; but it is during that quarter of a turn that the carriages are effected in the second and fourth rows. Since the bolts which drive the dials at the first, third, and fifth rows, have no mechanical connexion with the dials in the second and fourth rows, there is nothing in the process of shooting those bolts incompatible with that of moving the dials of the second and fourth rows: hence these two processes may both take place during the same quarter of a turn. But in order to equalize the resistance to the moving power, the same expedient is here adopted as that already described in the process of carrying. The arms which shoot the bolts of each row of dials are arranged *spirally*, so as to act successively throughout the quarter of a turn. There is, however, a contingency which, under certain circumstances, would here produce a difficulty which must be provided against. It is possible, and in fact does sometimes happen, that the process of carrying causes a dial to move under the index from 0 to 1. In that case, the bolt, preparatory to the next addition, ought not to be shot until after the carriage takes place; for if the arm which shoots it passes its point of action before the carriage takes place, the bolt will be moved out of its sphere of action, and will not be shot, which, as we have already explained, must always happen when 0 is at the index: therefore no addition would in this case take place during the next quarter of a turn of the axis; whereas, since 1 is brought to the index by the carriage, which immediately succeeds the passage of the arm which ought to bolt, 1 should be added during the next quarter of a turn. It is plain, accordingly, that the mechanism should be so arranged, that the action of the arms, which shoot the bolts successively, should immediately follow the action of those fingers which raise the carrying claws successively; and therefore either a separate quarter of a turn should be appropriated to each of those movements, or if they be executed in the same quarter of a turn, the mechanism must be so constructed, that the arms which shoot the bolts successively, shall severally follow immediately after those which raise the carrying claws successively. The latter object is attained by a mechanical arrangement of singular felicity, and partaking of that elegance which characterises all the details of this mechanism. Both sets of arms are spirally arranged on their respective axes, so as to be carried through their period in the same quarter of a turn; but the one spiral is shifted a few degrees, in angular position, behind the other, so that each pair of corresponding arms succeed each other in the most regular order,—equalizing the resistance, economizing time, harmonizing the mechanism, and giving to the whole mechanical action the utmost practical perfection.

The system of mechanical contrivances by which the results, here attempted to be described, are attained, form only one order of expedients adopted in this machinery; although such is the perfection of their action, that in any ordinary case they would be regarded as having attained the ends in view with an almost superfluous degree of precision. Considering, however, the immense importance of the purposes which the mechanism was destined to fulfil, its inventor determined that a higher order of expedients should be superinduced upon those already described; the purpose of which should be to obliterate all small errors or inequalities which might, even by remote possibility, arise either from defects in the original formation of the mechanism, from inequality of wear, from casual strain or derangement,—or, in short, from any other cause whatever. Thus the movements of the first and principal parts of the mechanism were regarded by him merely as a first, though extremely nice approximation, upon which a system of small corrections was to be subsequently made by suitable and independent mechanism. This supplementary system of mechanism is so contrived, that if one or more of the moving parts of the mechanism of the first order be slightly out of their places, they will be forced to their exact position by the action of the mechanical expedients of the second order to which we now allude. If a more considerable derangement were produced by any accidental disturbance, the consequence would be that the supplementary mechanism would cause the whole system to become locked, so that not a wheel would be capable of moving; the impelling power would necessarily lose all its energy, and the machine would stop. The consequence of this exquisite arrangement is, that the machine will either calculate rightly, or not at all.

The supernumerary contrivances which we now allude to, being in a great degree unconnected with each other, and scattered through the machinery to a certain extent, independent of the mechanical arrangement of the principal parts, we find it difficult to convey any distinct notion of their nature or form.

In some instances they consist of a roller resting between certain curved surfaces, which has but one position of stable equilibrium, and that position the same, however the roller or the curved surfaces may wear. A slight error in the motion of the principal parts would make this roller for the moment rest on one of the curves; but, being constantly urged by a spring, it would press on the curved surface in such a manner as to force the moving piece on which that curved surface is formed, into such a position that the roller may rest between the two surfaces; that position being the one which the mechanism should have. A greater derangement would bring the roller to the crest of the curve, on which it would rest in instable equilibrium; and the machine would either become locked, or the roller would throw it as before into its true position.

In other instances a similar object is attained by a solid cone being pressed into a conical seat: the position of the axis of the cone and that of its seat being necessarily invariable, however the cone may wear; and the action of the cone upon the seat being such, that it cannot rest in any position except that in which the axis of the cone coincides with the axis of its seat.

L 2

Having thus attempted to convey a notion, however inadequate, of the calculating section of the machinery, we shall proceed to offer some explanation of the means whereby it is enabled to print its calculations in such a manner as to preclude the possibility of error in any individual printed copy.

On the axle of each of the wheels which express the calculated number of the table T, there is fixed a solid piece of metal, formed into a curve, not unlike the wheel in a common clock, which is called the *snail*. This curved surface acts against the arm of a lever, so as to raise that arm to a higher or lower point according to the position of the dial with which the snail is connected. Without entering into a more minute description, it will be easily understood that the snail may be so formed that the arm of the lever shall be raised to ten different elevations, corresponding to the ten figures of the dial which may be brought under the index. The opposite arm of the lever here described puts in motion a solid arch, or sector, which carries ten punches; each punch bearing on its face a raised character of a figure, and the ten punches bearing the ten characters, 1, 2, 3, 4, 5, 6, 7, 8, 9, 0. It will be apparent from what has been just stated, that this *type sector* (as it is called) will receive ten different attitudes, corresponding to the ten figures which may successively be brought under the index of the dial-plate. At a point over which the type sector is thus moved, and immediately under a point through which it plays, is placed a frame, in which is fixed a plate of copper. Immediately over a certain point through which the type sector moves, is likewise placed a *bent lever*, which, being straightened, is forcibly pressed upon the punch which has been brought under it. If the type sector be moved, so as to bring under the bent lever one of the steel punches above mentioned, and be held in that position for a certain time, the bent lever, being straightened, acts upon the steel punch, and drives it against the face of the copper beneath, and thus causes a sunken impression of the character upon the punch to be left upon the copper. If the copper be now shifted slightly in its position, and the type sector be also shifted so as to bring another punch under the bent lever, another character may be engraved on the copper by straightening the bent lever, and pressing it on the punch as before. It will be evident, that if the copper were shifted from right to left through a space equal to two figures of a number, and, at the same time, the type sector so shifted as to bring the punches corresponding to the figures of the number successively under the bent lever, an engraved impression of the number might thus be obtained upon the copper by the continued action of the bent lever. If, when one line of figures is thus obtained, a provision be made to shift the copper in a direction at right angles to its former motion, through a space equal to the distance between two lines of figures, and at the same time to shift it through a space in the other direction equal to the length of an entire line, it will be evident that another line of figures might be printed below the first in the same manner.

The motion of the type sector, here described, is accomplished by the action of the snail upon the lever already mentioned. In the case where the number calculated is that expressed in fig. 1, the process would be as follows:—The snail of the wheel F^1, acting upon the lever, would throw the type sector into such an attitude, that the punch bearing the character 0 would come under the bent lever. The next turn of the moving axis would cause the bent lever to press on the tail of the punch, and the character 0 would be impressed upon the copper. The bent lever being again drawn up, the punch would recoil from the copper by the action of a spring; the next turn of the moving axis would shift the copper through the interval between two figures, so as to bring the point destined to be impressed with the next figure under the bent lever. At the same time, the snail of the wheel E would cause the type sector to be thrown into the same attitude as before, and the punch 0 would be brought under the bent lever; the next turn would impress the figure 0 beside the former one, as before described. The snail upon the wheel D would now come into action, and throw the type sector into that position in which the punch bearing the character 7 would come under the bent lever, and at the same time the copper would be shifted through the interval between two figures; the straightening of the lever would next follow, and the character 7 would be engraved. In the same manner, the wheels C, B, and A would successively act by means of their snails; and the copper being shifted, and the lever allowed to act, the number 007776 would be finally engraved upon the copper: this being accomplished, the calculating machinery would next be called into action, and another circulation would be made, producing the next number of the Table exhibited in fig. 5. During this process the machinery would be engaged in shifting the copper both in the direction of its length and its breadth, with a view to commence the printing of another line; and this change of position would be accomplished at the moment when the next calculation would be completed: the printing of the next number would go on like the former, and the operation of the machine would proceed in the same manner, calculating and printing alternately. It is not, however, at all necessary —though we have here supposed it, for the sake of simplifying the explanation—that the calculating part of the mechanism should have its action suspended while the printing part is in operation, or *vice versâ*: it is not intended, in fact, to be so suspended in the actual machinery. The same turn of the axis by which one number is printed, executes a part of the movements necessary for the succeeding calculation; so that the whole mechanism will be simultaneously and continuously in action.

Of the mechanism by which the position of the copper is shifted from figure to figure, from line to line, we shall not attempt any description. We feel that it would be quite vain. Complicated and difficult to describe as every other part of this machinery is, the mechanism for moving the copper is such as it would be quite impossible to render at all intelligible, without numerous illustrative drawings.

The engraved plate of copper obtained in the manner above described, is designed to be used as a mould from which a stereotyped plate may be cast; or, if deemed advisable, it may be used as the immediate means of printing. In the one case we should produce a table, printed from type, in the same manner as common letter-press printing; in the other an engraved table. If it be thought most advisable to print from the stereotyped plates, then as many stereotyped plates as may be required may be taken from the copper mould; so that when once a table has been calculated and engraved by the machinery, the whole world may be supplied with stereotyped plates to print it, and may continue to be so supplied for an unlimited period of time. There is no practical limit to the number of stereotyped plates which may be taken from the engraved copper; and there is scarcely any limit to the number of printed copies which may be taken from any single stereotyped plate. Not only, therefore, is the numerical table by these means engraved and stereotyped with infallible accuracy, but such stereotyped plates are producible in unbounded quantity. Each plate, when produced, becomes itself the means of producing printed copies of the table, in accuracy perfect, and in number without limit.

Unlike all other machinery, the calculating mechanism produces, not the object of consumption, but the machinery by which that object may be made. To say that it computes and prints with infallible accuracy, is to understate its merits:—it computes and fabricates *the means* of printing with absolute correctness and an unlimited abundance.

For the sake of clearness, and to render ourselves more easily intelligible to the general reader, we have in the preceding explanation thrown the mechanism into an arrangement somewhat different from that which is really adopted. The dials expressing the numbers of the tables of the successive differences are not placed, as we have supposed them, in horizontal rows, and read from right to left, in the ordinary way; they are, on the contrary, placed vertically, one below the other, and read from top to bottom. The number of the table occupies the first vertical column on the right, the units being expressed on the lowest dial, and the tens on the next above that, and so on. The first difference occupies the next vertical column on the left; and the numbers of the succeeding differences occupy vertical columns, proceeding regularly to the left; the constant difference being on the last vertical column. It is intended in the machine now in progress to introduce six orders of differences, so that there will be seven columns of dials; it is also intended that the calculations shall extend to eighteen places of figures: thus each column will have eighteen dials. We have referred to the dials as if they were inscribed upon the faces of wheels, whose axes are horizontal and planes vertical. In the actual machinery the axes are vertical and the planes horizontal, so that the edges of the *figure wheels*, as they are called, are presented to the eye. The figures are inscribed, not upon the dial-plate, but around the surface of a small cylinder or barrel, placed upon the axis of the figure wheel, which revolves with it; so that as the figure wheel revolves, the figures on the barrel are successively brought to the front, and pass under an index engraved upon a plate of metal immediately above the barrel. This arrangement has the obvious practical advantage, that, instead of each figure wheel having a separate axis, all the figure wheels of the same vertical column revolve on the same axis; and the same observation will apply to all the wheels with which the figure wheels are in mechanical connexion. This arrangement has the further mechanical advantage over that which has been assumed for the purposes of explanation, that the friction of the wheel-work on the axes is less in amount, and more uniformly distributed, than it could be if the axes were placed in the horizontal position.

A notion may therefore be formed of the front elevation of the calculating part of the mechanism, by conceiving seven steel axes erected, one beside another, on each of which shall be placed eighteen wheels,* five inches in diameter, having cylinders or barrels upon them an inch and a half in height, and inscribed, as already stated, with the ten arithmetical characters. The entire elevation of the machinery would occupy a space measuring ten feet broad, ten feet high, and five feet deep. The process of calculation would be observed by the alternate motion of the figure wheels on the several axes. During the first quarter of a turn, the wheels on the first, third, and fifth axes would turn, receiving their addition from the second, fourth, and sixth; during the second quarter of a turn, such of the wheels on the first, third, and fifth axes, to which carriages are due, would be moved forward one additional figure: the second, fourth, and sixth columns of wheels being all this time quiescent.

---

* The wheels, and every other part of the mechanism except the axes, springs, and such parts as are necessarily of steel, are formed of an alloy of copper with a small portion of tin.

During the third quarter of a turn, the second, fourth, and sixth columns would be observed to move, receiving their additions from the third, fifth, and seventh axes: and during the fourth quarter of a turn, such of these wheels to which carriages are due, would be observed to move forward one additional figure; the wheels of the first, third, and fifth columns being quiescent during this time.

It will be observed that the wheels of the seventh column are always quiescent in this process; and it may be asked, of what use they are, and whether some mechanism of a fixed nature would not serve the same purpose? It must, however, be remembered, that for different tables there will be different constant differences; and that when the calculation of a table is about to commence, the wheels on the seventh axis must be moved by the hand, so as to express the constant difference, whatever it may be. In tables, also, which have not a difference rigorously constant, it will be necessary, after a certain number of calculations, to change the constant difference by the hand; and in this case the wheels of the seventh axis must be moved when occasion requires. Such adjustment, however, will only be necessary at very distant intervals, and after a considerable extent of printing and calculation has taken place; and when it is necessary, a provision is made in the machinery by which notice will be given by the sounding of a bell, so that the machine may not run beyond the extent of its powers of calculations.

Immediately behind the seven axes on which the figure wheels revolve, are seven other axes; on which are placed, first, the wheels already described as driven by the figure wheels, and which bear upon them the wedge which withdraws the bolt immediately over these latter wheels, and on the same axis is placed the adding bolt. From the bottom of this bolt there projects downwards the pin, which acts upon the unbolting wedge by which the bolt is withdrawn: from the upper surface of the bolt proceeds a tooth, which, when the bolt is shot, enters between the teeth of the adding wheel, which turns on the same axis, and is placed immediately above the bolt: its teeth, on which the bolt acts, are like the teeth of a crown wheel, and are presented downwards. The bolt is fixed upon this axis, and turns with it; but the adding wheel above the bolt, and the unbolting wheel below it, both turn upon the axis, and independently of it. When the axis is made to revolve by the moving power, the bolt revolves with it; and so long as the tooth of the bolt remains inserted between those of the adding wheel, the latter is likewise moved; but when the lower pin of the bolt encounters the unbolting wedge on the lower wheel, the tooth of the bolt is withdrawn and the motion of the adding wheel is stopped. This adding wheel is furnished with spur teeth, besides the crown teeth just mentioned; and these spur teeth are engaged with those of that unbolting wheel which is in connexion with the adjacent figure wheel to which the addition is to be made. By such an arrangement it is evident that the revolution of the bolt will necessarily add to the adjacent figure wheel the requisite number.

It will be perceived, that upon the same axis are placed an unbolting wheel, a bolt, and an adding wheel, one above the other, for every figure wheel; and as there are eighteen figure wheels there will be eighteen tiers; each tier formed of an unbolting wheel, a bolt, and an adding wheel, placed one above the other; the wheels on this axis all revolving independent of the axis, but the bolts being all fixed upon it. The same observations, of course will apply to each of the seven axes.

At the commencement of every revolution of the adding axes, it is evident that the several bolts placed upon them must be shot in order to perform the various additions. This is accomplished by a third set of seven axes, placed at some distance behind the range of the wheels, which turn upon the adding axes: these are called *bolting axes*. On these bolting axes are fixed, so as to revolve with them, a bolting finger opposite to each bolt: as the bolting axis is made to revolve by the moving power, the bolting finger is turned, and as it passes near the bolt, it encounters the shoulder of a hammer or lever, which strikes the heel of the bolt, and presses it forward so as to shoot its tooth between the crown teeth of the adding wheel. The only exception to this action is the case in which 0 happens to be at the index of the figure wheel; in that case, the lever or hammer, which the bolting finger would encounter, is, as before stated, lifted out of the way of the bolting finger, so that it revolves without encountering it. It is on the bolting axes that the fingers are spirally arranged so as to equalize their action as already explained.

The same axes in front of the machinery on which the figure wheels turn are made to serve the purpose of *carrying*. Each of these bear a series of fingers which turn with them, and which encounter a carrying claw, already described, so as to make the carriage: these carrying fingers are also spirally arranged on their axes, as already described.

Although the absolute accuracy which appears to be ensured by the mechanical arrangements here described is such as to render further precautions nearly superfluous, still it may be right to state, that, supposing it were possible for an error to be produced in calculation, this error could be easily and speedily detected in the printed tables: it would only be necessary to calculate a number of the table taken at intervals, through which the mechanical action of the machine has not been suspended, and during which it has received no adjustment by the hand : if the computed number be found to agree with those printed, it may be taken for granted that all the intermediate numbers are correct ; because, from the nature of the mechanism, and the principle of computation, an error occurring in any single number of the table would be unavoidably entailed, in an increasing ratio, upon all the succeeding numbers.

We have hitherto spoken merely of the practicability of executing by the machinery, when completed, that which its inventor originally contemplated—namely, the calculating and printing of all numerical tables, derived by the method of differences from a constant difference. It has, however happened that the actual powers of the machinery greatly transcend those contemplated in its original design :—they not only have exceeded the most sanguine anticipations of its inventor, but they appear to have an extent to which it is utterly impossible, even for the most acute mathematical thinker, to fix a probable limit. Certain subsidiary mechanical inventions have, in the progress of the enterprise, been, by the very nature of the machinery, suggested to the mind of the inventor, which confer upon it capabilities which he had never foreseen. It would be impossible even to enumerate, within the limits of this article, much less to describe in detail, those extraordinary mechanical arrangements, the effects of which have not failed to strike with astonishment every one who has been favoured with an opportunity of witnessing them, and who has been enabled, by sufficient mathematical attainments, in any degree to estimate their probable consequences.

As we have described the mechanism, the axes containing the several differences are successively and regularly added one to another ; but there are certain mechanical adjustments, and these of a very simple nature, which being thrown into action, will cause a difference of any order to be added any number of times to a difference of any other order ; and that either proceeding backwards or forwards, from a difference of an inferior to one of a superior order, and *vice versâ*.*

Among other peculiar mechanical provisions in the machinery is one by which, when the table for any order of difference amounts to a certain number, a certain arithmetical change would be made in the constant difference. In this way a series may be tabulated by the machine, in which the constant difference is subject to periodical change ; or the very nature of the table itself may be subject to periodical change, and yet to one which has a regular law.

Some of these subsidiary powers are peculiarly applicable to calculations required in astronomy, and are therefore of eminent and immediate practical utility : others there are by which tables are produced, following the most extraordinary, and apparently capricious, but still regular laws. Thus a table will be computed, which, to any required extent, shall coincide with a given table, and which shall deviate from that table for a single term, or for any required number of terms, and then resume its course, or which shall permanently alter the law of its construction. Thus the engine has calculated a table which agreed precisely with a table of square numbers, until it attained the hundred and first term, which was not the square of 101, nor were any of the subsequent numbers squares. Again, it has computed a table which coincided with the series of natural numbers, as far as 100,000,001, but which subsequently followed another law. This result was obtained, not by working the engine through the whole of the first table, for that would have required an enormous length of time ; but by showing, from the arrangement of the mechanism, that it must continue to exhibit the succession of natural numbers, until it would reach 100,000,000. To save time, the engine was set by the hand to the number 99999995, and was then put in regular operation. It produced successively the following numbers† :—

$$99,999,996$$
$$99,999,997$$
$$99,999,998$$
$$99,999,999$$
$$100,000,000$$
$$100,010,002$$
$$100,030,003$$
$$100,060,004$$
$$100,100,005$$
$$100,160,006$$
&c., &c.

* The machine was constructed with the intention of tabulating the equation $\Delta^7 u_z = 0$, but, by the means above alluded to, it is capable of tabulating such equations as the following : $\Delta^7 u_z = a \ \Delta \ u_z$, $\Delta^7 u_z = a \ \Delta^3 u_z$, $\Delta^7 u =$ units figure of Δu.

† Such results as this suggest a train of reflection on the nature and operation of general laws, which would lead to very curious and interesting speculations. The natural philosopher and astronomer will be hardly less struck with them than the metaphysician and theologian.

Equations have been already tabulated by the portion of the machinery which has been put together, which are so far beyond the reach of the present power of mathematics, that no distant term of the table can be predicted, nor any function discovered capable of expressing its general law. Yet the very fact of the table being produced by mechanism of an invariable form, and including a distinct principle of mechanical action, renders it quite manifest that *some* general law must exist in every table which it produces. But we must dismiss these speculations: we feel it impossible to stretch the powers of our own mind, so as to grasp the probable capabilities of this splendid production of combined mechanical and mathematical genius; much less can we hope to enable others to appreciate them, without being furnished with such means of comprehending them as those with which we have been favoured. Years must in fact elapse, and many inquirers direct their energies to the cultivation of the vast field of research thus opened, before we can fully estimate the extent of this triumph of matter over mind. "Nor is it," says Mr. Colebrooke, "among the least curious results of this "ingenious device, that it affords a new opening for discovery, since it is applicable, as has been "shown by its inventor, to surmount novel difficulties of analysis. Not confined to constant differ- "ences, it is available in every case of differences that follow a definite law, reducible therefore to an "equation. An engine adjusted to the purpose being set to work will produce any distant term, or "succession of terms, required—thus presenting the numercial solution of a problem, even though "the analytical solution be yet undetermined." That the future path of some important branches of mathematical inquiry must now in some measure be directed by the dictates, of mechanism is sufficiently evident; for who would toil on in any course of analytical inquiry, in which he must ultimately depend on the expensive and fallible aid of human arithmetic, with an instrument in his hands, in which all the dull monotony of numerical computation is turned over to the untiring action and unerring certainty of mechanical agency.

It is worth notice, that each of the axes in front of the machinery on which the figure wheels revolve, is connected with a bell, the tongue of which is governed by a system of levers, moved by the several figure wheels; an adjustment is provided by which the levers shall be dismissed, so as to allow the hammer to strike against the bell, whenever any proposed number shall be exhibited on the axis. This contrivance enables the machine to give notice to its attendants at any time that an adjustment may be required.

Among a great variety of curious accidental properties (so to speak) which the machine is found to possess, is one by which it is capable of solving numerical equations which have rational roots. Such an equation being reduced (as it always may be) by suitable transformations to that state in which the roots shall be whole numbers, the values 0, 1, 2, 3, &c., are substituted for the unknown quantity, and the corresponding values of the equation ascertained. From these a sufficient number of differences being derived, they are set upon the machine. The machine being then put in motion, the table axis will exhibit the successive values of the formula, corresponding to the substitutions of the successive whole numbers for the unknown quantity: at length the number exhibited on the table axis will be 0, which will evidently correspond to a root of the equation. By previous adjust- ment, the bell of the table axis will in this case ring and give notice of the exhibition of the value of the root in another part of the machinery.

If the equation have imaginary roots, the formula being necessarily a maximum or minimum on the occurrence of such roots, the first difference will become nothing; and the dials of that axis will under such circumstances present 0 to the respective indices. By previous adjustment, the bell of this axis would here give notice of a pair of imaginary roots.

Mr. Colebrooke speculates on the probable extension of these powers of the machine: "It may "not therefore be deemed too sanguine an anticipation when I express the hope that an instrument "which, in its simpler form, attains to the extraction of roots of numbers, and approximates to the "roots of equations, may in a more advanced state of improvement, rise to the approximate solution "of algebraic equations of elevated degrees. I refer to solutions of such equations proposed by La "Grange, and more recently by other annalists, which involve operations too tedious and intricate "for use, and which must remain without efficacy, unless some mode be devised of abridging the "labour, or facilitating the means of its performance. In any case this engine tends to lighten the "excessive and accumulating burden of arithmetical application of mathematical formulæ, and to "relieve the progress of science from what is justly termed by the author of this invention, the over- "whelming encumbrance of numerical detail."

Although there are not more than eighteen figure wheels on each axis, and therefore it might be supposed that the machinery was capable of calculating only to the extent of eighteen decimal places; yet there are contrivances connected with it, by which, in two successive calcula- tions, it will be possible to calculate even to the extent of thirty decimal places. Its powers, therefore, in this respect, greatly exceed any which can be required in practical science. It is also remarkable, that the machinery is capable of producing the calculated results *true to the last figure.* We have already explained, that when the figure which would follow the last is greater than 4, then it would be necessary to increase the last figure by 1; since the excess of the calculated number above the true value would in such case be less than its defect from it would be, had the regularly computed final figure been adopted: this is a precaution necessary in all numerical tables, and it is one which would hardly have been expected to be provided for in the calculating machinery.

As might be expected in a mechanical undertaking of such complexity and novelty, many practical difficulties have since its commencement been encountered and surmounted. It might have been foreseen, that many expedients would be adopted and carried into effect, which farther experiments would render it necessary to reject: and thus a large source of additional expense could scarcely fail to be produced. To a certain extent this has taken place; but owing to the admirable system of mechanical drawings, which in every instance Mr. Babbage has caused to be made, and owing to his own profound acquaintance with the practical working of the most complicated mechanism, he has been able to predict in every case what the result of any contrivance would be, as perfectly from the drawing, as if it had been reduced to the form of a working model. The drawings, consequently, form a most extensive and essential part of the enterprise. They are executed with extraordinary ability and precision, and may be considered as perhaps the best specimens of mechanical drawings which have ever been executed. It has been on these, and on these only, that the work of invention has been bestowed. In these, all those progressive modifications suggested by consideration and study have been made; and it was not until the inventor was fully satisfied with the result of any contrivance, that he had it reduced to a working form. The whole of the loss which has been incurred by the necessarily progressive course of invention, has been the expense of rejected drawings. Nothing can perhaps more forcibly illustrate the extent of labour and thought which has been incurred in the production of this machinery, than the contemplation of the working drawings which have been executed previously to its construction: these drawings cover above a thousand square feet of surface, and many of them are of the most elaborate and complicated description.

One of the practical difficulties which presented themselves at a very early stage in the progress of this undertaking, was the impossibility of bearing in mind all the variety of motions propagated simultaneously through so many complicated trains of mechanism. Nothing but the utmost imaginable harmony and order among such a number of movements, could prevent obstructions arising from incompatible motions encountering each other. It was very soon found impossible, by a mere act of memory, to guard against such an occurrence; and Mr. Babbage found, that, without some effective expedient by which he could at a glance see what every moving piece in the machinery was doing at each instant of time, such inconsistencies and obstructions as are here alluded to must continually have occurred. This difficulty was removed by another invention of even a more general nature than the calculating machinery itself, and pregnant with results probably of higher importance. This invention consisted in the contrivance of a scheme of *mechanical notation* which is generally applicable to all machinery whatsoever; and which is exhibited on a table or plan consisting of two distinct sections. In the first is traced, by a peculiar system of signs, the origin of every motion which takes place throughout the machinery; so that the mechanist or inventor is able, by moving his finger along a certain line, to follow out the motion of every piece from effect to cause, until he arrives at the prime mover. The same sign which thus indicates the *source* of motion indicates likewise the *species* of motion, whether it be continuous or reciprocating, circular or progressive, &c. The same system of signs further indicates the nature of the mechanical connexion between the mover and the thing moved, whether it be permanent and invariable (as between the two arms of a lever), or whether the mover and the moved are separate and independent pieces, as is the case when a pinion drives a wheel; also whether the motion of one piece necessarily implies the motion of another; or when such motion in the one is interrupted, and in the other continuous, &c.

The second section of the table divides the time of a complete period of the machinery into any required number of parts; and it exhibits in a map, as it were, that which every part of the machine is doing at each moment of time. In this way, incompatibility in the motions of different parts is rendered perceptible at a glance. By such means the contriver of machinery is not merely prevented from introducing into one part of the mechanism any movement inconsistent with the simultaneous action of the other parts; but when he finds that the introduction of any particular movement is necessary for his purpose, he can easily and rapidly examine the whole range of the machinery during one of its periods, and can find by inspection whether there is any, and what portion of time, at which no motion exists incompatible with the desired one, and thus discover a *niche*, as it were, in which to place the required movement. A further and collateral advantage consists in placing it in the power of the contriver to exercise the utmost possible economy

M

of *time* in the application of his moving power. For example, without some instrument of mechanical inquiry equally powerful with that now described, it would be scarcely possible, at least in the first instance, so to arrange the various movements that they should be all executed in the least possible number of revolutions of the moving axis. Additional revolutions would almost inevitably be made for the purpose of producing movements and changes which it would be possible to introduce in some of the phases of previous revolutions; and there is no one acquainted with the history of mechanical invention who must not be aware, that in the progressive contrivance of almost every machine the earliest arrangements are invariably defective in this respect; and that it is only by a succession of improvements, suggested by long experience, that that arrangement is at length arrived at, which accomplishes all the necessary motions in the shortest possible time. By the application of the mechanical notation, however, absolute perfection may be arrived at in this respect; even before a single part of the machinery is constructed, and before it has any other existence than that which it obtains upon paper.

Examples of this class of advantages derivable from the notation will occur to the mind of every one acquainted with the history of mechanical invention. In the common suction-pump, for example, the effective agency of the power is suspended during the descent of the piston. A very simple contrivance, however, will transfer to the descent the work to be accomplished in the next ascent; so that the duty of four strokes of the piston may thus be executed in time of two. In the earlier applications of the steam-engine, that machine was applied almost exclusively to the process of pumping; and the power acted only during the descent of the piston, being suspended during its ascent. When, however, the notion of applying the engine to the general purposes of manufacture occurred to the mind of Watt, he saw that it would be necessary to cause it to produce a continued rotatory motion; and, therefore, that the intervals of intermission must be filled up by the action of the power. He first proposed to accomplish this by a second cylinder working alternately with the first; but it soon became apparent that the blank which existed during the up-stroke in the action of the power, might be filled up by introducing the steam at both ends of the cylinder alternately. Had Watt placed before him a scheme of mechanical notation such as we allude to, this expedient would have been so obtruded upon him that he must have adopted it from the first.

One of the circumstances from which the mechanical notation derives a great portion of its power as an instrument of investigation and discovery, is that it enables the inventor to dismiss from his thoughts, and to disencumber his imagination of the arrangement and connexion of the mechanism; which, when it is very complex (and it is in that case that the notation is most useful), can only be kept before the mind by an embarrassing and painful effort. In this respect the powers of the notation may not inaptly be illustrated by the facilities derived in complex and difficult arithmetical questions from the use of the language and notation of algebra. When once the peculiar conditions of the question are translated into algebraical signs, and "reduced to an equation," the computist dismisses from his thoughts all the circumstances of the question, and is relieved from the consideration of the complicated relations of the quantities of various kinds which may have entered it. He deals with the algebraical symbols, which are the representatives of those quantities and relations, according to certain technical rules of a general nature, the truth of which he has previously established; and, by a process almost mechanical, he arrives at the required result. What algebra is to arithmetic, the notation we now allude to is to mechanism. The various parts of the machinery under consideration being once expressed upon paper by proper symbols, the inquirer dismisses altogether from his thoughts the mechanism itself, and attends only to the symbols; the management of which is so extremely simple and obvious, that the most unpractised person, having once acquired an acquaintance with the signs, cannot fail to comprehend their use.

A remarkable instance of the power and utility of this notation occurred in a certain stage of the invention of the calculating machinery. A question arose as to the best method of producing and arranging a certain series of motions necessary to print and calculate a number. The inventor, assisted by a practical engineer of considerable experience and skill, had so arranged these motions, that the whole might be performed by twelve revolutions of the principal moving axis. It seemed, however, desirable, if possible, to execute these motions by a less number of revolutions. To accomplish this, the engineer sat down to study the complicated details of a part of the machinery which had been put together; the inventor at the same time applied himself to the consideration of the arrangement and connection of the symbols in his scheme of notation. After a short time, by some transposition of symbols, he caused the received motions to be completed by eight turns of the axis. This he accomplished by transferring the symbols which occupied the last four divisions of his scheme, into such blank spaces as he could discover in the first eight divisions; due care being taken that no symbols should express actions at once simultaneous and incompatible. Pushing his inquiry, however, still further, he proceeded to ascertain whether his scheme of symbols did not admit of a still more compact arrangement, and whether eight revolutions were not more than enough to accomplish what was required. Here the powers of the practical engineer completely broke down. By no effort could he bring before his mind such a view of the complicated mechanism as would enable him to decide upon any improved arrangement. The inventor, however, without any extraordinary mental exertion, and merely by sliding a bit of ruled pasteboard up and down his plan, in search of a vacancy where the different motions might be placed, at length contrived to *pack* all the motions, which had previously occupied eight turns of the handle, into five turns. The symbolic instrument with which he conducted the investigation, now informed him of the impossibility of reducing the action of the machine to a more condensed form. This appeared by the fulness of every space along the lines of compatible action. It was, however, still possible, by going back to the actual machinery, to ascertain whether movements, which, under existing arrangements, were incompatible, might not be brought into harmony. This he accordingly did, and succeeded in diminishing the number of incompatible conditions, and thereby rendered it possible to make actions simultaneous which were before necessarily successive. The notation was now again called into requisition, and a new disposition of the parts was made. At this point of the investigation, this extraordinary instrument of mechanical analysis put forth one of its most singular exertions of power. It presented to the eye of the engineer two currents of mechanical action, which, from their nature, could not be simultaneous ; and each of which occupied a complete revolution of the axis, except about a twentieth ; the one occupying the last nineteen-twentieths of a complete revolution of the axis, and the other occupying the first nineteen-twentieths of a complete revolution. One of these streams of action was, the successive picking up by the carrying fingers of the successive carrying claws ; and the other was, the successive shooting of nineteen bolts by the nineteen bolting fingers. The notation rendered it obvious, that as the bolting action commenced a small space below the commencement of the carrying, and ended an equal space below the termination of the carrying, the two streams of action could be made to flow after one another in one and the same revolution of the axis. He thus succeeded in reducing the period of completing the action to four turns of the axis ; when the notation again informed him that he had again attained a limit of condensed action, which could not be exceeded without a further change in the mechanism. To the mechanism he again recurred, and soon found that it was possible to introduce a change which would cause the action to be completed in three revolutions of the axis. An odd number of revolutions, however, being attended with certain practical inconveniences, it was considered more advantageous to execute the motions in four turns ; and here again the notation put forth its powers, by informing the inventor, *through the eye*, almost independent of his mind, what would be the most elegant, symmetrical, and harmonious disposition of the required motions in four turns. This application of an almost metaphysical system of abstract signs, by which the motion of the hand performs the office of the mind, and of profound practical skill in mechanics alternately, to the construction of a most complicated engine, forcibly reminds us of a parallel in another science, where the chemist with difficulty succeeds in dissolving a refractory mineral, by the alternate action of the most powerful acids, and the most caustic alkalies, repeated in long-continued succession.

This important discovery was explained by Mr. Babbage, in a short paper read before the Royal Society, and published in the Philosophical Transactions in 1826.* It is to us more a matter of regret than surprise, that the subject did not receive from scientific men in this country that attention to which its importance in every practical point of view so fully entitled it. To appreciate it would indeed have been scarcely possible, from the very brief memoir which its inventor presented, unaccompanied by any observations or arguments of a nature to force it upon the attention of minds unprepared for it by the nature of their studies or occupations. In this country, science has been generally separated from practical mechanics by a wide chasm. It will be easily admitted, that an assembly of eminent naturalists and physicians, with a sprinkling of astronomers, and one or two abstract mathematicians, were not precisely the persons best qualified to appreciate such an instrument of mechanical investigation as we have here described. We shall not therefore be understood as intending the slightest disrespect for these distinguished persons, when we express our regret, that a discovery of such paramount practical value, in a country pre-eminently conspicuous for the results of its machinery, should fall still-born and inconsequential through their hands, and be buried unhonoured and undiscriminated in their miscellaneous transactions. We trust that a more auspicious period is at hand ; that the chasm which has separated practical from scientific men will speedily close ; and that that combination of knowledge will be effected, which can only be obtained when we see the men of science more frequently extending their observant eye over the wonders of our factories, and our great practical manufacturers, with a reciprocal ambition, presenting themselves as active and useful members of our scientific associations. When this has taken place, an order of

* Phil. Trans. 1826, Part iii., p. 250, on a method of expressing by signs the action of machinery.

scientific men will spring up, which will render impossible an oversight so little creditable to the country as that which has been committed respecting the mechanical notation.* This notation has recently undergone very considerable extension and improvement. An additional section has been introduced into it ; designed to express the process of circulation in machines, through which fluids, whether liquid or gaseous, are moved. Mr. Babbage, with the assistance of a friend, who happened to be conversant with the structure and operation of the steam-engine, has illustrated it with singular felicity and success in its application to that machine. An eminent French surgeon, on seeing the scheme of notation thus applied, immediately suggested the advantages which must attend it as an instrument for expressing the structure, operation, and circulation of the animal system ; and we entertain no doubt of its adequacy for that purpose. Not only the mechanical connection of the solid members of the bodies of men and animals, but likewise the structure and operation of the softer parts, including the muscles, integuments, membranes, &c. ; the nature, motion, and circulation of the various fluids, their reciprocal effects, the changes through which they pass, the deposits which they leave in various parts of the system ; the functions of respiration, digestion, and assimilation— all would find appropriate symbols and representatives in the notation, even as it now stands, without those additions of which, however, it is easily susceptible. Indeed, when we reflect for what a very different purpose this scheme of symbols was contrived, we cannot refrain from expressing our wonder that it should seem, in all respects, as if it had been designed expressly for the purposes of anatomy and physiology.

Another of the uses which the slightest attention to the details of this notation irresistibly forces upon our notice, is to exhibit, in the form of a connected plan or map, the organization of an extensive factory, or any great public institution, in which a vast number of individuals are employed, and their duties regulated (as they generally are or ought to be) by a consistent and well-digested system. The mechanical notation is admirably adapted, not only to express such an organized connection of human agents, but even to suggest the improvements of which such organization is susceptible—to betray its weak and defective points, and to disclose, at a glance, the origin of any fault which may, from time to time, be observed in the working of the system. Our limits, however, preclude us from pursuing this interesting topic to the extent which its importance would justify. We shall be satisfied if the hints here thrown out should direct to the subject the attention of those who, being most interested in such an inquiry, are likely to prosecute it with greatest success.

One of the consequences which have arisen in the prosecution of the invention of the calculating machinery, has been the discovery of a multitude of mechanical contrivances, which have been elicited by the exigencies of the undertaking, and which are as novel in their nature as the purposes were novel which they were designed to attain. In some cases several different contrivances were devised for the attainment of the same end ; and that among them which was best suited for the purpose was finally selected : the rejected expedients—those overflowings or waste of the invention—were not, however, always found useless. Like the *waste* in various manufactures, they were soon converted to purposes of utility. These rejected contrivances have found their way, in many cases, into the mills of our manufacturers ; and we now find them busily effecting purposes, far different from any which their inventor dreamed of, in the spinning-frames of Manchester.†

Another department of mechanical art, which has been enriched by this invention, has been that of *tools*. The great variety of new forms which it was necessary to produce, created the necessity of contriving and constructing a vast number of novel and most valuable tools, by which, with the aid of the lathe, and that alone, the required forms could be given to the different parts of the machinery with all the requisite accuracy.

The idea of calculation by mechanism is not new. Arithmetical instruments, such as the calculating boards of the ancients, on which they made their computations by the aid of counters— the *Abacus*, an instrument for computing by the aid of balls sliding upon parallel rods—the method of calculation invented by Baron Napier, called by him *Rhabdology*, and since called *Napier's bones*— the Swan Pan of the Chinese—and other similar contrivances, among which more particularly may be mentioned the Sliding Rule, of so much use in practical calculations to modern engineers, will occur to every reader : these may more properly be called *arithmetical instruments*, partaking more or less of a mechanical character. But the earliest piece of mechanism to which the name of "calculating "machine" can fairly be given, appears to have been a machine invented by the celebrated Pascal. This philosopher and mathematician, at a very early age, being engaged with his father, who held an

* This discovery has been more justly appreciated by scientific men abroad. It was, almost immediately after its publication, adopted as the topic of lectures, in an institution on the Continent for the instruction of Civil Engineers.
† An eminent and wealthy retired manufacturer at Manchester assured us, that on the occasion of a visit to London, when he was favoured with a view of the calculating machinery, he found in it mechanical contrivances, which he subsequently introduced with the greatest advantage into his own spinning-machinery.

official situation in Upper Normandy, the duties of which required frequent numerical calculations, contrived a piece of mechanism to facilitate the performance of them. This mechanism consisted of a series of wheels, carrying cylindrical barrels, on which were engraved the ten arithmetical characters, in a manner not very dissimilar to that already described. The wheel which expressed each order of units was so connected with the wheel which expressed the superior order, that when the former passed from 9 to 0, the latter was necessarily advanced one figure ; and thus the process of carrying was executed by mechanism : when one number was to be added to another by this machine, the addition of each figure to the other was performed by the hand ; when it was required to add more than two numbers, the additions were performed in the same manner successively ; the second was added to the first, the third to their sum, and so on.

Subtraction was reduced to addition by the method of arithmetical complements ; multiplication was performed by a succession of additions ; and division by a succession of subtractions. In all cases, however, the operations were executed from wheel to wheel by the hand.*

This mechanism, which was invented about the year 1650, does not appear to have been brought into any practical use ; and seems to have speedily found its appropriate place in a museum of curiosities. It was capable of performing only particular arithmetical operations, and these subject to all the chances of error in manipulation ; attended also with little more expedition (if so much), as would be attained by the pen of an expert computer.

This attempt of Pascal was followed by various others, with very little improvement, and with no additional success. Polenus, a learned and ingenious Italian, invented a machine by which multiplication was performed, but which does not appear to have afforded any material facilities, nor any more security against error than the common process of the pen. A similar attempt was made by Sir Samuel Moreland, who is described as having transferred to wheel-work the figures of *Napier's bones*, and as having made some additions to the machine of Pascal.†

Grillet, a French mechanician, made a like attempt with as little success. Another contrivance for mechanical calculation was made by Saunderson. Mechanical contrivances for performing particular arithmetical processes were also made about a century ago by Depréne and Boitissendeau ; but they were merely modifications of Pascal's, without varying or extending its objects. But one of the most remarkable attempts of this kind which has been made since that of Pascal, was a machine invented by Leibnitz, of which we are not aware that any detailed or intelligible description was ever published. Leibnitz described its mode of operation, and its results, in the " Berlin Miscellany,"‡ but he appears to have declined any description of its details. In a letter addressed by him to Bernoulli, in answer to a request of the latter that he would afford a description of the machinery, he says, " Descriptionem ejus dare accuratam res non facilis foret. De effectu ex eo judicaveris quod ad " multiplicandum numerum sex figurarum, e.g., rotam quamdam tantum sexies gyrari necesse est, " nulla alia opera mentis, nullis additionibus intervenientibus ; quo facto, integrum absolutumque " productum oculis objicietur."§ He goes on to say that the process of division is performed independently of a succession of subtractions, such as that used by Pascal.

It appears that this machine was one of an extremely complicated nature, which would be attended with considerable expense of construction, and only fit to be used in cases where numerous and expensive calculations were necessary.‖ Leibnitz observes to his correspondent, who required whether it might not be brought into common use, " Non est facta pro his qui olera aut pisculos vendunt, sed " pro observatoriis aut cameris computorum, aut aliis, qui sumptus facile ferunt et multo calculo egent." Nevertheless, it does not appear that this contrivance, of which the inventor states that he caused two models to be made, was ever applied to any useful purpose ; nor indeed do the mechanical details of the invention appear ever to have been published.

Even had the mechanism of these machines performed all which their inventors expected from them, they would have been still altogether inapplicable for the purposes to which it is proposed that the calculating machinery of Mr. Babbage shall be applied. They were all constructed with a view to perform *particular arithmetical operations,* and in all of them the accuracy of the result depended more or less upon manipulation. The principle of the calculating machinery of Mr. Babbage is perfectly general in its nature, not depending on any particular arithmetical operation, and is equally

---

* See a description of this machine by Diderot, in the *Encyc. Method.*; also in the works of Pascal, tom. iv., p. 7 ; Paris, 1819.

† Equidem Morelandus in Anglia, tubæ stentoriæ author, Rhabdologiam ex baculis in cylindrulos transtulit, et additiones auxiliares peragit in adjuncta machina additionum Pascaliana.

‡ Tom. i., p. 317. § *Com. Epist.*, tom. i., p. 289.

‖ Sed machinam esse sumptuosam et multarum rotarum instar horologii : Huygenius aliquoties admonuit ut absolvi curarem ; quod non sine magno sumptu tædioque factum est, dum varie mihi cum opificibus fuit conflictandum.--*Com. Epist.*

applicable to numerical tables of every kind. This distinguishing characteristic was well expressed by Mr. Colebrooke in his address to the Astronomical Society on this invention. "The principle which " essentially distinguishes Mr. Babbage's invention from all these is, that it proposes to calculate a " series of numbers following any law, by the aid of differences, and that by setting a few figures at " the outset, a long series of numbers is readily produced by a mechanical operation. The method of " differences in a very wide sense is the mathematical principle of the contrivance. A machine to " add a number of arbitrary figures together is no economy of time or trouble, since each individual " figure must be placed in the machine; but it is otherwise when those figures follow some law. The " insertion of a few at first determines the magnitude of the next, and those of the succeeding. It is " this constant repetition of similar operations which renders the computation of tables a fit subject " for the application of machinery. Mr. Babbage's invention puts an engine in the place of the com- " puter; the question is set to the instrument, or the instrument is set to the question, and by simply " giving it motion the solution is wrought, and a string of answers is exhibited." But perhaps the greatest of its advantages is, that it prints what it calculates; and this completely precludes the possibility of error in these numerical results which pass into the hands of the public. "The useful- " ness of the instrument," says Mr. Colebrooke, "is thus more than doubled; for it not only saves " time and trouble in transcribing results into a tabular form, and setting types for the printing of the " table, but it likewise accomplishes the yet more important object of insuring accuracy, obviating " numerous sources of error through the careless hands of transcribers and compositors."

Some solicitude will doubtless be felt respecting the present state of the calculating machinery, and the probable period of its completion. In the beginning of the year 1829, Government directed the Royal Society to institute such inquiries as would enable them to report upon the state to which it had then arrived; and also whether the progress made in its construction confirmed them in the opinion which they had formerly expressed—that it would ultimately prove adequate to the important object which it was intended to obtain. The Royal Society, in accordance with these directions, appointed a Committee to make the necessary inquiry, and report. This Committee consisted of Mr. Davies Gilbert, then President, the Secretaries, Sir John Herschel, Mr. Francis Baily, Mr. Brunel, engineer, Mr. Donkin, engineer, Mr. G. Rennie, engineer, Mr. Barton, comptroller of the Mint, and Mr. Warburton, M.P. The voluminous drawings, the various tools, and the portion of the machinery then executed, underwent a close and elaborate examination by this Committee, who reported upon it to the Society.

They stated in their report, that they declined the consideration of the principle on which the practicability of the machinery depends, and of the public utility of the object which it proposes to attain; because they considered the former fully admitted, and the latter obvious to all who consider the immense advantage of accurate numerical tables in all matters of calculation, especially in those which relate to astronomy and navigation, and the great variety and extent of those which it is professedly the object of the machinery to calculate and print with perfect accuracy;—that absolute accuracy being one of the prominent pretensions of the undertaking, they had directed their attention especially to this point, by careful examination of the drawings and of the work already executed, and by repeated conferences with Mr. Babbage on the subject:—that the result of their inquiry was, that such precautions appeared to have been taken in every part of the contrivance, and so fully aware was the inventor of every circumstance which might by possibility produce error, that they had no hesita- tion in stating their belief that these precautions were effectual, and that whatever the machine would do, it would do truly.

They further stated, that the progress which Mr. Babbage had then made, considering the very great difficulties to be overcome in an undertaking of so novel a kind, fully equalled any expectations that could reasonably have been formed; and that although several years had elapsed since the com- mencement of the undertaking, yet when the necessity of constructing plans, sections, elevations, and working drawings of every part; of constructing, and in many cases inventing, tools and machinery of great expense and complexity, necessary to form with the requisite precision parts of the apparatus differing from any which had previously been introduced in ordinary mechanical works; of making many trials to ascertain the value of each proposed contrivance; of altering, improving, and simplify- ing the drawings;—that, considering all these matters, the Committee, instead of feeling surprise at the time which the work has occupied, felt more disposed to wonder at the possibility of accomplishing so much.

The Committee expressed their confident opinion of the adequacy of the machinery to work under all the friction and strain to which it can be exposed; of its durability, strength, solidity, and equilibrium; of the prevention of, or compensation for, wear by friction; of the accuracy of the various adjustments; and of the judgment and discretion displayed by the inventor, in his determination to admit into the mechanism nothing but the very best and most finished workmanship; as a contrary course would have been false economy, and might have led to the loss of the whole capital expended on it.

Finally, considering all that had come before them, and relying on the talent and skill displayed by Mr. Babbage as a mechanist in the progress of this arduous undertaking, not less for what remained, than on the matured and digested plan and admirable execution of what is completed, the Committee did not hesitate to express their opinion, that in the then state of the engine, they regarded it as likely to fulfil the expectations entertained of it by its inventor.

This report was printed in the commencement of the year 1829. From that time until the beginning of the year 1833, the progress of the work has been slow and interrupted. Meanwhile many unfounded rumours have obtained circulation as to the course adopted by Government in this undertaking; and as to the position in which Mr. Babbage stands with respect to it. We shall here state, upon authority on which the most perfect reliance may be placed, what have been the actual circumstances of the arrangement which has been made, and of the steps which have been already taken.

Being advised that the objects of the projected machinery were of paramount national importance to a maritime country, and that, from its nature, it could never be undertaken with advantage by any individual as a pecuniary speculation, Government determined to engage Mr. Babbage to construct the calculating engine for the nation. It was then thought that the work could be completed in two or three years; and it was accordingly undertaken on this understanding about the year 1821, and since then has been in progress. The execution of the workmanship was confided to an engineer by whom all the subordinate workmen were employed, and who supplied for the work the requisite tools and other machinery; the latter being his own property, and not that of Government. This engineer furnished, at intervals, his accounts, which were duly audited by proper persons appointed for that purpose. It was thought advisable—with a view, perhaps, to invest Mr. Babbage with a more strict authority over the subordinate agents—that the payments of these accounts of the engineer should pass through his hands. The amount was accordingly from time to time issued to him by the Treasury, and paid over to the engineer. This circumstance has given rise to reports, that he has received considerable sums of money as a remuneration for his skill and labour in inventing and constructing this machinery. Such reports are altogether destitute of truth. He has received, neither directly nor indirectly, any remuneration whatever;—on the contrary, owing to various official delays in the issues of money from the Treasury for the payment of the engineer, he has frequently been obliged to advance these payments himself, that the work might proceed without interruption. Had he not been enabled to do this from his private resources, it would have been impossible that the machinery could have arrived at its present advanced state.

It will be a matter of regret to every friend of science to learn, that, notwithstanding such assistance, the progress of the work has been suspended, and the workmen dismissed for more than a year and a half; nor does there at the present moment appear to be any immediate prospect of its being resumed. What the causes may be of a suspension so extraordinary, of a project of such great national and universal interest—in which the country has already invested a sum of such serious amount as £15,000—is a question which will at once suggest itself to every mind; and is one to which, notwithstanding frequent inquiries in quarters from which correct information might be expected, we have not been able to obtain any satisfactory answer. It is not true, we are assured, that the Government object to make the necessary payments, or even advances, to carry on the work. It is not true, we also are assured, that any practical difficulty has arisen in the construction of the mechanism;—on the contrary, the drawings of all the parts of it are completed, and may be inspected by any person appointed on the part of Government to examine them.* Mr. Babbage is known as a man of unwearied activity, and aspiring ambition. Why, then, it may be asked, is it that he, seeing his present reputation and future fame depending in so great a degree upon the successful issue of this undertaking, has nevertheless allowed it to stand still for so long a period, without distinctly pointing out to Government the course which they should adopt to remove the causes of delay? Had he done this (which we consider to be equally due to the nation and to himself), he would have thrown upon Government and its agents the whole responsibility for the delay and consequent loss; but we believe he has not done so. On the contrary, it is said that he has of late almost withdrawn from all interference on the

* Government has erected a fire-proof building, in which it is intended that the calculating machinery shall be placed when completed. In this building are now deposited the large collection of drawings, containing the designs, not only of the part of the machinery which has been already constructed, but what is of much greater importance, of those parts which have not yet been even modelled. It is gratifying to know that Government has shown a proper solicitude for the preservation of those precious but perishable documents, the loss or destruction of which would, in the event of the death of the inventor, render the completion of the machinery impracticable.

subject, either with the Government or the engineer. Does not Mr. Babbage perceive the inference which the world will draw from this course of conduct? Does he not see that they will impute it to a distrust of his own power, or even to a consciousness of his own inability to complete what he has begun? We feel assured that such is not the case; and we are anxious, equally for the sake of science, and for Mr. Babbage's own reputation, that the mystery—for such it must be regarded—should be cleared up; and that all obstructions to the progress of the undertaking should immediately be removed. Does this supineness and apparent indifference, so incompatible with the known character of Mr. Babbage, arise from any feeling of dissatisfaction at the existing arrangements between himself and the Government? If such be the actual cause of the delay (and we believe that, in some degree, it is so), we cannot refrain from expressing our surprise that he does not adopt the candid and straightforward course of declaring the grounds of his discontent, and explaining the arrangement which he desires to be adopted. We do not hesitate to say, that every reasonable accommodation and assistance ought to be afforded him. But if he will pertinaciously abstain from this, to our minds, obvious and proper course, then it is surely the duty of Government to appoint proper persons to inquire into and report upon the present state of the machinery; to ascertain the causes of its suspension; and to recommend such measures as may appear to be most effectual to insure its speedy completion. If they do not by such means succeed in putting the project in a state of advancement, they will at least shift from themselves all responsibility for its suspension.

EXCERPT FROM "PASSAGES FROM THE LIFE OF A PHILOSOPHER."

CHAPTER V.

DIFFERENCE ENGINE NO. I.

"Oh no! we never mention it,
Its name is never heard."

Difference Engine No. 1—First Idea at Cambridge, 1812—Plan for Dividing Astronomical Instruments—Idea of a Machine to calculate Tables by Differences—Illustrations by Piles of Cannon-balls.

CALCULATING MACHINES comprise various pieces of mechanism for assisting the human mind in executing the operations of arithmetic. Some few of these perform the whole operation without any mental attention when once the given numbers have been put into the machine.

Others require a moderate portion of mental attention: these latter are generally of much simpler construction than the former, and it may also be added, are less useful.

The simplest way of deciding to which of these two classes any calculating machine belongs is to ask its maker—Whether, when the numbers on which it is to operate are placed in the instrument, it is capable of arriving at its result by the mere motion of a spring, a descending weight, or any other constant force? If the answer be in the affirmative, the machine is really automatic; if otherwise, it is not self-acting.

Of the various machines I have had occasion to examine, many of those for Addition and Subtraction have been found

N

to be automatic. Of machines for Multiplication and Division, which have fully come under my examination, I cannot at present recall one to my memory as absolutely fulfilling this condition.

The earliest idea that I can trace in my own mind of calculating arithmetical Tables by machinery arose in this manner :—

One evening I was sitting in the rooms of the Analytical Society, at Cambridge, my head leaning forward on the Table in a kind of dreamy mood, with a Table of logarithms lying open before me. Another member, coming into the room, and seeing me half asleep, called out, " Well, Babbage, what are you dreaming about ?" to which I replied, " I am thinking that all these Tables (pointing to the logarithms) might be calculated by machinery."

I am indebted to my friend, the Rev. Dr. Robinson, the Master of the Temple, for this anecdote. The event must have happened either in 1812 or 1813.

About 1819 I was occupied with devising means for accurately dividing astronomical instruments, and had arrived at a plan which I thought was likely to succeed perfectly. I had also at that time been speculating about making machinery to compute arithmetical Tables.

One morning I called upon the late Dr. Wollaston, to consult him about my plan for dividing instruments. On talking over the matter, it turned out that my system was exactly that which had been described by the Duke de Chaulnes, in the Memoirs of the French Academy of Sciences, about fifty or sixty years before. I then mentioned my other idea of computing Tables by machinery, which Dr. Wollaston thought a more promising subject.

I considered that a machine to execute the mere isolated

operations of arithmetic, would be comparatively of little value, unless it were very easily set to do its work, and unless it executed not only accurately, but with great rapidity, whatever it was required to do.

On the other hand, the method of differences supplied a general principle by which *all* Tables might be computed through limited intervals, by one uniform process. Again, the method of differences required the use of mechanism for Addition only. In order, however, to insure accuracy in the printed Tables, it was necessary that the machine which computed Tables should also set them up in type, or else supply a mould in which stereotype plates of those Tables could be cast.

I now began to sketch out arrangements for accomplishing the several partial processes which were required. The arithmetical part must consist of two distinct processes—the power of adding one digit to another, and also of carrying the tens to the next digit, if it should be necessary.

The first idea was, naturally, to add each digit successively. This, however, would occupy much time if the numbers added together consisted of many places of figures.

The next step was to add all the digits of the two numbers each to each at the same instant, but reserving a certain mechanical memorandum, wherever a carriage became due. These carriages were then to be executed successively.

Having made various drawings, I now began to make models of some portions of the machine, to see how they would act. Each number was to be expressed upon wheels placed upon an axis; there being one wheel for each figure in the number operated upon.

Having arrived at a certain point in my progress, it became necessary to have teeth of a peculiar form cut upon these wheels. As my own lathe was not fit for this job, I took the wheels to a wheel-cutter at Lambeth, to whom I carefully conveyed my instructions, leaving with him a drawing as his guide.

These wheels arrived late one night, and the next morning I began putting them in action with my other mechanism, when, to my utter astonishment, I found they were quite unfit for their task. I examined the shape of their teeth, compared them with those in the drawings, and found they agreed perfectly; yet they could not perform their intended work. I had been so certain of the truth of my previous reasoning, that I now began to be somewhat uneasy. I reflected that, if the reasoning about which I had been so certain should prove to have been really fallacious, I could then no longer trust the power of my own reason. I therefore went over with my wheels to the artist who had formed the teeth, in order that I might arrive at some explanation of this extraordinary contradiction.

On conferring with him, it turned out that, when he had understood fully the peculiar form of the teeth of wheels, he discovered that his wheel-cutting engine had not got amongst its divisions that precise number which I had required. He therefore had asked me whether another number, which his machine possessed, would not equally answer my object. I had inadvertently replied in the affirmative. He then made arrangements for the precise number of teeth I required; and the new wheels performed their expected duty perfectly.

The next step was to devise means for printing the tables to be computed by this machine. My first plan was to make it put together moveable type. I proposed to make metal boxes, each containing 3,000 types of one of the ten digits. These types were to be made to pass out one by one from the bottom of their boxes, when required by the computing part of the machine.

But here a new difficulty arose. The attendant who put the types into the boxes might, by mistake, put a wrong type in one or more of them. This cause of error I removed in the following manner :—There are usually certain notches in the side of the type. I caused these notches to be so placed that all the types of any given digit possessed the same characteristic notches, which no other type had. Thus, when the boxes were filled, by passing a small wire down these peculiar notches, it would be impeded in its passage, if there were included in the row a single wrong figure. Also, if any digit were accidentally turned upside down, it would be indicated by the stoppage of the testing wire.

One notch was reserved as common to every species of type. The object of this was that, before the types which the Difference Engine had used for its computation were removed from the iron platform on which they were placed, a steel wire should be passed through this common notch, and remain there. The tables, composed of moveable types, thus interlocked, could never have any of their figures drawn out by adhesion to the inking-roller, and then by possibility be restored in an inverted order. A small block of such figures tied together by a bit of string, remained unbroken for several years, although it was rather roughly used as a plaything by my children. One such box was finished, and delivered its type satisfactorily.

Another plan for printing the tables, was to place the ordinary printing type round the edges of wheels. Then, as each successive number was produced by the arithmetical part, the type-wheels would move down upon a plate of soft composition, upon which the tabular number would be impressed. This mould was formed of a mixture of plaster-of-Paris with other materials, so as to become hard in the course of a few hours.

The first difficulty arose from the impression of one tabular number on the mould being distorted by the succeeding one.

I was not then aware that a very slight depth of impression from the type would be quite sufficient. I surmounted the difficulty by previously passing a roller, having longitudinal wedge-shaped projections, over the plastic material. This formed a series of small depressions in the matrix between each line. Thus the expansion arising from the impression of one line partially filled up the small depression or ditch which occurred between each successive line.

The various minute difficulties of this kind were successively overcome; but subsequent experience has proved that the depth necessary for stereotype moulds is very small, and that even thick paper, prepared in a peculiar manner, is quite sufficient for the purpose.

Another series of experiments were, however, made for the purpose of punching the computed numbers upon copper plate. A special machine was contrived and constructed, which might be called a co-ordinate machine, because it moved the copper plate and steel punches in the direction of three rectangular co-ordinates. This machine was afterwards found very useful for many other purposes. It was, in fact, a general shaping machine, upon which many parts of the Difference Engine were formed.

Several specimens of surface and copper-plate printing, as well as of the copper plates, produced by these means, were exhibited at the Exhibition of 1862.

I have proposed and drawn various machines for the purpose of calculating a series of numbers forming Tables

by means of a certain system called " The Method of Differences," which it is the object of this sketch to explain.

The first Difference Engine with which I am acquainted
comprised a few figures, and was made by myself, between
1820 and June 1822. It consisted of from six to eight
figures. A much larger and more perfect engine was subsequently commenced in 1823 for the Government.

It was proposed that this latter Difference Engine should
have six orders of differences, each consisting of about
twenty places of figures, and also that it should print the
Tables it computed.

The small portion of it which was placed in the International Exhibition of 1862 was put together nearly thirty
years ago. It was accompanied by various parts intended to
enable it to print the results it calculated, either as a single
copy on paper—or by putting together moveable types—or by
stereotype plates taken from moulds punched by the machine
—or from copper plates impressed by it. The parts necessary for the execution of each of these processes were made,
but these were not at that time attached to the calculating
part of the machine.

A considerable number of the parts by which the printing
was to be accomplished, as also several specimens of portions
of tables punched on copper, and of stereotype moulds, were
exhibited in a glass case adjacent to the Engine.

In 1834 Dr. Lardner published, in the " Edinburgh Review," *
a very elaborate description of this portion of the machine,
in which he explained clearly the method of Differences.

It is very singular that two persons, one resident in London,
the other in Sweden, should both have been struck, on reading
this review, with the simplicity of the mathematical principle

* " Edinburgh Review," No. cxx., July, 1834.

of differences as applied to the calculation of Tables, and should have been so fascinated with it as to have undertaken to construct a machine of the kind.

Mr. Deacon, of Beaufort House, Strand, whose mechanical skill is well known, made, for his own satisfaction, a small model of the calculating part of such a machine, which was shown only to a few friends, and of the existence of which I was not aware until after the Swedish machine was brought to London.

Mr. Scheutz, an eminent printer at Stockholm, had far greater difficulties to encounter. The construction of mechanism, as well as the mathematical part of the question, was entirely new to him. He, however, undertook to make a machine having four differences, and fourteen places of figures, and capable of printing its own Tables.

After many years' indefatigable labour, and an almost ruinous expense, aided by grants from his Government, by the constant assistance of his son, and by the support of many enlightened members of the Swedish Academy, he completed his Difference Engine. It was brought to London, and some time afterwards exhibited at the great Exhibition at Paris. It was then purchased for the Dudley Observatory at Albany by an enlightened and public-spirited merchant of that city, John F. Rathbone, Esq.

An exact copy of this machine was made by Messrs. Donkin and Co., for the English Government, and is now in use in the Registrar-General's Department at Somerset House. It is very much to be regretted that this specimen of English workmanship was not exhibited in the International Exhibition.

Explanation of the Difference Engine.

Those who are only familiar with ordinary arithmetic may, by following out with the pen some of the examples which will be given, easily make themselves acquainted with the simple principles on which the Difference Engine acts.

It is necessary to state distinctly at the outset, that the Difference Engine is not intended to answer special questions. Its object is to calculate and print a *series* of results formed according to given laws. These are called Tables—many such are in use in various trades. For example—there are collections of Tables of the amount of any number of pounds from 1 to 100 lbs. of butchers' meat at various prices per lb. Let us examine one of these Tables: viz.—the price of meat 5*d*. per lb., we find

Number. Lbs.	Table. Price.
	s. *d.*
1	0 5
2	0 10
3	1 3
4	1 8
5	2 1

There are two ways of computing this Table:—

1st. We might have multiplied the number of lbs. in each line by 5, the price per lb., and have put down the result in *l. s. d.*, as in the 2nd column: or,

2nd. We might have put down the price of 1 lb., which is 5*d*., and have added five pence for each succeeding lb.

Let us now examine the relative advantages of each plan. We shall find that if we had multiplied each number of lbs. in

o

the Table by 5, and put down the resulting amount, then every number in the Table would have been computed independently. If, therefore, an error had been committed, it would not have affected any but the single tabular number at which it had been made. On the other hand, if a single error had occurred in the system of computing by adding five at each step, any such error would have rendered the whole of the rest of the Table untrue.

Thus the system of calculating by differences, which is the easiest, is much more liable to error. It has, on the other hand, this great advantage: viz., that when the Table has been so computed, if we calculate its last term directly, and if it agree with the last term found by the continual addition of 5, we shall then be quite certain that every term throughout is correct. In the system of computing each term directly, we possess no such check upon our accuracy.

Now the Table we have been considering is, in fact, merely a Table whose first difference is constant and equal to five. If we express it in pence it becomes—

	Table.	1st Difference.
1	5	5
2	10	5
3	15	5
4	20	5
5	25	

Any machine, therefore, which could add one number to another, and at the same time retain the original number called the first difference for the next operation, would be able to compute all such Tables.

Let us now consider another form of Table which might readily occur to a boy playing with his marbles, or to a young lady with the balls of her solitaire board.

The boy may place a row of his marbles on the sand, at equal distances from each other, thus—

●　　　●　　　●　　　●　　　●

He might then, beginning with the second, place two other marbles under each, thus—

He might then, beginning with the third, place three other marbles under each group, and so on; commencing always one group later, and making the addition one marble more each time.   The several groups would stand thus arranged—

He will not fail to observe that he has thus formed a series of triangular groups, every group having an equal number of marbles in each of its three sides.   Also that the side of each successive group contains one more marble than that of its preceding group.

Now an inquisitive boy would naturally count the numbers in each group and he would find them thus—

1　　　3　　　6　　　10　　　15　　　21

He might also want to know how many marbles the thirtieth or any other distant group might contain.   Perhaps he might go to papa to obtain this information; but I much fear papa would snub him, and would tell him that it was nonsense—that it was useless—that nobody knew the number, and so forth.   If the boy is told by papa, that he is not able to answer the question, then I recommend him to pay careful attention to whatever that father may at any time say, for he has overcome two of the greatest obstacles to the acquisition

o 2

of knowledge—inasmuch as he possesses the consciousness that he does not know—and he has the moral courage to avow it.*

If papa fail to inform him, let him go to mamma, who will not fail to find means to satisfy her darling's curiosity. In the meantime the author of this sketch will endeavour to lead his young friend to make use of his own common sense for the purpose of becoming better acquainted with the triangular figures he has formed with his marbles.

In the case of the Table of the price of butchers' meat, it was obvious that it could be formed by adding the same *constant* difference continually to the first term. Now suppose we place the numbers of our groups of marbles in a column, as we did our prices of various weights of meat. Instead of adding a certain difference, as we did in the former case, let us subtract the figures representing each group of marbles from the figures of the succeeding group in the Table. The process will stand thus :—

	Table.	1st Difference.	2nd Difference.
Number of the Group.	Number of Marbles in each Group.	Difference between the number of Marbles in each Group and that in the next.	
1	1	1	1
2	3	2	1
3	6	3	1
4	10	4	1
5	15	5	1
6	21	6	
7	28	7	

It is usual to call the third column thus formed *the column of*

* The most remarkable instance I ever met with of the distinctness with which any individual perceived the exact boundary of his own knowledge, was that of the late Dr. Wollaston.

first differences. It is evident in the present instance that that column represents the natural numbers. But we already know that the first difference of the natural numbers is constant and equal to unity. It appears, therefore, that a Table of these numbers, representing the group of marbles, might be constructed to any extent by mere addition—using the number 1 as the first number of the Table, the number 1 as the first Difference, and also the number 1 as the second Difference, which last always remains constant.

Now as we could find the value of any given number of pounds of meat directly, without going through all the previous part of the Table, so by a somewhat different rule we can find at once the value of any group whose number is given.

Thus, if we require the number of marbles in the fifth group, proceed thus :—

Take the number of the group 5
Add 1 to this number, it becomes 6
———
Multiply these numbers together . . . 2)30
———
Divide the product by 2 15
This gives 15, the number of marbles in the 5th group.

If the reader will take the trouble to calculate with his pencil the five groups given above, he will soon perceive the general truth of this rule.

We have now arrived at the fact that this Table—like that of the price of butchers' meat—can be calculated by two different methods. By the first, each number of the Table is calculated independently: by the second, the truth of each number depends upon the truth of all the previous numbers.

Perhaps my young friend may now ask me, What is the use of such Tables? Until he has advanced further in his arithmetical studies, he must take for granted that they are of some use. The very Table about which he has been reasoning possesses a special name—it is called a Table of Triangular Numbers. Almost every general collection of Tables hitherto published contains portions of it of more or less extent.

Above a century ago, a volume in small quarto, containing the first 20,000 triangular numbers, was published at the Hague by E. De Joncourt, A.M., and Professor of Philosophy.* I cannot resist quoting the author's enthusiastic expression of the happiness he enjoyed in composing his celebrated work :

" The Trigonals here to be found, and nowhere else, are
" exactly elaborate. Let the candid reader make the best
" of these numbers, and feel (if possible) in perusing my work
" the pleasure I had in composing it."

" That sweet joy may arise from such contemplations
" cannot be denied. Numbers and lines have many charms,
" unseen by vulgar eyes, and only discovered to the unwearied
" and respectful sons of Art. In features the serpentine line
" (who starts not at the name) produces beauty and love ; and
" in numbers, high powers, and humble roots, give soft delight.

" Lo ! the raptured arithmetician ! Easily satisfied, he
" asks no Brussels lace, nor a coach and six. To calculate,
" contents his liveliest desires, and obedient numbers are
" within his reach."

I hope my young friend is acquainted with the fact—that he product of any number multiplied by itself is called the square of that number. Thus 36 is the product of 6 multiplied by 6, and 36 is called the square of 6. I would now recommend him to examine the series of square numbers

1, 4, 9, 16, 25, 36, 49, 64, &c.,

* ' On the Nature and Notable Use of the most Simple Trigonal Numbers.' By E. De Joncourt, at the Hague. 1762.

and to make, for his own instruction, the series of their first and second differences, and then to apply to it the same reasoning which has been already applied to the Table of Triangular Numbers.

When he feels that he has mastered that Table, I shall be happy to accompany mamma's darling to Woolwich or to Portsmouth, where he will find some practical illustrations of the use of his newly-acquired numbers. He will find scattered about in the Arsenal various heaps of cannon balls, some of them triangular, others square or oblong pyramids.

Looking on the simplest form—the triangular pyramid—he will observe that it exactly represents his own heaps of marbles placed each successively above one another until the top of the pyramid contains only a single ball.

The new series thus formed by the addition of his own triangular numbers is—

Number.	Table.	1st Difference.	2nd Difference.	3rd Difference.
1	1	3	3	1
2	4	6	4	1
3	10	10	5	1
4	20	15	6	
5	35	21		
6	56			

He will at once perceive that this Table of the number of cannon balls contained in a triangular pyramid can be carried to any extent by simply adding successive differences, the third of which is constant.

The next step will naturally be to inquire how any number in this Table can be calculated by itself. A little consideration will lead him to a fair guess; a little industry will enable him to confirm his conjecture.

It will be observed at p. 92 that in order to find independently any number of the Table of the price of butchers' meat, the following rule was observed :—

Take the number whose tabular number is required.

Multiply it by the first difference.

This product is equal to the required tabular number.

Again, at p. 53, the rule for finding any triangular number was :—

Take the number of the group	.	.	.	5
Add 1 to this number, it becomes		.	.	6
				——
Multiply these numbers together .		.	.	2)30
				——
Divide the product by 2	.	.	.	.	15

This is the number of marbles in the 5th group.

Now let us make a bold conjecture respecting the Table of cannon balls, and try this rule :—

Take the number whose tabular number is required, say	.	.	.	.	.	5
Add 1 to that number .	.	.	.	.	6
Add 1 more to that number .		.	.	.	7
						——
Multiply all three numbers together	.		.	2)210
						——
Divide by 2	.	.	.	.	.	105

The real number in the 5th pyramid is 35. But the number 105 at which we have arrived is exactly three times as great. If, therefore, instead of dividing by 2 we had divided by 2 and also by 3, we should have arrived at a true result in this instance.

The amended rule is therefore—

Take the number whose tabular number is
    required, say   .   .   .   .   . n
Add 1 to it .   .   .   .   .   . $n + 1$
Add 1 to this   .   .   .   .   . $n + 2$
Multiply these three numbers
    together   .   .   . $n \times (n + 1) \times (n + 2)$
Divide by $1 \times 2 \times 3$.

The result is   .   .   . $\dfrac{n \, (n + 1) \, (n + 2)}{6}$

This rule will, upon trial, be found to give correctly every tabular number.

By similar reasoning we might arrive at the knowledge of the number of cannon balls in square and rectangular pyramids. But it is presumed that enough has been stated to enable the reader to form some general notion of the method of calculating arithmetical Tables by differences which are constant.

It may now be stated that mathematicians have discovered that all the Tables most important for practical purposes, such as those relating to Astronomy and Navigation, can, although they may not possess any constant differences, still be calculated in detached portions by that method.

Hence the importance of having machinery to calculate by differences, which, if well made, cannot err; and which, if carelessly set, presents in the last term it calculates the power of verification of every antecedent term.

Of the Mechanical Arrangements necessary for computing Tables by the Method of Differences.

From the preceding explanation it appears that all Tables may be calculated, to a greater or less extent, by the method of Differences. That method requires, for its successful

P

execution, little beyond mechanical means of performing the arithmetical operation of Addition.   Subtraction can, by the aid of a well-known artifice, be converted into Addition.

The process of Addition includes two distinct parts— 1st. The first consists of the addition of any one digit to another digit; 2nd. The second consists in carrying the tens to the next digit above.

Let us take the case of the addition of the two following numbers, in which no carriages occur :—

$$\begin{array}{r} 6023 \\ 1970 \\ \hline 7993 \end{array}$$

It will be observed that, in making this addition, the mind acts by successive steps.   The person adding says to himself—

0 and 3 make three,
7 and 2 make nine,
9 and 0 make nine,
1 and 6 make seven.

In the following addition there are several carriages :—

$$\begin{array}{r} 2648 \\ 4564 \\ \hline 7212 \end{array}$$

The person adding says to himself—

4 and 8 make 12 : put down 2 and carry one.
1 and 6 are 7 and 4 make 11·: put down 1 and carry one.
1 and 5 are 6 and 6 make 12 : put down 2 and carry one.
1 and 4 are 5 and 2 make 7 : put down 7.

Now, the length of time required for adding one number to another is mainly dependent upon the number of figures to be added. If we could tell the average time required by the mind to add two figures together, the time required for adding any given number of figures to another equal number would be found by multiplying that average time by the number of digits in either number.

When we attempt to perform such additions by machinery we might follow exactly the usual process of the human mind. In that case we might take a series of wheels, each having marked on its edges the digits 0, 1, 2, 3, 4, 5, 6, 7, 8, 9. These wheels might be placed above each other upon an axis. The lowest would indicate the units' figure, the next above the tens, and so on, as in the Difference Engine at the Exhibition, a woodcut of which faces the title-page.

Several such axes, with their figure wheels, might be placed around a system of central wheels, with which the wheels of any one or more axes might at times be made to gear. Thus the figures on any one axis might, by means of those central wheels, be added to the figure wheels of any other axis.

But it may fairly be expected, and it is indeed of great importance that calculations made by machinery should not merely be exact, but that they should be done in a much shorter time than those performed by the human mind. Suppose there were no tens to carry, as in the first of the two cases; then, if we possessed mechanism capable of adding any one digit to any other in the units' place of figures, a similar mechanism might be placed above it to add the tens' figures, and so on for as many figures as might be required.

But in this case, since there are no carriages, each digit might be added to its corresponding digit at the same time. Thus, the time of adding by means of mechanism, any two numbers, however many figures they might consist of, would not exceed that of adding a single digit to another digit.　If this could be accomplished it would render additions and subtractions with numbers having ten, twenty, fifty, or any number of figures, as rapid as those operations are with single figures.

Let us now examine the case in which there were several carriages. Its successive stages may be better explained, thus—

		2648
		4584
Stages.		——
1 Add units' figure　4　.　.　.		2642
2 Carry　.　.　.　.　.　.		1
		——
		2652
3 Add tens' figure = 8　.　.　.		8
		——
		2632
4 Carry　.　.　.　.　.　.		1
		——
		2732
5 Add hundreds' figure = 5　.　.		5
		——
		2232
6 Carry　.　.　.　.　.　.		1
		——
		3232
7 Add thousands' figure = 4　.　.		4
		——
		7232
8 Carry 0.　There is no carr.		

Now if, as in this case, all the carriages were known, it would then be possible to make all the additions of digits at the same time, provided we could also record each carriage as it became due. We might then complete the addition by adding, at the same instant, each carriage in its proper place. The process would then stand thus:—

```
          2648
          4564
          ____
Stages  ⎰ 6102   Add each digit to the digit above.
  1     ⎱ 111    Record the carriages.
          ____
  2     { 7212   Add the above carriages.
```

Now, whatever mechanism is contrived for adding any one digit to any other must, of course, be able to add the largest digit, nine, to that other digit. Supposing, therefore, one unit of number to be passed over in one second of time, it is evident that any number of pairs of digits may be added together in nine seconds, and that, when all the consequent carriages are known, as in the above case, it will cost one second more to make those carriages. Thus, addition and carriage would be completed in ten seconds, even though the numbers consisted each of a hundred figures.

But, unfortunately, there are multitudes of cases in which the carriages that become due are only known in successive periods of time. As an example, add together the two following numbers :—

```
                                              8473
                                              1528
Stages                                        ____
  1  Add all the digits      .    .    .   .  9991
  2  Carry on tens and warn next car.    .      1
                                              ____
                                              9901
  3  Carry on hundreds, and ditto    .   .      1
                                              ____
                                              9001
  4  Carry on thousands, and ditto   .   .      1
                                              ____
                                              00001
  5  Carry on ten thousands   .    .     .      1
                                              ____
                                              10001
```

In this case the carriages only become known successively, and they amount to the number of figures to be added ; consequently, the mere addition of two numbers, each of fifty places of figures, would require only nine seconds of time, whilst the possible carriages would consume fifty seconds.

The mechanical means I employed to make these carriages bears some slight analogy to the operation of the faculty of memory. A toothed wheel had the ten digits marked upon its edge; between the nine and the zero a projecting tooth was placed. Whenever any wheel, in receiving addition, passed from nine to zero, the projecting tooth pushed over a certain lever. Thus, as soon as the nine seconds of time required for addition were ended, every carriage which had *become due* was indicated by the altered position of its lever. An arm now went round, which was so contrived that the act of replacing that lever caused the carriage which its position indicated to be made to the next figure above. But this figure might be a nine, in which case, in passing to zero, it would put over its lever, and so on. By placing the arms spirally round an axis, these successive carriages were accomplished.

Multitudes of contrivances were designed, and almost endless drawings made, for the purpose of economizing the time and simplifying the mechanism of carriage. In that portion of the Difference Engine in the Exhibition of 1862 the time of carriage has been reduced to about one-fourth part of what was at first required.

At last having exhausted, during years of labour, the principle of successive carriages, it occurred to me that it might be possible to teach mechanism to accomplish another mental process, namely—to foresee. This idea occurred to me in October, 1834. It cost me much thought, but the principle was arrived at in a short time. As soon as that was attained, the next step was to teach the mechanism which could foresee to act upon that foresight. This was not so difficult : certain mechanical means were soon devised which, although very far from simple, were yet sufficient to demonstrate the possibility of constructing such machinery.

The process of simplifying this form of carriage occupied me, at intervals, during a long series of years. The demands of the Analytical Engine, for the mechanical execution of arithmetical operations, were of the most extensive kind. The multitude of similar parts required by the Analytical Engine, amounting in some instances to upwards of fifty thousand, rendered any, even the simplest, improvement of each part a matter of the highest importance, more especially as regarded the diminished amount of expenditure for its construction.

Description of the existing portion of Difference Engine No. 1.

That portion of Difference Engine, No. 1, which during the last twenty years has been in the museum of King's College, at Somerset House, is represented in the woodcut opposite the title page.

It consists of three columns; each column contains six cages; each cage contains one figure-wheel.

The column on the right hand has its lowest figure-wheel covered by a shade which is never removed, and to which the reader's attention need not be directed.

The figure-wheel next above may be placed by hand at any one of the ten digits. In the woodcut it stands at zero.

The third, fourth, and fifth cages are exactly the same as the second.

The sixth cage contains exactly the same as the four just

described. It also contains two other figure-wheels, which with a similar one above the frame, may also be dismissed from the reader's attention. Those wheels are entirely unconnected with the moving part of the engine, and are only used for memoranda.

It appears, therefore, that there are in the first column on the right hand five figure-wheels, each of which may be set by hand to any of the figures 0, 1, 2, 3, 4, 5, 6, 7, 8, 9.

The lowest of these figure-wheels represents the unit's figure of any number; the next above the ten's figure, and so on. The highest figure-wheel will therefore represent tens of thousands.

Now, as each of these figure-wheels may be set by hand to any digit, it is possible to place on the first column any number up to 99999. It is on these wheels that the Table to be calculated by the engine is expressed. This column is called the Table column, and the axis of the wheels the Table axis.

The second or middle column has also six cages, in each of which a figure-wheel is placed. It will be observed that in the lowest cage, the figure on the wheel is concealed by a shade. It may therefore be dismissed from the attention. The five other figure-wheels are exactly like the figure-wheels on the Table axis, and can also represent any number up to 99999.

This column is called the First Difference column, and the axis is called the First Difference axis.

The third column, which is that on the left hand, has also six cages, in each of which is a figure-wheel capable of being set by hand to any digit.

The mechanism is so contrived that whatever may be the numbers placed respectively on the figure-wheels of each of the three columns, the following succession of operations will take place as long as the handle is moved:—

1st. Whatever number is found upon the column of first differences will be added to the number found upon the Table column.

2nd. The same first difference remaining upon its own column, the number found upon the column of second differences will be added to that first difference.

It appears, therefore, that with this small portion of the Engine any Table may be computed by the method of differences, provided neither the Table itself, nor its first and second differences, exceed five places of figures.

If the whole Engine had been completed it would have had six orders of differences, each of twenty places of figures, whilst the three first columns would each have had half a dozen additional figures.

This is the simplest explanation of that portion of the Difference Engine No. 1, at the Exhibition of 1862. There are, however, certain modifications in this fragment which render its exhibition more instructive, and which even give a mechanical insight into those higher powers with which I had endowed it in its complete state.

As a matter of convenience in exhibiting it, there is an arrangement by which the *three* upper figures of the second difference are transformed into a small engine which counts the natural numbers.

By this means it can be set to compute any Table whose second difference is constant and less than 1000, whilst at the same time it thus shows the position in the Table of each tabular number.

In the existing portion there are three bells; they can be respectively ordered to ring when the Table, its first difference

Q

and its second difference, pass from positive to negative. Several weeks after the machine had been placed in my drawing-room, a friend came by appointment to test its power of calculating Tables. After the Engine had computed several Tables, I remarked that it was evidently finding the root of a quadratic equation; I therefore set the bells to watch it. After some time the proper bell sounded twice, indicating, and giving the two positive roots to be 28 and 30. The Table thus calculated related to the barometer and really involved a quadratic equation, although its maker had not previously observed it. I afterwards set the Engine to tabulate a formula containing impossible roots, and of course the other bell warned me when it had attained those roots. I had never before used these bells, simply because I did not think the power it thus possessed to be of any practical utility.

Again, the lowest cages of the Table, and of the first difference, have been made use of for the purpose of illustrating three important faculties of the finished engine.

1st. The portion exhibited can calculate any Table whose third difference is constant and less than 10.

2nd. It can be used to show how much more rapidly astronomical Tables can be calculated in an engine in which there is no constant difference.

3rd. It can be employed to illustrate those singular laws which might continue to be produced through ages, and yet after an enormous interval of time change into other different laws; each again to exist for ages, and then to be superseded by new laws. These views were first proposed in the "Ninth Bridgewater Treatise."

Amongst the various questions which have been asked respecting the Difference Engine, I will mention a few of the most remarkable:—One gentleman addressed me thus:

" Pray, Mr. Babbage, can you explain to me in two words " what is the principle of this machine?" Had the querist possessed a moderate acquaintance with mathematics I might in four words have conveyed to him the required information by answering, " The method of differences." The question might indeed have been answered with six characters thus—

$$\Delta^7 u_x = 0.$$

but such information would have been unintelligible to such inquirers.

On two occasions I have been asked,—" Pray, Mr. Babbage, " if you put into the machine wrong figures, will the right " answers come out?" In one case a member of the Upper, and in the other a member of the Lower, House put this question. I am not able rightly to apprehend the kind of confusion of ideas that could provoke such a question. I did, however, explain the following property, which might in some measure approach towards an answer to it.

It is possible to construct the Analytical Engine in such a manner that after the question is once communicated to the engine, it may be stopped at any turn of the handle and set on again as often as may be desired. At each stoppage every figure-wheel throughout the Engine, which is capable of being moved without breaking, may be moved on to any other digit. Yet after all these apparent falsifications the engine will be found to make the next calculation with perfect truth.

The explanation is very simple, and the property itself useless. The whole of the mechanism ought of course to be enclosed in glass, and kept under lock and key, in which case the mechanism necessary to give it the property alluded to would be useless.

CHAPTER VI.

Statement relative to the Difference Engine, drawn up by the late
Sir H. Nicolas from the Author's Papers.

THE following statement was drawn up by the late Sir
Harris Nicolas, G.S.M. & G., from papers and documents in
my possession relating to the Difference Engine. I believe
every paper I possessed at all bearing on the subject was in
his hands for several months.

For some time previous to 1822, Mr. Babbage had been
engaged in contriving machinery for the execution of extensive arithmetical operations, and in devising mechanism by
which the machine that made the calculations might also
print the results.

On the 3rd of July, 1822, he published a letter to Sir
Humphry Davy, President of the Royal Society, containing
a statement of his views on that subject; and more particularly describing an Engine for calculating astronomical, nautical, and other Tables, by means of the "method of differences." In that letter it is stated that a small Model, consisting of six figures, and capable of working two orders of
differences, had been constructed; and that it performed its
work in a satisfactory manner.

The concluding paragraph of that letter is as follows :—

" Whether I shall construct a larger Engine of this kind, and bring to perfection the others I have described, will, in a great measure, depend on the nature of the encouragement I may receive.

"Induced, by a conviction of the great utility of such Engines, to withdraw, for some time, my attention from a subject on which it has been engaged during several years, and which possesses charms of a higher order, I have now arrived at a point where success is no longer doubtful. It must, however, be attained at a very considerable expense, which would not probably be replaced, by the works it might produce, for a long period of time; and which is an undertaking I should feel unwilling to commence, as altogether foreign to my habits and pursuits."

The Model alluded to had been shown to a large number of Mr. Babbage's acquaintances, and to many other persons; and copies of his letter having been given to several of his friends, it is probable that one of the copies was sent to the Treasury.

On the 1st of April, 1823, the Lords of the Treasury referred that Letter to the Royal Society, requesting—

"The opinion of the Royal Society on the merits and utility of this invention."

On the 1st of May the Royal Society reported to the Treasury, that—

"Mr. Babbage has displayed great talent and ingenuity in the construction of his Machine for Computation, which the Committee think fully adequate to the attainment of the objects proposed by the inventor; and they consider Mr. Babbage as highly deserving of public encouragement, in the prosecution of his arduous undertaking." *

On the 21st of May these papers were ordered to be printed by the House of Commons.

In July, 1823, Mr. Babbage had an interview with the Chancellor of the Exchequer (Mr. Robinson†), to ascertain if it was the wish of the Government that he should construct a large Engine of the kind, which would also print the results it calculated.

* Parliamentary Paper, No. 370, printed 22nd May, 1823.
† Afterwards Lord Goderich, now Earl of Ripon.

From the conversation which took place on that occasion, Mr. Babbage apprehended that such was the wish of the Government. The Chancellor of the Exchequer remarked that the Government were in general unwilling to make grants of money for any inventions, however meritorious; because, if they really possessed the merit claimed for them, the sale of the article produced would be the best, as well as largest reward of the inventor: but that the present case was an *exception*; it being apparent that the construction of such a Machine could not be undertaken with a view to profit from the sale of its produce; and that, as mathematical Tables were peculiarly valuable for nautical purposes, it was deemed a fit object of encouragement by the Government.

The Chancellor of the Exchequer mentioned two modes of advancing money for the construction:—either through the recommendation of a Committee of the House of Commons, or by taking a sum from the Civil Contingencies: and he observed that, as the Session of Parliament was near its termination, the latter course might, perhaps, be the most convenient.

Mr. Babbage thinks the Chancellor of the Exchequer also made some observation, indicating that the amount of money taken from the Civil Contingencies would be smaller than that which might be had by means of a Committee of the House of Commons: and he then proposed to take 1,000*l.* as a commencement from the Civil Contingencies Fund. To this Mr. Babbage replied, in words which he distinctly remembers, " Would it be too much, in the first instance, to take 1,500*l.* ?" The Chancellor of the Exchequer immediately answered, that 1,500*l.* should be advanced.

Mr. Babbage's opinion at that time was, that the Engine would be completed in two, or at the most in three years; and that by having 1,500*l.* in the first instance, he would be enabled to advance, from his own private funds, the residue of the 3,000*l.*, or even 5,000*l.*, which he then imagined the Engine might possibly cost; so that he would not again have occasion to apply to Government until it was completed. Some observations were made by the Chancellor of the Exchequer about the mode of accounting for the money received, as well as about its expenditure; but it seemed to be admitted that it was not possible to prescribe any very definite system, and that much must be left to Mr. Babbage's own judgment.

Very unfortunately, no Minute of that conversation was made at the time, nor was any sufficiently distinct understanding between the parties arrived at. Mr. Babbage's conviction was, that whatever might be the labour and difficulty of the undertaking, the Engine itself would, of course, become the property of the Government, which had paid for its construction.

Soon after this interview with the Chancellor of the Exchequer, a letter was sent from the Treasury to the Royal Society, informing that body that the Lords of the Treasury

" Had directed the issue of 1,500*l.* to Mr. Babbage, to enable him to bring his invention to perfection, in the manner recommended."

These latter words, "*in the manner recommended,*" can only refer to the previous recommendation of the Royal Society; but it does not appear, from the Report of the Royal Society, that *any plan, terms,* or *conditions* had been pointed out by that body.

Towards the end of July, 1823, Mr. Babbage took measures for the construction of the present *Difference* Engine,* and it was regularly proceeded with for four years.

* See Note on next page.

In October, 1827, the expense incurred had amounted to 3,475*l*.; and Mr. Babbage having suffered severe domestic affliction, and being in a very ill state of health, was recommended by his medical advisers to travel on the Continent. He left, however, sufficient drawings to enable the work to be continued, and gave an order to his own banker to advance 1,000*l*. during his absence: he also received, from time to time, drawings and inquiries relating to the mechanism, and returned instructions to the engineer who was constructing it.

As it now appeared probable that the expense would much exceed what Mr. Babbage had originally anticipated, he thought it desirable to inform the Government of that fact, and to procure a further grant. As a preliminary step, he wrote from Italy to his brother-in-law, Mr. Wolryche Whitmore, to request that he would see Lord Goderich upon the subject of the interview in July, 1823; but it is probable that he did not sufficiently inform Mr. Whitmore of all the circumstances of the case.

Mr. Whitmore, having had some conversation with Lord Goderich on the subject, addressed a letter, dated on the 29th of February, 1828, to Mr. Babbage, who was then at Rome, stating that

"That interview was unsatisfactory; that Lord Goderich did not like to admit that there was any understanding, at the time the 1,500*l*. was advanced, that more would be given by Government."

On Mr. Babbage's return to England, towards the end of

NOTE.—It will be convenient to distinguish between—

1. The small *Model* of the original or Difference Engine.
2. The *Difference* Engine itself, belonging to the Government, a part only of which has been put together.
3. The designs for another *Engine*, which in this Statement is called the Analytical Engine.

1828, he waited in person upon Lord Goderich, who admitted that the understanding of 1823 was not very definite. He then addressed a statement to the Duke of Wellington, as the head of the Government, explaining the previous steps in the affair ; stating the reasons for his inferences from what took place at the interview with the Chancellor of the Exchequer in July, 1823 ; and referring his Grace for further information to Lord Goderich, to whom also he sent a copy of that statement.

The Duke of Wellington, in consequence of this application, requested the Royal Society to inquire—

" Whether the progress of the Machine confirms them in their former opinion, that it will ultimately prove adequate to the important object it was intended to attain."

The Royal Society reported, in February, 1829, that—

" They had not the slightest hesitation in pronouncing their decided opinion in the affirmative."

The Royal Society also expressed their hope that—

" Whilst Mr. Babbage's mind is intensely occupied in an undertaking likely to do so much honour to his country, he may be relieved, as much as possible, from all other sources of anxiety."

On the 28th of April, 1829, a Treasury Minute directed a further payment to Mr. Babbage of

" 1,500l. to enable him to complete the Machine by which such important benefit to Science might be expected."

At that time the sum expended on the Engine amounted to 6,697l. 12s., of which 3,000l. had been received from the Treasury ; so that Mr. Babbage had provided 3,697l. 12s. from his own private funds.

Under these circumstances, by the advice of Mr. Wolryche Whitmore, a meeting of Mr. Babbage's personal friends was held on the 12th of May, 1829. It consisted of—

THE DUKE OF SOMERSET,
LORD ASHLEY,
SIR JOHN FRANKLIN,
MR. WOLRYCHE WHITMORE,
DR. FITTON,
MR. FRANCIS BAILY,
MR. (now SIR JOHN) HERSCHEL.

Being satisfied, upon inquiry, of the following facts, they came to the annexed resolutions :—

"1st. That Mr. Babbage was originally induced to take up the work, on its present extensive scale, by an understanding on his part that it was the wish of Government that he should do so, and by an advance of 1,500*l*, at the outset; with a full impression on his mind, that such further advances would be made as the work might require.

" 2nd. That Mr. Babbage's expenditure had amounted to nearly 7,000*l*., while the whole sum advanced by Government was 3,000*l*.

"3rd. That Mr. Babbage had devoted the most assiduous and anxious attention to the progress of the Engine, to the injury of his health, and the neglect and refusal of other profitable occupations.

" 4th. That a very large expense remained to be incurred; and that his private fortune was not such as would justify his completing the Engine, without further and effectual assistance from Government.

" 5th. That a personal application upon the subject should be made to the Duke of Wellington.

"6th. That if such application should be unsuccessful in procuring effectual and adequate assistance, they must regard Mr. Babbage (considering the great pecuniary and personal sacrifices he will then have made; the entire expenditure of all he had received from the public on the subject of its destination; and the moral certainty of completing it, to which it was, by his exertions, reduced) as no longer called on to proceed with an undertaking which might destroy his health, and injure, if not ruin, his fortune.

" 7th. That Mr. Wolryche Whitmore and Mr. Herschel should request an interview with the Duke of Wellington, to state to his Grace these opinions on the subject."

Mr. Whitmore and Mr. Herschel accordingly had an interview with the Duke of Wellington; and some time after they were informed by the Chancellor of the Exchequer, to whom they had applied for his Grace's answer, that the Duke of

Wellington intended to see the portion of the Engine which had been then made.

In November, 1829, the Duke of Wellington, accompanied by the Chancellor of the Exchequer (Mr. Goulburn) and Lord Ashley, saw the *Model* of the Engine, the drawings, and the parts in progress. On the 23rd of that month Mr. Babbage received a note from Mr. Goulburn, dated on the 20th, informing him that the Duke of Wellington and himself had recommended the Treasury to make a further payment towards the completion of the Machine; and that their Lordships had in consequence directed a payment of 3,000*l.* to be made to him. This letter also contained a suggestion about separating the Calculating from the Printing part of the Machine, which was repeated in the letter from the Treasury of the 3rd of December, 1829, communicating officially the information contained in Mr. Goulburn's private note, and stating that directions had been given—

" To pay to you the further sum of 3,000*l.*, to enable you to complete the Machine which you have invented for the calculation of various tables; but I have to intimate to you that, in making this additional payment, my Lords think it extremely desirable that the Machine should be so constructed, that, if any failure should take place in the attempt to print by it, the calculating part of the Machine may nevertheless be perfect and available for that object."

Mr. Babbage inferred from this further grant, that Government had adopted his view of the arrangement entered into with the Chancellor of the Exchequer in July, 1823; but, to prevent the recurrence of difficulty from any remaining indistinctness, he wrote to Mr. Goulburn, stating that, before he received the 3,000*l.*, he wished to propose some general arrangements for expediting the completion of the Engine, further notes of which he would shortly submit to him. On the 25th of November, 1829, he addressed a letter to Lord

Ashley, to be communicated to the Chancellor of the Exchequer, stating the grounds on which he thought the following arrangements desirable :—

1st. That the Engine should be considered as the property of Government.

2nd. That professional engineers should be appointed by Government to examine the charges made for the work already executed, as well as for its future progress ; and that such charges should be defrayed by Government.

3rd. That under this arrangement he himself should continue to direct the construction of the Engine, as he had hitherto done.

Mr. Babbage also stated that he had been obliged to suspend the work for nearly nine months ; and that such delay risked the final completion of the Engine.

In reply to these suggestions, Mr. Goulburn wrote to Lord Ashley, stating—

" That we (the Government) could not adopt the course which Mr. Babbage had pointed out, consistently with the principle on which we have rendered him assistance in the construction of his Machine, and without considerable inconvenience. The view of the Government was, to assist an able and ingenious man of science, whose zeal had induced him to exceed the limits of prudence, in the construction of a work which would, if successful, redound to his honour, and be of great public advantage. We feel ourselves, therefore, under the necessity of adhering to our original intention, as expressed in the Minute of the Treasury, which granted Mr. Babbage the last 3,000*l.*, and in the letter in which I informed him of that grant."

Mr. Goulburn's letter was enclosed by Lord Ashley to Mr. Babbage, with a note, in which his Lordship observed, with reference to Mr. Goulburn's opinion, that it was

" A wrong view of the position in which Mr. Babbage was placed, after his conference with Lord Goderich—which must be explained to him (Mr. Goulburn)."

" *The original intention* " of the Government is here stated to have been communicated to Mr. Babbage, both in the letter from the Treasury of the 3rd of December, 1829, granting the 3,000*l.*, and also in Mr. Goulburn's private letter of the 20th of November, 1829. These letters have been just given; and it certainly does not appear from either of them, that the " original intention " was then in any degree more apparent than it was at the commencement of the undertaking in July, 1823.

On the 16th of December, 1829, Mr. Babbage wrote to Lord Ashley, observing, that Mr. Goulburn seemed to think that he [Mr. Babbage] had commenced the machine on his own account; and that, pursuing it zealously, he had expended more than was prudent, and had then applied to Government for aid. He remarked, that a reference to papers and dates would confirm his own positive declaration, that this was never for one moment, in *his* apprehension, the ground on which the matter rested; and that the following facts would prove that it was absolutely impossible it could have been so:—

1stly. Mr. Babbage referred to the passage* (already quoted) in his letter to Sir Humphry Davy, in which he had expressed his opinion as *decidedly adverse* to the plan of making a larger Machine, on his own account.

2ndly. Mr. Babbage stated that the small Model of the Machine seen by the Duke of Wellington and Mr. Goulburn, was completed *before* his interview with Lord Goderich in July, 1823; for it was alluded to in the Report of the Royal Society, of the 1st of May, 1823.

3rdly. That the interview with Lord Goderich having taken place in July, 1823; the present Machine (*i. e.* the *Difference*

* See page 111.

Engine) was commenced in *consequence* of that interview; and *after* Mr. Babbage had received the first grant of 1,500*l.* on the 7th of August, 1823.

Having thus shown that the light in which Mr. Goulburn viewed these transactions was founded on a misconception, Mr. Babbage requested Lord Ashley to inquire whether the facts to which he had called Mr. Goulburn's attention might not induce him to reconsider the subject. And in case Mr. Goulburn should decline revising his opinion, then he wished Lord Ashley to ascertain the opinion of Government, upon the contingent questions which he enclosed; viz.—

1. Supposing Mr. Babbage received the 3,000*l.* now directed to be issued, what are the claims which Government will have on the Engine, or on himself?

2. Would Mr. Babbage owe the 6,000*l.*, or any part of that sum to the Government?

If this question be answered in the negative,

3. Is the portion of the Engine now made, as completely Mr. Babbage's property as if it had been entirely paid for with his own money?

4. Is it expected by Government that Mr. Babbage should continue to construct the Engine at his own private expense; and, if so, to what extent in money?

5. Supposing Mr. Babbage should decline resuming the construction of the Engine, to whom do the drawings and parts already made belong?

The following statement was also enclosed:—

Expenses up to 9th May, 1829, when the work ceased . .		* £6,628
Two grants of 1,500*l.* each, amounting to	£3,000	
By Treasury Minute, Nov. 1829, but not yet received .	3,000	
	———	6,000
		£628

In January, 1830, Mr. Babbage wrote to Lord Goderich,

---

* The difference between this sum and 6,697*l.* 12*s.* mentioned in page 115, seems to have arisen from the fact of the former sum having included the estimated amount of a bill which, when received, was found to be less than had been anticipated.

stating that the Chancellor of the Exchequer (Mr. Goulburn) would probably apply to his Lordship respecting the interview in July, 1823. He therefore recalled some of the circumstances attending it to Lord Goderich, and concluded thus :—

"The matter was, as you have justly observed on another occasion, left, in a certain measure, indefinite; and I have never contended that any promise was made to me. My subsequent conduct was founded upon the impression left on my mind by that interview. I always considered that, whatever difficulties I might encounter, it could never happen that I should ultimately suffer any pecuniary loss.

"I understand that Mr. Goulburn wishes to ascertain from your Lordship whether, from the nature of that interview, it was reasonable that I should have such expectation."

In the mean time Mr. Babbage had encountered difficulties of another kind. The Engineer who had been constructing the Engine under Mr. Babbage's direction had delivered his bills in such a state that it was impossible to judge how far the charges were just and reasonable; and although Mr. Babbage had paid several thousand pounds, yet there remained a considerable balance, which he was quite prepared and willing to pay, as soon as the accounts should be examined, and the charges approved of by professional engineers.

The delay in deciding whether the Engine was the property of Government, added greatly to this embarrassment. Mr. Babbage, therefore, wrote to Lord Ashley on the 8th of February, to mention these difficulties; and to point out the serious inconvenience which would arise, in the future progress of the Engine, from any dispute between the Engineer and himself relative to payments.

On the 24th of February, 1830, Mr. Babbage called on Lord Ashley, to request he would represent to the Duke of Wellington the facts of the case, and point out to his Grace the importance of a decision. In the afternoon of the same day, he again saw Lord Ashley, who communicated to him the decision of the Government; to the following effect:—

1st. Although the Government would not pledge themselves to COMPLETE *the. Machine, they were willing to declare it their property.*

2nd. That professional Engineers should be appointed to examine the bills.

3rd. That the Government were willing to advance 3,000l. more than the sum (6,000l.) already granted.

4th. That, when the Machine was completed, the Government would be willing to attend to any claim of Mr. Babbage to remuneration, either by bringing it before the Treasury, or the House of Commons.

Thus, after considerable discussion, the doubts arising from the indefiniteness of the understanding with the Chancellor of the Exchequer, in July, 1823, were at length removed. Mr. Babbage's impression of the original arrangement entered into between Lord Goderich and himself was thus formally adopted in the first three propositions: and the Government voluntarily added the expression of their disposition to attend to any claim of his for remuneration when the Engine should be completed.

When the arrangements consequent upon this decision were made, the work of the Engine was resumed, and continued to advance.

After some time, the increasing amount of costly drawings, and of parts of the Engine already executed. remaining exposed to destruction from fire and from other casualties became a source of some anxiety.

These facts having been represented to Lord Althorp (then Chancellor of the Exchequer), an experienced surveyor was directed to find a site adapted for a building for the reception of the Engine in the neighbourhood of Mr. Babbage's residence.

On the 19th of January the Surveyor's reports were forwarded to Lord Althorp (the Chancellor of the Exchequer), who referred the case to a committee of practical Engineers for their opinion. This committee reported strongly in favour of the removal, on the grounds of security, and of economy in completing the Engine; and also recommended the site which had been previously selected by the Surveyor. The Royal Society, also, to whom Lord Althorp had applied, examined the question, and likewise reported strongly to the same effect.

A lease of some property, adjacent to Mr. Babbage's residence, was therefore subsequently granted by him to the Government; and a fire-proof building, capable of containing the Engine, with its drawings, and workshops necessary for its completion, were erected.

With respect to the expenses of constructing the Engine, the following plan was agreed upon and carried out :—The great bulk of the work was executed by the Engineer under the direction of Mr. Babbage. When the bills were sent in, they were immediately forwarded by him to two eminent Engineers, Messrs. Donkin and Field, who, at the request of Government, had undertaken to examine their accuracy. On these gentlemen certifying those bills to be correct, Mr. Babbage transmitted them to the Treasury; and after the usual forms, a warrant was issued directing the payment of the respective sums to Mr. Babbage. This course, however, required considerable time; and the Engineer having represented that he was unable to pay his workmen without more immediate advances, Mr. Babbage, to prevent delay in completing the Engine, did himself, from time to time, advance from his own funds several sums of money; so that he was, in fact, usually in advance from 500*l.* to 1,000*l.* Those sums were, of course, repaid when the Treasury warrants were issued.

Early in the year 1833, an event of great importance in the history of the Engine occurred. Mr. Babbage had directed a portion of it, consisting of sixteen figures, to be put together. It was capable of calculating Tables having two or three orders of differences; and, to some extent, of forming other Tables. The action of this portion completely justified the expectations raised, and gave a most satisfactory assurance of its final success.

The fire-proof building and workshops having been completed, arrangements were made for the removal of the Engine. Mr. Babbage finding it no longer convenient to make payments in advance, informed the Engineer that he should in future not pay him until the money was received from the Treasury. Upon receiving this intimation, the Engineer immediately discontinued the construction of the Engine, and dismissed the workmen employed on it; which fact Mr. Babbage immediately communicated to the Treasury.

In this state of affairs it appeared, both to the Treasury and to Mr. Babbage, that it would be better to complete the removal of the drawings, and all the parts of the Engine to the fire-proof building; and then make such arrangements between the Treasury and the Engineer, respecting the future payments, as might prevent further discussion on that subject.

After much delay and difficulty the whole of the drawings, and parts of the Engine, were at length removed to the fire-proof building in East-street, Manchester-square. Mr. Babbage wrote, on the 16th of July, 1834, to the Treasury, informing their Lordships of the fact;—adding that no advance had been made in its construction for above a year and a quarter; and requesting further instructions on the subject.

Mr. Babbage received a letter from the Treasury, expressing their Lordships' satisfaction at learning that the drawings, and parts of the Calculating Engine were removed to the fire-proof building, and stating that as soon as Mr. Clement's Accounts should be received and examined, they would

"Take into consideration what further proceedings may be requisite with a view to its completion."

A few weeks afterwards Mr. Babbage received a letter from the Treasury, conveying their Lordships' authority to proceed with the construction of the Engine.

During the time which had elapsed since the Engineer had ceased to proceed with the construction of the Engine, Mr. Babbage had been deprived of the use of his own drawings. Having, in the meanwhile, naturally speculated upon the general principles on which machinery for calculation might be constructed, *a principle of an entirely new kind* occurred to him, the power of which over the most complicated arithmetical operations seemed nearly unbounded. On re-examining his drawings when returned to him by the Engineer, the new principle appeared to be limited only by the extent of the mechanism it might require. The invention of simpler mechanical means for executing the elementary operations of the Engine now derived a far greater importance than it had hitherto possessed; and should such simplifications be discovered, it seemed difficult to anticipate, or even to over-estimate, the vast results which might be attained. In the Engine for calculating by differences, such simplifications affected only about a hundred and twenty

s 2

similar parts, whilst in the new or *Analytical* Engine, they would affect a great many thousand. The *Difference* Engine might be constructed with more or less advantage by employing various mechanical modes for the operation of addition: the *Analytical* Engine could not *exist* without inventing for it a method of mechanical addition possessed of the utmost simplicity. In fact, it was not until upwards of twenty different mechanical modes for performing the operation of addition had been designed and drawn, that the necessary degree of simplicity required for the Analytical Engine was ultimately attained. Hence, therefore, the powerful motive for simplification.

These new views acquired additional importance, from their bearings upon the Engine already partly executed for the Government. For, if such simplifications should be discovered, it might happen that the *Analytical* Engine would execute more rapidly the calculations for which the *Difference* Engine was intended; or, that the *Difference* Engine would itself be superseded by a far simpler mode of construction. Though these views might, perhaps, at that period have appeared visionary, both have subsequently been completely realized.

To withhold those new views from the Government, and under such circumstances to have allowed the construction of the Engine to be resumed, would have been improper; yet the state of uncertainty in which those views were then necessarily involved rendered any *written* communication respecting their probable bearing on the Difference Engine a matter of very great difficulty. It appeared to Mr. Babbage that the most straightforward course was to ask for an interview on the subject with the Head of the Government, and to communicate to him the exact state of the case.

Had that interview taken place, the First Lord of the Treasury might have ascertained from his inquiries, in a manner quite impracticable by any written communications, the degree of importance which Mr. Babbage attached to his new inventions, and his own opinion of their probable effect, in superseding the whole or any part of the original, or *Difference*, Engine. The First Lord of the Treasury would then have been in a position to decide, either on the immediate continuation and completion of the original design, or on its temporary suspension, until the character of the new views should be more fully developed by further drawings and examination.

There was another, although a far less material point, on which also it was desirable to obtain the opinion of the Government: the serious impediments to the progress of the Engine, arising from the Engineer's conduct, as well as the consequent great expense, had induced Mr. Babbage to consider, whether it might not be possible to employ some other person as his agent for constructing it. His mind had gradually become convinced of the practicability of that measure ; but he was also aware that however advantageous it might prove to the Government, from its greater economy, yet that it would add greatly to his own personal labour, responsibility, and anxiety.

On the 26th of September, 1834, Mr. Babbage therefore requested an interview with Lord Melbourne, for the purpose of placing before him these views. Lord Melbourne acceded to the proposed interview, but it was then postponed; and soon after, the Administration of which his Lordship was the Head went out of Office, without the interview having taken place.

For the same purpose, Mr. Babbage applied in December,

1834, for an interview with the Duke of Wellington, who, in reply, expressed his wish to receive a written communication on the subject. He accordingly addressed a statement to his Grace, pointing out the only plans which, in his opinion, could be pursued for terminating the questions relative to the *Difference* Engine ; namely,

1st. The Government might desire Mr. Babbage to continue the construction of the Engine, in the hands of the person who has hitherto been employed in making it.

2ndly. The Government might wish to know whether any other person could be substituted for the Engineer at present employed to continue the construction ;—a course which was possible.

3rdly. The Government might (although he did not presume that they would) substitute some person to superintend the completion of the Engine instead of Mr. Babbage himself.

4thly. The Government might be disposed to give up the undertaking entirely.

He also stated to the Duke of Wellington, the circumstances which had led him to the invention of a *new* Engine, of far more extensive powers of calculation; which he then observed did not supersede the former one, but added greatly to its utility.

At this period, the impediments relating to the *Difference* Engine had been partially and temporarily removed. The chief difficulty would have been either the formation of new arrangements with the Engineer, or the appointment of some other person to supply his place. This latter alternative, which was of great importance for economy as well as for its speedy completion, Mr. Babbage had carefully examined, and was then prepared to point out means for its accomplishment.

The duration of the Duke of Wellington's Administration was short; and no decision on the subject of the *Difference* Engine was obtained.

On the 15th of May the *Difference* Engine was alluded to in the House of Commons; when the Chancellor of the Exchequer did Mr. Babbage the justice to state distinctly, that the whole of the money voted had been expended in paying the workmen and for the materials employed in constructing it, and that not one shilling of it had ever gone into his own pocket.

About this time several communications took place between the Chancellor of the Exchequer and Mr. Babbage, respecting a reference to the Royal Society for an opinion on the subject of the Engine.

A new and serious impediment to the possibility of executing one of the plans which had been suggested to the Duke of Wellington for completing the *Difference* Engine arose from these delays. The draftsman whom Mr. Babbage had, at his own expense, employed, both on the *Difference* and on the *Analytical* Engine, received an offer of a very liberal salary, if he would enter into an engagement abroad, which would occupy many years. His assistance was indispensable, and his services were retained only by Mr. Babbage considerably increasing his salary.

On the 14th of January, 1836, Mr. Babbage received a communication from the Chancellor of the Exchequer (Mr. Spring Rice*), expressing his desire to come to some definite result on the subject of the Calculating Engine, in which he remarked, that the conclusion to be drawn from Mr. Babbage's statement to the Duke of Wellington was, that he

* The present Lord Monteagle.

(Mr. Babbage) having invented a new machine, of far greater powers than the former one, wished to be informed if the Government would undertake to defray the expense of this *new* Engine.

The Chancellor of the Exchequer then pointed out reasons why he should feel himself bound to look to the completion of the first machine, before he could propose to Parliament to enter on the consideration of the second : and he proposed to refer to the Royal Society for their opinion, authorizing them, if they thought fit, to employ any practical mechanist or engineer to assist them in their inquiries. The Chancellor of the Exchequer concluded with expressing his readiness to communicate with Mr. Babbage respecting the best mode of attaining that result.

From these statements it is evident that Mr. Babbage had failed in making his own views distinctly understood by the Chancellor of the Exchequer. His first anxiety, when applying to Lord Melbourne, had been respecting the question, whether the *Discoveries* with which he was then advancing might not ultimately supersede the work already executed. His second object had been to point out a possible arrangement, by which great expense might be saved in the mechanical construction of the *Difference* Engine.

So far was Mr. Babbage from having proposed to the Government to defray the expenses of the *new* or *Analytical* Engine, that though he expressly pointed out in the statement to the Duke of Wellington* four courses which it was possible for the Government to take,—yet in no one of them was the construction of the *new* Engine alluded to.

Those views of improved machinery for making calculations

* See page 128.

which had appeared in but faint perspective in 1834, as likely to lead to important consequences, had, by this time, assumed a form and distinctness which fully justified the anticipations then made. By patient inquiry, aided by extensive drawings and notations, the projected *Analytical* Engine had acquired such powers, that it became necessary, for its further advancement, to simplify the elements of which it was composed. In the progress of this inquiry, Mr. Babbage had gradually arrived at simpler mechanical modes of performing those arithmetical operations on which the action of the *Difference* Engine depended ; and he felt it necessary to communicate these new circumstances, as well as their consequences, to the Chancellor of the Exchequer.

On the 20th of January, 1836, Mr. Babbage wrote, in answer to the communication from the Chancellor of the Exchequer, that he did not, on re-examining the statement addressed to the Duke of Wellington, perceive that it contained *any application to take up the new or Analytical Engine ;* and he accompanied this reply by a statement relative to the progress of the *Analytical* Engine, and its bearing upon the *Difference* Engine belonging to the Government. The former, it was said,

"Is not only capable of accomplishing all those other complicated calculations which I had intended, but it also performs all calculations which were peculiar to the Difference Engine, both in less time, and to a greater extent : in fact, it completely supersedes the Difference Engine."

The Reply then referred to the statement laid before the Duke of Wellington in July, 1834, in which it was said,

"That all the elements of the Analytical were essentially different from those of the Difference Engine;"

and that the mechanical simplicity to which its elements had now been reduced was such, that it would probably cost more

T

to finish the *old Difference* Engine on its original plan than to construct a *new* Difference Engine with the simplified elements devised for the *Analytical* Engine.

It then proceeded to state that—

" The fact of a *new* superseding an *old* machine, in a very few years, is one of constant occurrence in our manufactories ; and instances might be pointed out in which the advance of invention has been so rapid, and the demand for machinery so great, that half-finished machines have been thrown aside as useless before their completion.

" It is now nearly fourteen years since I undertook for the Government to superintend the making of the Difference Engine. During nearly four years its construction has been absolutely stopped, and, instead of being employed in overcoming the physical impediments, I have been harassed by what may be called the moral difficulties of the question. It is painful to reflect that, in the time so employed, the first Difference Engine might, under more favourable circumstances, have been completed.

" In making this Report, I wish distinctly to state, that I do not entertain the slightest doubt of the success of the Difference Engine ; *nor do I intend it as any application to finish the one or to construct the other ;* but I make it from a conviction that the information it contains ought to be communicated to those who must decide the question relative to the Difference Engine."

The reference to the Royal Society, proposed by the Chancellor of the Exchequer, in his letter of the 14th of January, 1836,* did not take place ; and during more than a year and a half no further measures appear to have been adopted by the Government respecting the Engine.

It was obviously of the greatest importance to Mr. Babbage that a final decision should be made by the Government. When he undertook to superintend the construction of the *Difference* Engine for the Government, it was, of course, understood that he would not leave it unfinished. He had now been engaged fourteen years upon an object which he

* See page 130.

had anticipated would not require more than two or three ; and there seemed no limit to the time his engagement with the Government might thus be supposed to endure, unless some steps were taken to terminate it.   Without such a decision Mr. Babbage felt that he should be impeded in any plans he might form, and liable to the most serious interruption, if he should venture to enter upon the execution of them.   He therefore most earnestly pressed, both by his personal applications and by those of his friends, for the settlement of the question.   Mr. Wolryche Whitmore, in particular, repeatedly urged upon the Chancellor of the Exchequer, personally, as well as by letter, the injustice of keeping Mr. Babbage so very long in a state of suspense.

Time, however, passed on, and during nearly two years the question remained in the same state.   Mr. Babbage, wearied with this delay, determined upon making a last effort to obtain a decision.   He wrote to the First Lord of the Treasury (Lord Melbourne) on the 26th of July, 1838, recalling to his Lordship's attention the frequency of his applications on this subject, and urging the necessity of a final decision upon it.   He observed, that if the question had become more difficult, because he had invented superior mechanism, which had superseded that which was already partly executed, this consequence had arisen from the very delay against which he had so repeatedly remonstrated.   He then asked, for the last time, not for any favour, but for that which it was an injustice to withhold—a decision.

On the 16th of August Mr. Spring Rice (the Chancellor of the Exchequer) addressed a note to Mr. Babbage, in reference to his application to Lord Melbourne.   After recapitulating his former statement of the subject, which had been shown to be founded on a misapprehension, viz., that Mr. Babbage

T 2

had made an *application* to the Government to construct for them the *Analytical* Engine, the Chancellor of the Exchequer inquired whether he was solicitous that steps should be taken for the completion of the old, or for the commencement of a new machine,—and what he considered would be the cost of the one proceeding, and of the other ?

Being absent on a distant journey, Mr. Babbage could not reply to this note until the 21st of October. He then reminded the Chancellor of the Exchequer of his previous communication of the 20th of January, 1836 (see p. 89), in which it was expressly stated that he did *not* intend to make any application to construct a *new* machine ; but that the communication to the Duke of Wellington and the one to himself were made, simply because he thought it would be unfair to conceal such important facts from those who were called upon to decide on the continuance or discontinuance of the construction of the *Difference Engine.*

With respect to the expense of either of the courses pointed out by the Chancellor of the Exchequer, Mr. Babbage observed that, not being a professional Engineer, and his past experience having taught him not to rely upon his own judgment on matters of that nature, he should be very reluctant to offer any opinion upon the subject.

In conclusion, Mr. Babbage stated that the question he wished to have settled was—

Whether the Government required him to superintend the completion of the Difference Engine, which had been suspended during the last five years, according to the original plan and principles ; or whether they intended to discontinue it altogether ?

In November, 1841, Mr. Babbage, on his return from the Continent, finding that Sir Robert Peel had become First

Lord of the Treasury, determined upon renewing his application for a decision of the question. With this view the previous pages of this Statement were drawn up, and a copy of it was forwarded to him, accompanied by a letter from Mr. Babbage, in which he observed—

" Of course, when I undertook to give the invention of the Calculating Engine to the Government, and to superintend its construction, there must have been an implied understanding that I should carry it on to its termination. I entered upon that understanding, believing that two or at the utmost that three years would complete it. The better part of my life has now been spent on that machine, and no progress whatever having been made since 1834, that understanding may possibly be considered by the Government as still subsisting : I am therefore naturally **very** anxious that this state of uncertainty should be put an end to as soon as possible."

Mr. Babbage, in reply, received a note from Sir George Clerk (Secretary to the Treasury), stating that Sir Robert Peel feared that it would not be in his power to turn his attention to the subject for some days, but that he hoped, as soon as the great pressure of business previous to the opening of the session of Parliament was over, he might be able to determine on the best course to be pursued.

The session of Parliament closed in August, and Mr. Babbage had received no further communication on the subject. Having availed himself of several private channels for recalling the question to Sir Robert Peel's attention without effect, Mr. Babbage, on the 8th of October, 1842, again wrote to him, requesting an early decision.

On the 4th of November, 1842, a note from Sir Robert Peel explained to Mr. Babbage that some delay had arisen, from his wish to communicate personally with the Chancellor of the Exchequer, who would shortly announce to him their joint conclusion on the subject.

On the same day Mr. Babbage received a letter from Mr.

Goulburn (the Chancellor of the Exchequer), who stated that he had communicated with Sir Robert Peel, and that they both regretted the necessity of abandoning the completion of a machine, on which so much scientific labour had been bestowed. He observed, that the expense necessary for rendering it either satisfactory to Mr. Babbage or generally useful appeared, on the lowest calculation, so far to exceed what they should be justified in incurring, that they considered themselves as having no other alternative.

Mr. Goulburn concluded by expressing their hope, that by the Government withdrawing all claim to the machine as already constructed, and placing it entirely at Mr. Babbage's disposal, they might in some degree assist him in his future exertions in the cause of Science.

On the 6th of November, 1842, Mr. Babbage wrote to Sir Robert Peel and the Chancellor of the Exchequer, acknowledging the receipt of their decision, thanking them for the offer of the machine as already constructed, but, under all the circumstances, declining to accept it.*

On the 11th of November Mr. Babbage obtained an interview with Sir Robert Peel, and stated, that having given the original Invention to the Government—having superintended for them its construction—having demonstrated the possibility of the undertaking by the completion of an important portion of it—and that the non-completion of the design arose neither from his fault nor his desire, but was the act of the Government itself, he felt that he had some claims on their consideration.

He rested those claims upon the sacrifices he had made,

* The part of the *Difference* Engine already constructed, together with all the Drawings relating to the whole machine, were, in January, 1843 (by the direction of the Government), deposited in the Museum of King's College, London.

both personal and pecuniary, in the advancement of the Mechanical Arts and of Science—on the anxiety and the injury he had experienced by the delay of eight years in the decision of the Government on the subject, and on the great annoyance he had constantly been exposed to by the prevailing belief in the public mind that he had been amply remunerated by large grants of public money. Nothing, he observed, but some public act of the Government could ever fully refute that opinion, or repair the injustice with which he had been treated.

The result of this interview was entirely unsatisfactory. Mr. Babbage went to it prepared, had his statement produced any effect, to have pointed out two courses, by either of which it was probable that not only a Difference Engine, but even the Analytical Engine, might in a few years have been completed. The state of Sir Robert Peel's information on the subject, and the views he took of Mr. Babbage's services and position, prevented Mr. Babbage from making any allusion to either of those plans.

Thus finally terminated an engagement, which had existed upwards of twenty years. During no part of the last eight of those years does there appear to have been any reason why the same decision should not have been arrived at by the Government as was at last actually pronounced.

It was during this last period that all the great principles on which the *Analytical* Engine rests were discovered, and that the mechanical contrivances in which they might be embodied were invented. The establishment which Mr. Babbage had long maintained in his own house, and at his own expense, was now directed with increased energy to the new inquiries required for its perfection.

In this Statement the heavy sacrifices, both pecuniary and

personal, which the invention of these machines has entailed upon their author, have been alluded to as slightly as possible. Few can imagine, and none will ever know their full extent. Some idea of those sacrifices must nevertheless have occurred to every one who has read this Statement. During upwards of twenty years Mr. Babbage has employed, in his own house, and at his own expense, workmen of various kinds, to assist him in making experiments necessary for attaining a knowledge of every art which could possibly tend to the perfection of those Engines; and with that object he has frequently visited the manufactories of the Continent, as well as our own.

Since the discontinuance of the Difference Engine belonging to the Government, Mr. Babbage has himself maintained an establishment for making drawings and descriptions demonstrating the nature and power of the *Analytical* Engine, and for its construction at some future period, when its value may be appreciated.

To these remarks it will only be added, that at an early stage of the construction of the *Difference* Engine he refused more than one highly desirable and profitable situation, in order that he might give his whole time and thoughts to the fulfilment of the engagement which he considered himself to have entered into with the Government.

August, 1843.

CHAPTER VII.

DIFFERENCE ENGINE NO. II.

Difference Engine No. 2—The Earl of Rosse, President of the Royal Society, proposed to the Government a Plan by which the Difference Engine No. 2 might have been executed—It was addressed to the Earl of Derby, and rejected by his Chancellor of the Exchequer.

It was not until 1848, when I had mastered the subject of the Analytical Engine, that I resolved on making a complete set of drawings of the Difference Engine No. 2. In this I proposed to take advantage of all the improvements and simplifications which years of unwearied study had produced for the Analytical Engine.

In 1852, the Earl of Rosse, who, from its commencement, had looked forward with the greatest interest to the application of mechanism to purposes of calculation, and who was well acquainted with the drawings and notations of the Difference Engine No. 2, inquired of me whether I was willing to give them to the Government, provided they would have the Engine constructed. My feeling was, after the sad experience of the past, that I ought not to think of sacrificing any further portion of my life upon the subject. If, however, they chose to have the Difference Engine made, I was ready to give them the whole of the drawings, and also the notations by which it was demonstrated that such a machine could be constructed, and that when made it would necessarily do the work prescribed for it.

U

My much-valued friend, the late Sir Benjamin Hawes, had also been consulted, and it was agreed that the draft of a letter to Lord Derby, who was then prime minister, should be prepared ; in which I should make this offer.   Lord Rosse proposed to place my letter in Lord Derby's hands, with his own statement of a plan by which the whole question might be determined.

Lord Rosse's suggestion was, that the Government should apply to the President of the Institution of Civil Engineers to ascertain,

> 1st. Whether it was possible, from the drawings and notations, to make an estimate of the cost of constructing the machine ?
>
> 2ndly. In case this question was answered in the affirmative—then, could a Mechanical Engineer be found who would undertake to construct it, and at what expense ?

The Institution of Civil Engineers was undoubtedly the highest authority upon the first question.   That being decided in the affirmative, no other body had equal power to find out those mechanical engineers who might be willing to undertake the contract.

Supposing both these questions, or even the latter only, answered in the negative, the proposition, of course, fell to the ground.   But if they were both answered in the affirmative, then there would have arisen a further question for the consideration of the Government :   namely, Whether the object to be obtained was worthy of the expenditure ?

The final result of this eminently *practical* plan was communicated to the Royal Society by their President, in his address at their anniversary on the 30th November, 1854. The following is an extract :—

" The progress of the work was suspended: there was a
" change of Government.   Science was weighed against gold
" by a new standard, and it was resolved to proceed no
" further.   No enterprise could have had its beginning under
" more auspicious circumstances: the Government had taken
" the initiative—they had called for advice, and the adviser
" was the highest scientific authority in this country;—your
" Council; guided by such men as Davy, Wollaston, and
" Herschel.   By your Council the undertaking was inaugu-
" rated,—by your Council it was watched over in its progress.
" That the first great effort to employ the powers of calcu-
" lating mechanism, in aid of the human intellect, should
" have been suffered in this great country to expire fruitless,
" because there was no tangible evidence of immediate profit,
" as a British subject I deeply regret, and as a Fellow my
" regret is accompanied with feelings of bitter disappoint-
" ment.   Where a question has once been disposed of, suc-
" ceeding Governments rarely reopen it, still I thought I
" should not be doing my duty if I did not take some oppor-
" tunity of bringing the facts once more before Government.
" Circumstances had changed, mechanical engineering had
" made much progress; the tools required and trained work-
" men were to be found in the workshops of the leading
" mechanists, the founder's art was so advanced that casting
" had been substituted for cutting, in making the change
" wheels, even of screw-cutting engines, and therefore it was
" very probable that persons would be found willing to under-
" take to complete the Difference Engine for a specific sum.
" That finished, the question would then have arisen, how
" far it was advisable to endeavour, by the same means, to
" turn to account the great labour which had been expended
" under the guidance of inventive powers the most original,

U 2

" controlled by mathematics of a very high order; and which
" had been wholly devoted for so many years to the great
" task of carrying the powers of calculating machinery to its
" utmost limits. Before I took any step I wrote to several
" very eminent men of science, inquiring whether, in their
" opinion, any great scientific object would be gained if Mr.
" Babbage's views, as explained in Ménabrèa's little essay,
" were completely realized. The answers I received were
" strongly in the affirmative. As it was necessary the subject
" should be laid before Government in a form as practical
" as possible, I wrote to one of our most eminent mechanical
" engineers to inquire whether I should be safe in stating
" to Government that the expense of the Calculating Engine
" had been more than repaid in the improvements in me-
" chanism directly referable to it; he replied,—unquestion-
" ably. Fortified by these opinions, I submitted this propo-
" sition to Government:—that they should call upon the
" President of the Society of Civil Engineers to report
" whether it would be practicable to make a contract for the
" completion of Mr. Babbage's Difference Engine, and if so,
" for what sum. This was in 1852, during the short admi-
" nistration of Lord Derby, and it led to no result. The
" time was unfortunate; a great political contest was impend-
" ing, and before there was a lull in politics, so that the
" voice of Science could be heard, Lord Derby's government
" was at an end."

The following letter was then drawn up, and placed in
Lord Derby's hands by Lord Rosse:—

My Lord,   *June* 8, 1852.

    I TAKE the liberty of drawing your Lordship's attention
to the subject of the construction of a Difference Engine, for

calculating and printing Astronomical and Nautical Tables, which was brought under the notice of the Government so far back as the year 1823, and upon which the Government of that day desired the opinion of the Royal Society.

I annex a copy of the correspondence which took place at that time, and which your Lordship will observe was laid before Parliament.

The Committee of the Royal Society, to which the subject was referred, reported generally that the invention was one " fully adequate to the attainment of the objects proposed by " the inventor, and that they considered Mr. Babbage as highly " deserving of public encouragement in the prosecution of his " arduous undertaking."—*Report of Royal Society, 1st May,* 1823. *Parliamentary Paper,* 370, 22nd May, 1823.

And in a subsequent and more detailed Report, which I annex also, they state :—

" The Committee have no intention of entering into any " consideration of the abstract mathematical principle on which " the practicability of such a machine as Mr. Babbage's relies, " nor of its public utility when completed. They consider " the former as not only sufficiently clear in itself, but as " already admitted and acted on by the Council in their former " proceedings. The latter they regard as obvious to every one " who considers the immense advantage of accurate numerical " Tables in all matters of calculation, especially in those which " relate to Astronomy and Navigation, and the great variety " and extent of those which it is the object and within the " compass of Mr. Babbage's Engine to calculate and print " with perfect accuracy."—*Report of Committee of Royal Society, 12th Feb.,* 1829.

Upon the first of these Reports, the Government determined to construct the machine, under my personal superintendence and direction. The Engine was accordingly commenced and partially completed. Tables of figures were calculated, limited in extent only by the number of wheels put together.

Delays, from various causes arose in the progress of the work, and great expenses were incurred. The machine was altogether new in design and construction, and required the utmost mechanical skill which could be obtained for its execution. "It involved," to quote again from the Report of the Committee of the Royal Society, "the necessity of construct-
" ing, and in many instances inventing, tools and machinery
" of great expense and complexity (and in many instances of
" ingenious contrivances likely to prove useful for other pur-
" poses hereafter), for forming with the requisite precision
" parts of the apparatus dissimilar to any used in ordinary
" mechanical works ; that of making many previous trials to
" ascertain the validity of proposed movements ; and that of
" altering, improving, and simplifying those already contrived
" and reduced to drawings. Your Committee are so far from
" being surprised at the time it has occupied to bring it to its
" present state, that they feel more disposed to wonder it has
" been possible to accomplish so much." The true explanation both of the slow progress and of the cost of the work is clearly stated in this passage ; and I may remark in passing, that the tools which were invented for the construction of the machine were afterwards found of utility, and that this anticipation of the Committee has been realized, as some of our most eminent mechanical engineers will readily testify.

Similar circumstances will, I apprehend, always attend and prolong the period of bringing to perfection inventions which have no parallel in the previous history of mechanical construction. The necessary science and skill specially acquired in executing such works must also, as experience is gained, suggest deviations from, and improvements in, the original plan of those works; and the adoption or rejection of such changes, especially under circumstances similar to those in which I was placed, often involves questions of the greatest difficulty and anxiety.

From whatever cause, however, the delays and expenses arose, the result was that the Government was discouraged, and declined to proceed further with the work.

Mr. Goulburn's letter, intimating this decision to me, in 1842, will be found in the accompanying printed Statement. And that the impediments to the completion of the engine, described by the Royal Society, were those which influenced the Government in the determination they came to, I infer from the reason assigned by Mr. Goulburn for its discontinuance, viz., " the expense which would be necessary in order to " render it either satisfactory to yourself or generally useful." I readily admit that the work could not have been rendered satisfactory to myself unless I was free to introduce every improvement which experience and thought could suggest. But that even with this additional source of expense its general usefulness would have been impaired, I cannot assent to, for I believe, in the words of the Report I have already quoted, the " immense advantage of accurate Numerical " Tables in all matters of calculation, especially in those which " relate to Astronomy and Navigation, cannot, within any " reasonable limits, be over-estimated." As to the expense actually incurred upon the first Difference Engine, that of the Government was about 17,000*l.* On my own part, and out of my own private resources, I have sacrificed upon this and other works of science upwards of 20,000*l.*

From the date of Mr. Goulburn's letter, nothing has been done towards the further completion of the Difference Engine by the Government or myself. So much of it as was completed was deposited in the Museum of King's College, where it now remains.

Three consequences have, however, resulted from my subsequent labours, to which I attach great importance.

First, I have been led to conceive the most important elements of another Engine upon a new principle (the details of which are reduced accurately to paper), the power of which over the most complicated analytical operations appears nearly unlimited; but no portion of which is yet commenced. I have called this engine, in contradistinction to the other, the Analytical Engine.

Secondly, I have invented and brought to maturity a system of signs for the explanation of machinery, which I have called Mechanical Notation, by means of which the drawings, the times of action, and the trains for the transmission of force, are expressed in a language at once simple and concise. Without the aid of this language I could not have invented the Analytical Engine; nor do I believe that any machinery of equal complexity can ever be contrived without the assistance of that or of some other equivalent language. The Difference Engine No. 2, to which I shall presently refer, is entirely described by its aid.

Thirdly, in labouring to perfect this Analytical Machine of greater power and wider range of computation, I have discovered the means of simplifying and expediting the mechanical processes of the first or Difference Engine.

After what has passed, I cannot expect the Government to undertake the construction of the Analytical Engine, and I do not offer it for that purpose. It is not so matured as to enable any other person, without long previous training and application, even to attempt its execution; and on my own part, to superintend its construction would demand an amount of labour, anxiety, and time which could not, after the treatment I have received, be expected from me. I therefore make no such offer.

But that I may fulfil to the utmost of my power the original expectation that I should be able to complete, for the Government, an Engine capable of calculating astronomical and nautical Tables with perfect accuracy, such as that which is described in the Reports of the Royal Society, I am willing to place at the disposal of Government (if they will undertake to execute a new Difference Engine) all those improvements which I have invented and have applied to the Analytical Engine. These comprise a complete series of drawings and explanatory notations, finished in 1849, of the Difference Engine No. 2,—an instrument of greater power as well as of greater simplicity than that formerly commenced, and now in the possession of the Government.

I have sacrificed time, health, and fortune, in the desire to complete these Calculating Engines. I have also declined several offers of great personal advantage to myself. But, notwithstanding the sacrifice of these advantages for the purpose of maturing an engine of almost intellectual power, and after expending from my own private fortune a larger sum than the Government of England has spent on that machine, the execution of which it only commenced, I have received neither an acknowledgment of my labours, nor even the offer of those honours or rewards which are allowed to fall within the reach of men who devote themselves to purely scientific investigations. I might, perhaps, advance some claims to consideration, founded on my works and contribu-

x

tions in aid of various departments of industrial and physical science,—but it is for others to estimate those services.

I now, however, simply ask your Lordship to do me the honour to consider this statement and the offer I make. I prefer no claim to the distinctions or the advantages which it is in the power of the Crown or the Government to bestow. I desire only to discharge whatever *imagined* obligation may be supposed to rest upon me, in connexion with the original undertaking of the Difference Engine; though I cannot but feel that whilst the public has already derived advantage from my labours, I have myself experienced only loss and neglect.

If the work upon which I have bestowed so much time and thought were a mere triumph over mechanical difficulties, or simply curious, or if the execution of such engines were of doubtful practicability or utility, some justification might be found for the course which has been taken; but I venture to assert that no mathematician who has a reputation to lose will ever *publicly* express an opinion that such a machine would be useless if made, and that no man distinguished as a Civil Engineer will venture to declare the construction of such machinery impracticable. The names appended to the Report of the Committee of the Royal Society fully justify my expressing this opinion, which I apprehend will not be disputed.

And at a period when the progress of physical science is obstructed by that exhausting intellectual and manual labour, indispensable for its advancement, which it is the object of the Analytical Engine to relieve, I think the application of machinery in aid of the most complicated and abstruse calculations can no longer be deemed unworthy of the attention of the country. In fact, there is no reason why mental as well as bodily labour should not be economized by the aid of machinery.

With these views I have addressed your Lordship, as the head of the Government; and whatever may be my sense of the injustice that has hitherto been done me, I feel, in laying this representation before your Lordship, and in making the offer I now make, that I have discharged to the utmost limit every implied obligation I originally contracted with the country.

I have the honour to be,

&c., &c., &c.,

CHARLES BABBAGE.

Dorset Street, Manchester Square.
    June 8, 1852.

As this question was one of finance and of calculation, the sagacious Premier adroitly turned it over to his Chancellor of the Exchequer—that official being, from his office, *supposed* to be well versed in both subjects.

The opinion pronounced by the novelist and financier was, " That Mr. Babbage's projects appear to be so indefinitely " expensive, the ultimate success so problematical, and the " expenditure certainly so large and so utterly incapable of " being calculated, that the Government would not be justified " in taking upon itself any further liability."—*Extract from the Reply of Earl Derby to the application of the Earl of Rosse, K.P., President of the Royal Society.*

The answer of Lord Derby to Lord Rosse was in substance—

That he had consulted the Chancellor of the Exchequer, who pronounced Mr. Babbage's project as—

x 2

1. " Indefinitely expensive."
2. " The ultimate success problematical."
3. " The expenditure utterly incapable of being calculated."

1. With regard to the "indefinite expense." Lord Rosse had proposed to refer this question to the President of the Institution of Civil Engineers, who would have given his opinion after a careful examination of the drawings and notations. These had not been seen by the Chancellor of the Exchequer; and, if seen by him, would not have been comprehended.

The objection that its success was "problematical" may refer either to its mechanical construction or to its mathematical principles.

Who, possessing one grain of common sense, could look upon the unrivalled workmanship of the then existing portion of the Difference Engine No. 1, and doubt whether a simplified form of the same engine could be executed?

As to any doubt of its mathematical principles, this was excusable in the Chancellor of the Exchequer, who was himself too practically acquainted with the fallibility of his own figures, over which the severe duties of his office had stultified his brilliant imagination. Far other figures are dear to him—those of speech, in which it cannot be denied he is indeed pre-eminent.

Any junior clerk in his office might, however, have told him that the power of computing Tables by differences merely required a knowledge of simple addition.

As to the impossibility of ascertaining the expenditure, this merges into the first objection; but a poetical brain must be pardoned when it repeats or amplifies. I will recall to the ex-Chancellor of the Exchequer what Lord Rosse really pro-

posed, namely, that the Government should take the opinion
of the President of the Institution of Civil Engineers upon
the question, whether a contract could be made for constructing the Difference Engine, and if so, for what sum.

But the very plan proposed by Lord Rosse and refused by
Lord Derby, for the construction of the *English* Difference
Engine, was adopted some few years after by another administration for the *Swedish* Difference Engine. Messrs.
Donkin, the eminent Engineers, *made an estimate*, and a
contract was in consequence executed to construct for Government a fac-simile of the *Swedish* Difference Engine, which
is now in use in the department of the Registrar-General,
at Somerset House. There were far greater mechanical difficulties in the production of that machine than in the one the
drawings of which I had offered to the Government.

From my own experience of the cost of executing such
works, I have no doubt, although it was highly creditable to
the skill of the able firm who constructed it, but that it must
have been commercially unprofitable. Under such circumstances, surely it was harsh on the part of the Government to
refuse Messrs. Donkin permission to exhibit it as a specimen
of English workmanship at the Exhibition of 1862.

But the machine upon which everybody could calculate,
had little chance of fair play from the man on whom nobody
could calculate.

If the Chancellor of the Exchequer had read my letter to
Lord Derby, he would have found the opinion of the Committee of the Royal Society expressed in these words :—

" They consider the former [the abstract mathematical
" principle] as not only sufficiently clear in itself, but as
" already admitted and acted on by the Council in their
" former proceedings.

" The latter [its public utility] they consider as obvious to
" every one who considers the immense advantage of accurate
" numerical tables in all matters of calculation, especially in
" those which relate to astronomy and navigation."—*Report
of the Royal Society*, 12*th Feb.*, 1829.

Thus it appears :—

> 1st. That the Chancellor of the Exchequer presumed to
> set up his *own idea* of the utility of the Difference
> Engine in direct opposition to that of the Royal
> Society.
>
> 2nd. That he *refused* to take the opinion of the highest
> mechanical authority in the country on its pro-
> bable cost, and even *to be informed* whether a con-
> tract for its construction at a definite sum might
> not be attainable : he then boldly pronounced
> the expense to be "utterly incapable of being
> " calculated."

This much-abused Difference Engine is, however, like its
prouder relative the Analytical, a being of sensibility, of
impulse, and of power.

It can not only calculate the millions the ex-Chancellor of
the Exchequer squandered, but it can deal with the smallest
quantities ; nay, it feels even for zeros.*   It is as conscious
as Lord Derby himself is of the presence of a *negative quantity*, and it is not beyond the ken of either of them to foresee
the existence of *impossible ones*.†

Yet should any unexpected course of events ever raise the

* It discovers the roots of equations by feeling whether all the figures in
a certain column are *zeros*.

† It may be necessary to explain to the unmathematical reader and
to the ex-Chancellor of the Exchequer that *impossible quantities* in algebra
are something like *mare's-nests* in ordinary life.

ex-Chancellor of the Exchequer to his former dignity, I am sure he will be its *friend* as soon as he is convinced that it can be made *useful* to him.

It may possibly enable him to un-muddle even his own financial accounts, and to——

But as I have no wish to crucify him, I will leave his name in obscurity.

The Herostratus of Science, if he escape oblivion, will be linked with the destroyer of the Ephesian Temple.

CHAPTER VIII.

OF THE ANALYTICAL ENGINE.

Man wrongs, and Time avenges.
BYRON.—*The Prophecy of Dante.*

Built Workshops for constructing the Analytical Engine—Difficulties about carrying the Tens—Unexpectedly solved—Application of the Jacquard Principle—Treatment of Tables — Probable Time required for Arithmetical Operations—Conditions it must fulfil—Unlimited in Number of Figures, or in extent of Analytical Operations—The Author invited to Turin in 1840—Meetings for Discussion—Plana, Menabrea, MacCullagh, Mosotti—Difficulty proposed by the latter—Observations on the Errata of Astronomical Tables—Suggestions for a Reform of Analytical Signs.

THE circular arrangement of the axes of the Difference Engine round large central wheels led to the most extended prospects. The whole of arithmetic now appeared within the grasp of mechanism. A vague glimpse even of an Analytical Engine at length opened out, and I pursued with enthusiasm the shadowy vision. The drawings and the experiments were of the most costly kind. Draftsmen of the highest order were necessary to economize the labour of my own head; whilst skilled workmen were required to execute the experimental machinery to which I was obliged constantly to have recourse.

In order to carry out my pursuits successfully, I had purchased a house with above a quarter of an acre of ground in a very quiet locality. My coach-house was now converted into a forge and a foundry, whilst my stables were transformed into a workshop. I built other extensive workshops myself, and had a fire-proof building for my drawings and draftsmen. Having myself worked with a variety of tools, and having studied the art of constructing each of them, I at length laid it down as a principle—that, except in rare cases, I would never do anything myself if I could afford to hire another person who could do it for me.

The complicated relations which then arose amongst the various parts of the machinery would have baffled the most tenacious memory. I overcame that difficulty by improving and extending a language of signs, the Mechanical Notation, which in 1826 I had explained in a paper printed in the "Phil. Trans." By such means I succeeded in mastering trains of investigation so vast in extent that no length of years ever allotted to one individual could otherwise have enabled me to control. By the aid of the Mechanical Notation, the Analytical Engine became a reality: for it became susceptible of demonstration.

Such works could not be carried on without great expenditure. The fluctuations in the demand and supply of skilled labour were considerable. The railroad mania withdrew from other pursuits the most intellectual and skilful draftsmen. One who had for some years been my chief assistant was tempted by an offer so advantageous that in justice to his own family he could scarcely have declined it. Under these circumstances I took into consideration the plan of advancing his salary to one guinea per day. Whilst this was in abeyance, I consulted my venerable surviving parent. When I had fully explained the circumstances, my excellent mother replied : " My dear son, you have advanced

Y

far in the accomplishment of a great object, which is worthy of your ambition. You are capable of completing it. My advice is—pursue it, even if it should oblige you to live on bread and cheese."

This advice entirely accorded with my own feelings. I therefore retained my chief assistant at his advanced salary.

The most important part of the Analytical Engine was undoubtedly the mechanical method of carrying the tens. On this I laboured incessantly, each succeeding improvement advancing me a step or two. The difficulty did not consist so much in the more or less complexity of the contrivance as in the reduction of the *time* required to effect the carriage. Twenty or thirty different plans and modifications had been drawn. At last I came to the conclusion that I had exhausted the principle of successive carriage. I concluded also that nothing but teaching the Engine to foresee and then to act upon that foresight could ever lead me to the object I desired, namely, to make the whole of any unlimited number of carriages in one unit of time. One morning, after I had spent many hours in the drawing-office in endeavouring to improve the system of successive carriages, I mentioned these views to my chief assistant, and added that I should retire to my library, and endeavour to work out the new principle. He gently expressed a doubt whether the plan was *possible*, to which I replied that, not being able to prove its impossibility, I should follow out a slight glimmering of light which I thought I perceived.

After about three hours' examination, I returned to the drawing-office with much more definite ideas upon the subject. I had discovered a principle that proved the possibility, and I had contrived mechanism which, I thought, would accomplish my object.

I now commenced the explanation of my views, which I soon found were but little understood by my assistant; nor was this surprising, since in the course of my own attempt at explanation, I found several defects in my plan, and was also led by his questions to perceive others. All these I removed one after another, and ultimately terminated at a late hour my morning's work with the conviction that *anticipating* carriage was not only within my power, but that I had devised one mechanism at least by which it might be accomplished.

Many years after, my assistant, on his return from a long residence abroad, called upon me, and we talked over the progress of the Analytical Engine. I referred back to the day on which I had made that most important step, and asked him if he recollected it. His reply was that he perfectly remembered the circumstance; for that on retiring to my library, he seriously thought that my intellect was beginning to become deranged. The reader may perhaps be curious to know how I spent the rest of that remarkable day.

After working, as I constantly did, for ten or eleven hours a day, I had arrived at this satisfactory conclusion, and was revising the rough sketches of the new contrivance, when my servant entered the drawing-office, and announced that it was seven o'clock—that I dined in Park Lane—and that it was time to dress. I usually arrived at the house of my friend about a quarter of an hour before the appointed time, in order that we might have a short conversation on subjects on which we were both much interested. Having mentioned my recent success, in which my host thoroughly sympathized, I remarked that it had produced an exhilaration of the spirits which not even his excellent champagne could rival. Having enjoyed the society of Hallam, of Rogers, and of some few others of that delightful circle, I retired, and joined one or perhaps two much more extensive reunions. Having thus forgotten science, and enjoyed society for four or five hours, I returned home. About one o'clock I was asleep in my bed, and thus continued for the next five hours.

This new and rapid system of carrying the tens when two numbers are added together, reduced the actual time of the addition of any number of digits, however large, to nine units of time for the addition, and one unit for the carriage. Thus in ten's units of time, any two numbers, however large, might be added together. A few more units of time, perhaps five or six, were required for making the requisite previous arrangements.

Having thus advanced as nearly as seemed possible to the minimum of time requisite for arithmetical operations, I felt renewed power and increased energy to pursue the far higher object I had in view.

To describe the successive improvements of the Analytical Engine would require many volumes. I only propose here to indicate a few of its more important functions, and to give to those whose minds are duly prepared for it some information which will remove those vague notions of wonder, and even of its impossibilty, with which it is surrounded in the minds of some of the most enlightened.

To those who are acquainted with the principles of the Jacquard loom, and who are also familiar with analytical formulæ, a general idea of the means by which the Engine executes its operations may be obtained without much difficulty. In the Exhibition of 1862 there were many splendid examples of such looms.

It is known as a fact that the Jacquard loom is capable of

weaving any design which the imagination of man may conceive. It is also the constant practice for skilled artists to be employed by manufacturers in designing patterns. These patterns are then sent to a peculiar artist, who, by means of a certain machine, punches holes in a set of pasteboard cards in such a manner that when those cards are placed in a Jacquard loom, it will then weave upon its produce the exact pattern designed by the artist.

Now the manufacturer may use, for the warp and weft of his work, threads which are all of the same colour; let us suppose them to be unbleached or white threads. In this case the cloth will be woven all of one colour; but there will be a damask pattern upon it such as the artist designed.

But the manufacturer might use the same cards, and put into the warp threads of any other colour. Every thread might even be of a different colour, or of a different shade of colour; but in all these cases the *form* of the pattern will be precisely the same—the colours only will differ.

The analogy of the Analytical Engine with this well-known process is nearly perfect.

The Analytical Engine consists of two parts :—

1st. The store in which all the variables to be operated upon, as well as all those quantities which have arisen from the result of other operations, are placed.

2nd. The mill into which the quantities about to be operated upon are always brought.

Every formula which the Analytical Engine can be required to compute consists of certain algebraical operations to be performed upon given letters, and of certain other modifications depending on the numerical value assigned to those letters.

There are therefore two sets of cards, the first to direct the nature of the operations to be performed—these are called operation cards : the other to direct the particular variables on which those cards are required to operate—these latter are called variable cards. Now the symbol of each variable or constant, is placed at the top of a column capable of containing any required number of digits.

Under this arrangement, when any formula is required to be computed, a set of operation cards must be strung together, which contain the series of operations in the order in which they occur. Another set of cards must then be strung together, to call in the variables into the mill, the order in which they are required to be acted upon. Each operation card will require three other cards, two to represent the variables and constants and their numerical values upon which the previous operation card is to act, and one to indicate the variable on which the arithmetical result of this operation is to be placed.

But each variable has below it, on the same axis, a certain number of figure-wheels marked on their edges with the ten digits : upon these any number the machine is capable of holding can be placed. Whenever variables are ordered into the mill, these figures will be brought in, and the operation indicated by the preceding card will be performed upon them. The result of this operation will then be replaced in the store.

The Analytical Engine is therefore a machine of the most general nature. Whatever formula it is required to develop, the law of its development must be communicated to it by two sets of cards. When these have been placed, the engine is special for that particular formula. The numerical value of its constants must then be put on the columns of wheels below them, and on setting the Engine in motion it will calculate and print the numerical results of that formula.

Every set of cards made for any formula will at any future time recalculate that formula with whatever constants may be required.

Thus the Analytical Engine will possess a library of its own. Every set of cards once made will at any future time reproduce the calculations for which it was first arranged. The numerical value of its constants may then be inserted.

It is perhaps difficult to apprehend these descriptions without a familiarity both with analytical forms and mechanical structures. I will now, therefore, confine myself to the mathematical view of the Analytical Engine, and illustrate by example some of its supposed difficulties.

An excellent friend of mine, the late Professor MacCullagh, of Dublin, was discussing with me, at breakfast, the various powers of the Analytical Engine. After a long conversation on the subject, he inquired what the machine could do if, in the midst of algebraic operations, it was required to perform logarithmic or trigonometric operations.

My answer was, that whenever the Analytical Engine should exist, all the developments of formula would be directed by this condition—that the machine should be able to compute their numerical value in the shortest possible time. I then added that if this answer were not satisfactory. I had provided means by which, with equal accuracy, it might compute by logarithmic or other Tables.

I explained that the Tables to be used must, of course, be computed and punched on cards by the machine, in which case they would undoubtedly be correct. I then added that when the machine wanted a tabular number, say the logarithm of a given number, that it would ring a bell and then stop itself. On this, the attendant would look at a certain part of the machine, and find that it wanted the logarithm of a given number, say of 2303. The attendant would then go to the drawer containing the pasteboard cards representing its table of logarithms. From amongst these he would take the required logarithmic card, and place it in the machine. Upon this the engine would first ascertain whether the assistant had or had not given him the correct logarithm of the number; if so, it would use it and continue its work. But if the engine found the attendant had given him a wrong logarithm, it would then ring a louder bell, and stop itself. On the attendant again examining the engine, he would observe the words, "Wrong tabular number," and then discover that he really had given the wrong logarithm, and of course he would have to replace it by the right one.

Upon this, Professor MacCullagh naturally asked why, if the machine could tell whether the logarithm was the right one, it should have asked the attendant at all? I told him that the means employed were so ridiculously simple that I would not at that moment explain them; but that if he would come again in the course of a few days, I should be ready to explain it. Three or four days after, Bessel and Jacobi, who had just arrived in England, were sitting with me, inquiring about the Analytical Engine, when fortunately my friend MacCullagh was announced. The meeting was equally agreeable to us all, and we continued our conversation. After some time Bessel put to me the very same question which MacCullagh had previously asked. On this Jacobi remarked that he, too, was about to make the same inquiry when Bessel had asked the question. I then explained to them the following very simple means by which that verification was accomplished.

Besides the sets of cards which direct the nature of the operations to be performed, and the variables or constants which are to be operated upon, there is another class of cards called number cards. These are much less general in their uses than the others, although they are necessarily of much larger size.

Any number which the Analytical Engine is capable of using or of producing can, if required, be expressed by a card with certain holes in it; thus—

NUMBER.				TABLE.						
2	3	0	3	3	6	2	2	9	3	9
●	●	○	●	●	●	●	●	●	●	●
●	●	○	●	●	●	●	●	●	●	●
○	●	○	●	●	●	○	○	●	●	●
○	○	○	○	○	●	○	○	●	○	●
○	○	○	○	○	●	○	○	●	○	●
○	○	○	○	○	●	○	○	●	○	●
○	○	○	○	○	○	○	○	●	○	●
○	○	○	○	○	○	○	○	●	○	●
○	○	○	○	○	○	○	○	●	○	●

The above card contains eleven vertical rows for holes, each row having nine or any less number of holes. In this example the tabular number is 3 6 2 2 9 3 9, whilst its number in the order of the table is 2 3 0 3. In fact, the former number is the logarithm of the latter.

The Analytical Engine will contain,

1st. Apparatus for printing on paper, one, or, if required, two copies of its results.

2nd. Means for producing a stereotype mould of the tables or results it computes.

3rd. Mechanism for punching on blank pasteboard cards or metal plates the numerical results of any of its computations.

Of course the Engine will compute all the Tables which

z

it may itself be required to use.   These cards will therefore be entirely free from error.   Now when the Engine requires a tabular number, it will stop, ring a bell, and ask for such number.   In the case we have assumed, it asks for the logarithm of 2 3 0 3.

When the attendant has placed a tabular card in the Engine, the first step taken by it will be to verify the *number* of the card given it by subtracting its number from 2 3 0 3, the number whose logarithm it asked for.   If the remainder is zero, then the engine is certain that the logarithm must be the right one, since it was computed and punched by itself.

Thus the Analytical Engine first computes and punches on cards its own tabular numbers.   These are brought to it by its attendant when demanded.   But the engine itself takes care that the *right* card is brought to it by verifying the *number* of that card by the number of the card which it demanded.   The Engine will always reject a wrong card by continually ringing a loud bell and stopping itself until supplied with the precise intellectual food it demands.

It will be an interesting question, which time only can solve, to know whether such tables of cards will ever be required for the Engine.   Tables are used for saving the time of continually computing individual numbers.   But the computations to be made by the Engine are so rapid that it seems most probable that it will make shorter work by computing directly from proper formulæ than by having recourse even to its own Tables.

The Analytical Engine I propose will have the power of expressing every number it uses to fifty places of figures.   It will multiply any two such numbers together, and then, if required, will divide the product of one hundred figures by number of fifty places of figures.

Supposing the velocity of the moving parts of the Engine to be not greater than forty feet per minute, I have no doubt that

> Sixty additions or subtractions may be completed and printed in one minute.
>
> One multiplication of two numbers, each of fifty figures, in one minute.
>
> One division of a number having 100 places of figures by another of 50 in one minute.

In the various sets of drawings of the modifications of the mechanical structure of the Analytical Engines, already numbering upwards of thirty, two great principles were embodied to an unlimited extent.

1st. The entire control over *arithmetical* operations, however large, and whatever might be the number of their digits.

2nd. The entire control over the *combinations* of algebraic symbols, however lengthened those processes may be required. The possibility of fulfilling these two conditions might reasonably be doubted by the most accomplished mathematician as well as by the most ingenious mechanician.

The difficulties which naturally occur to those capable of examining the question, as far as they relate to arithmetic, are these,—

> (*a*). The number of digits in *each constant* inserted in the Engine must be without limit.
>
> (*b*). The number of constants to be inserted in the Engine must also be without limit.
>
> (*c*). The number of operations necessary for arithmetic is only four, but these four may be repeated an *unlimited* number of times.
>
> (*d*). These operations may occur in any order, or follow an *unlimited* number of laws.

z 2

The following conditions relate to the algebraic portion of the Analytical Engine :—

(*e*). The number of *literal* constants must be *unlimited*.

(*f*). The number of *variables* must be *without limit*.

(*g*). The combinations of the algebraic signs must *be unlimited*.

(*h*). The number of *functions* to be employed must be *without limit*.

This enumeration includes eight conditions, each of which is absolutely *unlimited* as to the number of its combinations.

Now it is obvious that no *finite* machine can include infinity. It is also certain that no question *necessarily* involving infinity can ever be converted into any other in which the idea f infinity under some shape or other does not enter.

It is impossible to construct machinery occupying unlimited space ; but it is possible to construct finite machinery, and to use it through unlimited time. It is this substitution of the *infinity of time* for the *infinity of space* which I have made use of, to limit the size of the engine and yet to retain its unlimited power.

(*a*). I shall now proceed briefly to point out the means by which I have effected this change.

Since every calculating machine must be constructed for the calculation of a definite number of figures, the first datum must be to fix upon that number. In order to be somewhat in advance of the greatest number that may ever be required, I chose fifty places of figures as the standard for the Analytical Engine. The intention being that in such a machine two numbers, each of fifty places of figures, might be multiplied together and the resultant product of one hundred places might then be divided by another number of fifty places. It seems to me probable that a long period must elapse before the demands of science will exceed this limit. To this it may be added that the addition and subtraction of numbers in an engine constructed for n places of figures would be equally rapid whether n were equal to five or five thousand digits. With respect to multiplication and division, the time required is greater :—

Thus if $a \cdot 10^{50} + b$ and $a' \cdot 10^{50} + b'$ are two numbers each of less than a hundred places of figures, then each can be expressed upon two columns of fifty figures, and a, b, a', b' are each less than fifty places of figures : they can therefore be added and subtracted upon any column holding fifty places of figures.

The product of two such numbers is—

$$a a' \, 10^{100} + (a b' + a' b) \, 10^{50} + b b'.$$

This expression contains four pair of factors, $a a'$, $a b'$, $a b'$, $b b'$, each factor of which has less than fifty places of figures. Each multiplication can therefore be executed in the Engine. The time, however, of multiplying two numbers, each consisting of any number of digits between fifty and one hundred, will be nearly four times as long as that of two such numbers of less than fifty places of figures.

The same reasoning will show that if the numbers of digits of each factor are between one hundred and one hundred and fifty, then the time required for the operation will be nearly nine times that of a pair of factors having only fifty digits.

Thus it appears that whatever may be the number of digits the Analytical Engine is capable of holding, if it is required to make all the computations with k times that number of digits, then it can be executed by the same Engine, but in an amount of time equal to k^2 times the former. Hence the condition (*a*), or the unlimited number of digits contained in each constant employed, is fulfilled.

It must, however, be admitted that this advantage is gained at the expense of diminishing the number of the constants the Engine can hold. An engine of fifty digits, when used as one of a hundred digits, can only contain half the number of variables. An engine containing m columns, each holding n digits, if used for computations requiring kn digits, can only hold $\frac{m}{k}$ constants or variables.

(*b*). The next step is therefore to prove (*b*), viz.: to show that a finite engine can be used as if it contained an unlimited number of constants. The method of punching cards for tabular numbers has already been alluded to. Each Analytical Engine will contain one or more apparatus for printing any numbers put into it, and also an apparatus for punching on pasteboard cards the holes corresponding to those numbers. At another part of the machine a series of number cards, resembling those of Jacquard, but delivered to and computed by the machine itself, can be placed. These can be called for by the Engine itself in any order in which they may be placed, or according to *any law* the Engine may be directed to use. Hence the condition (*b*) is fulfilled, namely: an *unlimited number of constants* can be inserted in the machine in an *unlimited* time.

I propose in the Engine I am constructing to have places for only a thousand constants, because I think it will be more than sufficient. But if it were required to have ten, or even a hundred times that number, it would be quite possible to make it, such is the simplicity of the structure of that portion of the Engine.

(*c*). The next stage in the arithmetic is the number of times the four processes of addition, subtraction, multiplication, and division can be repeated. It is obvious that four different cards thus punched

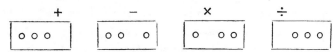

would give the orders for the four rules of arithmetic.

Now there is no limit to the number of such cards which may be strung together according to the nature of the operations required. Consequently the condition (c) is fulfilled.

(d). The fourth arithmetical condition (d), that the order of succession in which these operations can be varied, is itself *unlimited,* follows as a matter of course.

The four remaining conditions which must be fulfilled, in order to render the Analytical Engine as general as the science of which it is the powerful executive, relate to algebraic quantities with which it operates.

The thousand columns, each capable of holding any number of less than fifty-one places of figures, may each represent a constant or a variable quantity. These quantities I have called by the comprehensive title of variables, and have denoted them by V_n, with an index below. In the machine I have designed, n may vary from 0 to 999. But after any one or more columns have been used for variables, if those variables are not required afterwards, they may be printed upon paper, and the columns themselves again used for other variables. In such cases the variables must have a new index; thus, $^mV^n$. I propose to make n vary from 0 to 99. If more variables are required, these may be supplied by Variable Cards, which may follow each other in unlimited succession. Each card will cause its symbol to be printed with its proper indices.

For the sake of uniformity, I have used V with as many indices as may be required throughout the Engine. This, however, does not prevent the printed result of a development from being represented by any letters which may be thought to be more convenient. In that part in which the results are printed, type of any form may be used, according to the taste of the proposer of the question.

It thus appears that the two conditions, (e) and (f), which require that the number of constants and of variables should be unlimited, are both fulfilled.

The condition (g) requiring that the number of combinations of the four algebraic signs shall be unlimited, is easily fulfilled by placing them on cards in any order of succession the problem may require.

The last condition (h), namely, that the number of functions to be employed must be without limit, might seem at first sight to be difficult to fulfil. But when it is considered that any function of any number of operations performed upon any variables is but a combination of the four simple signs of operation with various quantities, it becomes apparent that any function whatever may be represented by two groups of cards, the first being signs of operation, placed in the order in which they succeed each other, and the second group of cards representing the variables and constants placed in the order of succession in which they are acted upon by the former.

Thus it appears that the whole of the conditions which enable a *finite* machine to make calculations of *unlimited* extent are fulfilled in the Analytical Engine. The means I have adopted are uniform. I have converted the infinity of space, which was required by the conditions of the problem, into the infinity of time. The means I have employed are in daily use in the art of weaving patterns. It is accomplished by systems of cards punched with various holes strung together to any extent which may be demanded. Two large boxes, the one empty and the other filled with perforated cards, are placed before and behind a polygonal prism, which revolves at intervals upon its axis, and advances through a short space, after which it immediately returns.

A card passes over the prism just before each stroke of the shuttle ; the cards that have passed hang down until they reach the empty box placed to receive them, into which they arrange themselves one over the other. When the box is full, another empty box is placed to receive the coming cards, and a new full box on the opposite side replaces the one just emptied. As the suspended cards on the entering side are exactly equal to those on the side at which the others are delivered, they are perfectly balanced, so that whether the formulæ to be computed be excessively complicated or very simple, the force to be exerted always remains nearly the same.

In 1840 I received from my friend M. Plana a letter pressing me strongly to visit Turin at the then approaching meeting of Italian philosophers. In that letter M. Plana stated that he had inquired anxiously of many of my country-men about the power and mechanism of the Analytical Engine. He remarked that from all the information he could collect the case seemed to stand thus :—

" Hitherto the *legislative* department of our analysis has been all powerful—the *executive* all feeble.

" Your engine seems to give us the same control over the executive which we have hitherto only possessed over the legislative department."

Considering the exceedingly limited information which

2 A

could have reached my friend respecting the Analytical Engine, I was equally surprised and delighted at his exact prevision of its powers. Even at the present moment I could not express more clearly, and in fewer terms, its real object. I collected together such of my models, drawings, and notations as I conceived to be best adapted to give an insight into the principles and mode of operating of the Analytical Engine. On mentioning my intention to my excellent friend the late Professor MacCullagh, he resolved to give up a trip to the Tyrol, and join me at Turin.

We met at Turin at the appointed time, and as soon as the first bustle of the meeting had a little abated, I had the great pleasure of receiving at my own apartments, for several mornings, Messrs. Plana, Menabrea, Mossotti, MacCullagh, Plantamour, and others of the most eminent geometers and engineers of Italy.

Around the room were hung the formula, the drawings, notations, and other illustrations which I had brought with me. I began on the first day to give a short outline of the idea. My friends asked from time to time further explanations of parts I had not made sufficiently clear. M. Plana had at first proposed to make notes, in order to write an outline of the principles of the engine. But his own laborious pursuits induced him to give up this plan, and to transfer the task to a younger friend of his, M. Menabrea, who had already established his reputation as a profound analyst.

These discussions were of great value to me in several ways. I was thus obliged to put into language the various views I had taken, and I observed the effect of my explanations on different minds. My own ideas became clearer, and I profited by many of the remarks made by my highly-gifted friends.

One day Mosotti, who had been unavoidably absent from the previous meeting, when a question of great importance had been discussed, again joined the party. Well aware of the acuteness and rapidity of my friend's intellect, I asked my other friends to allow me five minutes to convey to Professor Mosotti the substance of the preceding sitting. After putting a few questions to Mosotti himself, he placed before me distinctly his greatest difficulty.

He remarked that he was now quite ready to admit the power of mechanism over numerical, and even over algebraical relations, to any extent. But he added that he had no conception how the machine could perform the act of judgment sometimes required during an analytical inquiry, when two or more different courses presented themselves, especially as the proper course to be adopted could not be known in many cases until all the previous portion had been gone through.

I then inquired whether the solution of a numerical equation of any degree by the usual, but very tedious proceeding of approximation would be a type of the difficulty to be explained. He at once admitted that it would be a very eminent one.

For the sake of perspicuity and brevity I shall confine my present explanation to possible roots.

I then mentioned the successive stages :—

Number of Operation
  Cards used.

1 *a*. Ascertain the number of possible roots by applying Sturm's theorem to the coefficients.

2 *b*. Find a number greater than the greatest root.

3 *c*. Substitute the powers of ten (commencing with that next greater than the greatest root, and

2 A 2

diminishing the powers by unity at each step) for the value of x in the given equation.

Continue this until the sign of the resulting number changes from positive to negative.

The index of the last power of ten (call it n), which is positive, expresses the number of digits in that part of the root which consists of whole numbers. Call this index $n + 1$.

4 *d.* Substitute successively for x in the original equation 0×10^n, 1×10^n, 2×10^n, 3×10^n, ... 9×10^n, until a change of sign occurs in the result. The digit previously substituted will be the first figure of the root sought.

5 *e.* Transform the original equation into another whose roots are less by the number thus found.

The transformed equation will have a real root, the digit, less than 10^n.

6 *f.* Substitute $1 \times 10^{n-1}$, $2 \times 10^{n-1}$, $3 \times 10^{n-1}$, &c., successively for the root of this equation, until a change of sign occurs in the result, as in process 4.

This will give the second figure of the root.

This process of alternately finding a new figure in the root, and then transforming the equation into another (as in process 4 and 5), must be carried on until as many figures as are required, whether whole numbers or decimals, are arrived at.

7 *g.* The root thus found must now be used to reduce the original equation to one dimension lower.

8 *h*. This new equation of one dimension lower must now be treated by sections 3, 4, 5, 6, and 7, until the new root is found.

9 *i*. The repetition of sections 7 and 8 must go on until all the roots have been found.

Now it will be observed that Professor Mosotti was quite ready to admit at once that each of these different processes could be performed by the Analytical Machine through the medium of properly-arranged sets of Jacquard cards.

His real difficulty consisted in teaching the engine to know when to change from one set of cards to another, and back again repeatedly, at intervals not known to the person who gave the orders.

The dimensions of the algebraic equation being known, the number of arithmetical processes necessary for Sturm's theorem is consequently known. A set of operation cards can therefore be prepared. These must be accompanied by a corresponding set of variable cards, which will represent the columns in the store, on which the several coefficients of the given equation, and the various combinations required amongst them, are to be placed.

The next stage is to find a number greater than the greatest root of the given equation. There are various courses for arriving at such a number. Any one of these being selected, another set of operation and variable cards can be prepared to execute this operation.

Now, as this second process invariably follows the first, the second set of cards may be attached to the first set, and the engine will pass on from the first to the second process, and again from the second to the third process.

But here a difficulty arises : successive powers of ten are to be substituted for x in the equation, until a certain event happens. A set of cards may be provided to make the substitution of the highest power of ten, and similarly for the others; but on the occurrence of a certain event, namely, the change of a sign from $+$ to $-$, this stage of the calculation is to terminate.

Now at a very early period of the inquiry I had found it necessary to teach the engine to know when any numbers it might be computing passed through zero or infinity.

The passage through zero can be easily ascertained, thus : Let the continually-decreasing number which is being computed be placed upon a column of wheels in connection with a carrying apparatus. After each process this number will be diminished, until at last a number is subtracted from it which is greater than the number expressed on those wheels.

Thus let it be	.	00000,00000,00000,00423
Subtract . . .		00000,00000,00000,00511

$$99999,99999,99999,99912$$

Now in every case of a carriage becoming due, a certain lever is transferred from one position to another in the cage next above it.

Consequently in the highest cage of all (say the fiftieth in the Analytical Engine), an arm will be moved or not moved accordingly as the carriages do or do not run up beyond the highest wheel.

This arm can, of course, make any change which has previously been decided upon. In the instance we have been considering it would order the cards to be turned on to the next set.

If we wish to find when any number, which is increasing,

exceeds in the number of its digits the number of wheels on the columns of the machine, the same carrying arm can be employed. Hence any directions may be given which the circumstances require.

It will be remarked that this does not actually prove, even in the Analytical Engine of fifty figures, that the number computed has passed through infinity; but only that it has become greater than any number of fifty places of figures.

There are, however, methods by which any machine made for a given number of figures may be made to compute the same formulæ with double or any multiple of its original number. But the nature of this work prevents me from explaining that method.

It may here be remarked that in the process, the cards employed to make the substitutions of the powers of ten are *operation* cards. They are, therefore, quite independent of the numerical values substituted. Hence the same set of operation cards which order the substitutions 1×10^n will, if backed, order the substitution of 2×10^n, &c. We may, therefore, avail ourselves of mechanism for backing these cards, and call it into action whenever the circumstances themselves require it.

The explanation of M. Mosotti's difficulty is this:—Mechanical means have been provided for backing or advancing the operation cards to any extent. There exist means of expressing the conditions under which these various processes are required to be called into play. It is not even necessary that two courses only should be possible. Any number of courses may be possible at the same time; and the choice of each may depend upon any number of conditions.

It was during these meetings that my highly valued friend, M. Menabrea, collected the materials for that lucid and admirable description which he subsequently published in the Bibli. Univ. de Genève, t. xli. Oct. 1842.

The elementary principles on which the Analytical Engine rests were thus in the first instance brought before the public by General Menabrea.

Some time after the appearance of his memoir on the subject in the "Bibliothèque Universelle de Genève," the late Countess of Lovelace * informed me that she had translated the memoir of Menabrea. I asked why she had not herself written an original paper on a subject with which she was so intimately acquainted? To this Lady Lovelace replied that the thought had not occurred to her. I then suggested that she should add some notes to Menabrea's memoir; an idea which was immediately adopted.

We discussed together the various illustrations that might be introduced: I suggested several, but the selection was entirely her own. So also was the algebraic working out of the different problems, except, indeed, that relating to the numbers of Bernouilli, which I had offered to do to save Lady Lovelace the trouble. This she sent back to me for an amendment, having detected a grave mistake which I had made in the process.

The notes of the Countess of Lovelace extend to about three times the length of the original memoir. Their author has entered fully into almost all the very difficult and abstract questions connected with the subject.

These two memoirs taken together furnish, to those who are capable of understanding the reasoning, a complete demonstration—*That the whole of the developments and operations of analysis are now capable of being executed by machinery.*

There are various methods by which these developments

* Ada Augusta, Countess of Lovelace, only child of the Poet Byron.

are arrived at:—1. By the aid of the Differential and Integral Calculus. 2. By the Combinatorial Analysis of Hindenburg. 3. By the Calculus of Derivations of Arbogast.

Each of these systems professes to expand any function according to any laws. Theoretically each method may be admitted to be perfect; but practically the time and attention required are, in the greater number of cases, more than the human mind is able to bestow. Consequently, upon several highly interesting questions relative to the Lunar theory, some of the ablest and most indefatigable of existing analysts are at variance.

The Analytical Engine is capable of executing the laws prescribed by each of these methods. At one period I examined the Combinatorial Analysis, and also took some pains to ascertain from several of my German friends, who had had far more experience of it than myself, whether it could be used with greater facility than the Differential system. They seemed to think that it was more readily applicable to all the usual wants of analysis.

I have myself worked with the system of Arbogast, and if I were to decide from my own limited use of the three methods, I should, for the purposes of the Analytical Engine, prefer the Calcul des Derivations.

As soon as an Analytical Engine exists, it will necessarily guide the future course of the science. Whenever any result is sought by its aid, the question will then arise—By what course of calculation can these results be arrived at by the machine in the *shortest time?*

In the drawings I have prepared I proposed to have a thousand variables, upon each of which any number not having more than fifty figures can be placed. This machine would multiply 50 figures by other 50, and print the product

2 B

of 100 figures. Or it would divide any number having 100 figures by any other of 50 figures, and print the quotient of 50 figures. Allowing but a moderate velocity for the machine, the time occupied by either of these operations would be about one minute.

The whole of the *numerical* constants throughout the works of Laplace, Plana, Le Verrier, Hansen, and other eminent men whose indefatigable labours have brought astronomy to its present advanced state, might easily be recomputed. They are but the numerical coefficients of the various terms of functions developed according to certain series. In all cases in which these numerical constants can be calculated by more than one method, it might be desirable to compute them by several processes until frequent practice shall have confirmed our belief in the infallibility of mechanism.

The great importance of having accurate Tables is admitted by all who understand their uses; but the multitude of errors really occurring is comparatively little known. Dr. Lardner, in the " Edinburgh Review," has made some very instructive remarks on this subject.

I shall mention two within my own experience : these are selected because they occurred in works where neither care nor expense were spared on the part of the Government to insure perfect accuracy. It is, however, but just to the eminent men who presided over the preparation of these works for the press to observe, that the real fault lay not in them but in *the nature of things.*

In 1828 I lent the Government an original MS. of the table of Logarithmic Sines, Cosines, &c., computed to every second of the quadrant, in order that they might have it compared with Taylor's Logarithms, 4to., 1792, of which they possessed a considerable number of copies. Nineteen

errors were thus detected, and a list of these errata was published in the Nautical Almanac for 1832: these may be
called

Nineteen errata of the first order . . 1832

An error being detected in one of these errata, in the following Nautical Almanac we find an

Erratum of the errata in N. Alm. 1832 . . 1833

But in this very erratum of the second order a new mistake
was introduced larger than any of the original mistakes.
In the year next following there ought to have been found

Erratum in the erratum of the errata in N. Alm.
1832 1834

In the "Tables de la Lune," by M. P. A. Hansen, 4to, 1857,
published at the expense of the English Government, under
the direction of the Astronomer Royal, is to be found a list
of errata amounting to 155. In the 21st of these original
errata there have been found *three* mistakes. These are
duly noted in a newly-printed list of errata discovered during
computations made with them in the "Nautical Almanac;"
so that we now have the errata of an erratum of the original
work.

This list of errata from the office of the "Nautical Almanac"
is larger than the original list. The total number of errors
at present (1862) discovered in Hansen's "Tables of the
Moon" amounts to above three hundred and fifty. In
making these remarks I have no intention of imputing the
slightest blame to the Astronomer Royal, who, like other
men, cannot avoid submitting to inevitable fate. The only
circumstance which is really extraordinary is that, when it
was demonstrated that all tables are capable of being computed by machinery, and even when a machine existed which

computed certain tables, that the Astronomer Royal did not become the most enthusiastic supporter of an instrument which could render such invaluable service to his own science.

In the Supplementary Notices of the Astronomical Society, No. 9, vol. xxiii., p. 259, 1863, there occurs a Paper by M. G. de Ponteculant, in which forty-nine numerical coefficients relative to the Longitude, Latitude, and Radius vector of the Moon are given as computed by Plana, Delaunay, and Ponteculant. The computations of Plana and Ponteculant agree in thirteen cases; those of Delaunay and Ponteculant in two; and in the remaining thirty-four cases they all three differ.

I am unwilling to terminate this chapter without reference to another difficulty now arising, which is calculated to impede the progress of Analytical Science. The extension of analysis is so rapid, its domain so unlimited, and so many inquirers are entering into its fields, that a variety of new symbols have been introduced, formed on no common principles. Many of these are merely new ways of expressing well-known functions. Unless some philosophical principles are generally admitted as the basis of all notation, there appears a great probability of introducing the confusion of Babel into the most accurate of all languages.

A few months ago I turned back to a paper in the Philosophical Transactions, 1844, to examine some analytical investigations of great interest by an author who has thought deeply on the subject. It related to the separation of symbols of operation from those of quantity, a question peculiarly interesting to me, since the Analytical Engine contains the embodiment of that method. There was no ready, sufficient, and simple mode of distinguishing letters which represented quantity from those which indicated operation. To understand the results the author had arrived at, it became necessary to read the whole Memoir.

Although deeply interested in the subject, I was obliged, with great regret, to give up the attempt; for it not only occupied much time, but placed too great a strain on the memory.

Whenever I am thus perplexed it has often occurred to me that the very simple plan I have adopted in my *Mechanical Notation* for lettering drawings might be adopted in analysis.

On the geometrical drawings of machinery every piece of matter which represents framework is invariably denoted by an *upright* letter; whilst all letters indicating moveable parts are marked by *inclined* letters.

The analogous rule would be—

Let all letters indicating operations or modifications be expressed by *upright* letters;

Whilst all letters representing quantity should be represented by *inclined* letters.

The subject of the principles and laws of notation is so important that it is desirable, before it is too late, that the scientific academies of the world should each contribute the results of their own examination and conclusions, and that some congress should assemble to discuss them. Perhaps it might be still better if each academy would draw up its own views, illustrated by examples, and have a sufficient number printed to send to all other academies.

CHAPTER IX.

OF THE MECHANICAL NOTATION.

Art of Lettering Drawings—Of expressing the Time and Duration of Action of every Part—A New Demonstrative Science—Royal Medals of 1826

SOON after I had commenced the Difference Engine, my attention was strongly directed to the imperfection of all known modes of explaining and demonstrating the construction of machinery. It soon became apparent that my progress would be seriously impeded unless I could devise more rapid means of understanding and recalling the interpretation of my own drawings.

By a new system of very simple signs I ultimately succeeded in rendering the most complicated machine capable of explanation almost without the aid of words.

In order thoroughly to understand the action of any machine, we must have full information upon the following subjects, and it is of the greatest importance that this information should be acquired in the shortest possible time.

I. The actual shape and relative position of every piece of matter of which the machine is composed.

This can be accomplished by the ordinary mechanical drawings. Such drawings usually have letters upon them for the sake of reference in the description of the machine. Hitherto such letters were chosen without any principle, and in fact gave no indication of anything except the mere spot upon the paper on which they were written.

I then laid down rules for the selection of letters. I shall only mention one or two of them :—

1. All upright letters, as a, c, d, e, A, B, represent framing.

2. All inclined letters, as *a, c, d, e, A, B,* represent moveable parts.

3. All small letters represent working points. One of the most obvious advantages of these rules is that they enable the attention to be more easily confined to the immediate object sought.

By other rules it is rendered possible, when looking at a plan of any complicated machine, to perceive the *relative order* of super-position of any number of wheels, arms, &c., without referring to the elevation or end view.

II. The actual time and duration of every motion throughout the action of any machine can be ascertained almost instantly by a system of signs called the Notations of Periods.

It possesses equal facilities for ascertaining every contemporaneous as well as for every successive system of movements.

III. The actual connection of each moveable piece of the machine with every other on which it acts. Thus, taking from any special part of the drawing the indicating letter, and looking for it on a certain diagram, called the trains, the whole course of its movements may be traced, up to the prime mover, or down to the final result.

I have called this system of signs the Mechanical Notation. By its application to geometrical drawing it has given us a new demonstrative science, namely, that of proving that any given machine can or cannot exist; and if it can exist, that it will accomplish its desired object.

It is singular that this addition to human knowledge should have been made just about the period when it was beginning to be felt by those most eminently skilled in analysis that the time has arrived when many of its conclusions rested only on probable evidence. This state of things arose chiefly from the enormous extent to which the developments were necessarily carried in the lunar and planetary theories.

After employing this language for several years, it was announced, in December 1825, that King William IV. had founded two medals of fifty guineas each, to be given annually by the Royal Society according to rules to be laid down by the Council.

On the 26th January 1826, it was resolved,

"That it is the opinion of the Council that the medals be awarded for the most important discoveries or series of investigations, completed and made known to the Royal Society in the year preceding the day of the award."

This rule reduced the number of competitors to a very few. Although I had had some experience as to the mode in which medals were awarded, and therefore valued them accordingly, I was simple enough to expect that the Council of the Royal Society would not venture upon a fraud on the very first occasion of exercising the royal liberality. I had also another motive for taking a ticket in this philosophical lottery of medals.

In 1824, the Astronomical Society did me the honour to award to me the first gold medal they ever bestowed. It was rendered still more grateful by the address of that eminent man, the late Henry Thomas Colebrooke, the President, who in a spirit of prophecy anticipated the results of years, at that period, long future.

"It may not, therefore, be deemed too sanguine an anti-
" cipation, when I express the hope that an instrument which
" in its simpler form attains to the extraction of the roots of
" numbers, and approximates to the roots of equations, may,
" in a more advanced state of improvement, rise to the
" approximate solutions of algebraic equations of elevated
" degrees. I refer to solutions of such equations proposed
" by Lagrange, and more recently by other analysts, which
" involve operations too tedious and intricate for use, and
" which must remain without efficacy, unless some mode be
" devised of abridging the labour or facilitating the means of
" performance."*

I felt, therefore, that the *first* Royal Medal might fairly
become an object of ambition, whatever might be the worth
of subsequent ones.

In order to qualify myself for this chance, I carefully drew
up a paper, "On a Method of expressing by Signs the
Action of Machinery," which I otherwise should not have
published at that time.

This Memoir was read at the Royal Society on the 16th
March, 1826. To the system of signs which it first ex-
pounded I afterwards gave the name of " Mechanical Notation." It had been used in England and in Ireland, although
not taught in its schools. It applies to the description of
a combat by sea or by land. It can assist in representing
the functions of animal life; and I have had both from the
Continent and from the United States, specimens of such
applications. Finally, to whatever degree of simplicity I
may at last have reduced the Analytical Engine, the course

---

* "Discourse of the President on delivering the first Gold Medal of the Astronomical Society to Charles Babbage, Esq." " Memoirs of the Astronomical
Society," vol. i. p. 509.

through which I arrived at it was the most entangled and perplexed which probably ever occupied the human mind. Through the aid of the Mechanical Notation I examined numberless plans and systems of computing, and I am sure, from the nature of its self-necessary verifications that it is impossible I can have been deceived.

On the 16th November, 1826, that very Council of the Royal Society which had made the law took the earliest opportunity to violate it by awarding the two Royal Medals, the first to Dalton, whose great discovery had been made nearly twenty years before, and the other to Ivory, for a paper published in their "Transactions" three years before. The history of their proceedings will be found in the "Decline of Science in England," p. 115, 1830.

CHAPTER X.

THE EXHIBITION OF 1862.

"En administration, toutes les sottises sont mères."—*Maximes*, par M. G. De Levis.

" An abject worship of princes and an unaccountable appetite for knighthood are probably unavoidable results of placing second-rate men in prominent positions."—*Saturday Review*, January 16, 1864.

" Whose fault is this? But tallow, toys, and sweetmeats evidently stand high in the estimation of Her Majesty's Commissioners."—*The Times*, August 13, 1862.

Mr. Gravatt suggests to King's College the exhibition of the Difference Engine No. 1, and offers to superintend its Transmission and Return—Place allotted to it most unfit—Not Exhibited in 1851—Its Loan refused to New York—Refused to the Dublin Exhibition in 1847—Not sent to the great French Exhibition in 1855—Its Exhibition in 1862 entirely due to Mr. Gravatt—Space for its Drawings refused—The Payment of Six Shillings a Day for a competent person to explain it refused by the Commissioners—Copy of Swedish Difference Engine made by English Workmen not exhibited—Loan of various other Calculating Machines offered—Anecdote of Count Strzelecki's—The Royal Commissioners' elaborate taste for Children's Toys—A plan for making such Exhibitions profitable—Extravagance of the Commissioners to their favourite—Contrast between his Treatment and that of Industrious Workmen—The Inventor of the Difference Engine publicly insulted by his Countrymen in the Exhibition of 1862.

Circumstances connected with the Exhibition of the Difference Engine No. 1 in the International Exhibition of 1862.

WHEN the construction of the Difference Engine No. 1 was abandoned by the Government in 1842, I was consulted respecting the place in which it should be deposited. Well aware of the unrivalled perfection of its workmanship, and

2 c 2

conscious that it formed the first great step towards reducing the whole science of number to the absolute control of mechanism, I wished it to be placed wherever the greatest number of persons could see it daily.

With this view, I advised that it should be placed in one of the much-frequented rooms of the British Museum. Another locality was, however, assigned to it, and it was confided by the Government to the care of King's College, Somerset House. It remained in safe custody within its glass case in the Museum of that body for twenty years. It is remarkable that during that long period no person should have studied its structure, and, by explaining its nature and use, have acquired an amount of celebrity which the singularity of that knowledge would undoubtedly have produced.

The College authorities did justice to their charge. They put it in the place of honour, in the centre of their Museum, and would, no doubt have given facilities to any of their members or to other persons who might have wished to study it.

But the system quietly pursued by the Government, of ignoring the existence of the Difference Engine and its inventor doubtlessly exercised its deadening influence * on those who were inclined, by taste or acquirements, to take such a course.

---

* An illustration fell under my notice a few days after this paragraph was printed. A *new* work on Geometrical Drawing, commissioned by the Committee of Council on Education, was published by Professor Bradley. I have not been able to find in it a single word concerning " Mechanical Notation," not even the very simplest portion of that science, namely, the Art of Lettering Drawings. It would seem impossible that any *Professor* of so limited a subject could be ignorant of the existence of such an important addition to its powers.

I shall enumerate a few instances.

1. In 1850, the Government appointed a Commission to organize the Exhibition of 1851.

The name of the author of the *Economy of Manufactures* was not thought worthy by the Government to be placed on that Commission.

2. In 1851, the Commissioners of the International Exhibition did not think proper to exhibit the Difference Engine, although it was the property of the nation. They were as insensible to the greatest mechanical as to, what has been regarded by some, the greatest intellectual triumph of their country.

3. When it was decided by the people of the United States to have an Exhibition at New York, they sent a Commissioner to Europe to make arrangement for its success. He was authorized to apply for the loan of the Difference Engine for a few months, and was empowered to give any pecuniary guarantee which might be required for its safe return.

That Commissioner, on his arrival, applied to me on the subject. I explained to him the state of the case, and advised him to apply to the Government, whose property it was. I added that, if his application was successful, I would at my own expense put the machine in good working order, and give him every information requisite for its safe conveyance and use. His application was, however, unsuccessful.

4. In 1847, Mr. Dargan nobly undertook at a vast expense to make an Exhibition in Dublin to aid in the relief of his starving countrymen. It was thought that the exhibition of the Difference Engine would be a great attraction. I was informed at the time that an application was made to the Government for its loan, and that it was also unsuccessful.

5. In 1855 the great French Exhibition occurred. Previously to its opening, our Government sent Commissioners to arrange and superintend the English department.

These Commissioners reported that the English contribution was remarkably deficient in what in France are termed "instruments de précision," a term which includes a variety of instruments for scientific purposes. They recommended that "a Committee should be appointed who could represent to the producers of Philosophical Instruments how necessary it was that they should, upon an occasion of this kind, maintain their credit in the eyes of Europe." The Government also applied to the Royal Society for advice; but neither did the Royal Society advise, nor the Government propose, to exhibit the Difference Engine.

6. The French Exhibition of 1855 was remarkable beyond all former ones for the number and ingenuity of the machines which performed arithmetical operations.

Pre-eminently above all others stood the Swedish Machine for calculating and printing mathematical Tables. It is honourable to France that its highest reward was deservedly given to the inventor of that machine; whilst it is somewhat remarkable that the English Commissioners appointed to report upon the French Exhibition omitted all notice of these Calculating Machines.

The appearance of the finished portion of the unfinished Difference Engine No. 1 at the Exhibition of 1862 is entirely due to Mr. Gravatt. That gentleman had a few years before paid great attention to the Swedish Calculating Engine of M. Scheutz, and was the main cause of its success in this country.

Being satisfied that it was possible to calculate and print all Tables by machinery, Mr. Gravatt became convinced that the time must arrive when no Tables would ever be calculated or printed except by machines.   He felt that it was of great importance to accelerate the arrival of that period, more especially as numerical Tables, which are at present the most expensive kind of printing, would then become the cheapest.

In furtherance of this idea, Mr. Gravatt wrote to Dr. Jelf, the Principal of King's College, Somerset House, to suggest that the Difference Engine of Mr. Babbage, which had for so many years occupied a prominent place in the museum, should be exhibited in the International Exhibition of 1862. He at the same time offered his assistance in the removal and reinstatement of that instrument.

The authorities of the College readily acceded to this plan.   On further inquiry, it appeared that the Difference Engine belonged to the Government, and was only deposited with the College.   It was then found necessary to make an application to the Treasury for permission to exhibit it, which was accordingly done by the proper authorities.

The Government granted the permission, and referred it to the Board of Works to superintend its placement in the building.

The Board of Works sent to me a copy of the correspondence relative to this matter, asking my opinion whether any danger might be apprehended for the safety of the machine during its transport, and also inquiring whether I had any other suggestion to make upon the subject.

Knowing the great strength of the work, I immediately answered that I did not anticipate the slightest injury from its transport, and that, under the superintendence of Mr. Gravatt, I considered it might be removed with perfect safety.   The only suggestion I ventured to offer was, that as the Government possessed in the department of the Registrar-General a copy, made by English workmen, of the Swedish Difference Engine, that it should be exhibited by the side of mine: and that both the Engines should be kept constantly working with a very slow motion.

By a subsequent communication I was informed that the Swedish Machine could not be exhibited, because it was then in constant use, computing certain Tables relating to the values of lives. I regretted this very much. I had intended to alter the handle of my own Engine in order to make it moveable circularly by the same catgut which I had hoped might have driven both. The Tables which the Swedish Machine was employed in printing were *not* of any pressing necessity, and their execution could, upon such an occasion, have been postponed for a few months without loss or inconvenience.

Besides, if the Swedish Engine had, as I proposed, been placed at work, its superintendent might have continued his table-making with but little delay, and the public would have been highly gratified by the sight.

He could also have given information to the public by occasional explanations of its principles; thus might Her Majesty's Commissioners have gratified thousands of her subjects who came, with intense curiosity, prepared to be pleased and instructed, and whom they sent away amazed and disappointed.

From the experience I had during the first week of the Exhibition, I am convinced that if a fit place had been provided for the two Calculating Machines, so that the public might have seen them both in constant but slow motion, and if the superintendent had occasionally given a short explanation of the principles on which they acted, they would have been one of the greatest attractions within the building.

On Mr. Gravatt applying to the Commissioners for space, it was stated that the Engine must be placed amongst philosophical instruments, Class XIII.

The only place offered for its reception was a small hole, 4 feet 4 inches in front by 5 feet deep. On one side of this was the *only* passage to the office of the superintendent of the class. The opposite side was occupied by a glass case in which I placed specimens of the separate parts of the unfinished engine. These, although executed by English workmen above thirty years ago, were yet, in the opinion of the most eminent engineers, unsurpassed by any work the building of 1862 contained. The back of this recess was closed in and dark, and only allowed a space on the wall of about five feet by four, on which to place the *whole* of the drawings and illustrations of the Difference Engine. Close above the top of the machine was a flat roof, which deprived the drawings and the work itself of much light.

The public at first flocked to it : but it was so placed that only three persons could conveniently see it at the same time. When Mr. Gravatt kindly explained and set it in motion, he was continually interrupted by the necessity of moving away in order to allow access to the numerous persons whose business called them to the superintendent's office. At a very early period various representations were made to the Commissioners by the Jury, the superintendent, and very strongly by the press, of the necessity of having some qualified person to explain the machine to the public. I was continually informed by the attendants that hundreds of persons had, during my absence asked, when they could get an opportunity of seeing the machine in motion.

Admiring the earnestness of purpose and the sagacity with which Mr. Gravatt had steadily followed out the convictions of

2 D

his own mind relative to the abolition of all tables except those made and stereotyped by machinery, I offered all the assistance in my power to accelerate the accomplishment of his task.

I lent him for exhibition numerous specimens of the unfinished portions of the Difference Engine No. 1. These I had purchased on the determination of the Government to abandon its construction in 1842.

I proposed also to lend him the Mechanical Notations of the Difference Engine, which had been made at my own expense, and were finished by myself and my eldest son, Mr. B. Herschel Babbage.

I had had several applications from foreigners* for some account of my system of Mechanical Notation, and great desire was frequently expressed to see the illustrations of the method itself, and of its various applications.

These, however, were so extensive that it was impossible, without very great inconvenience, to exhibit them even in my own house.

I therefore wrote to Mr. Gravatt to offer him the loan of the following property for the Exhibition :—

1. A small Calculating Machine of the simplest order for adding together any number of separate sums of money, provided the total was under 100,000*l.*, by Sir Samuel Morland. 1666.

2. A very complete and well-executed Machine for answering all questions in plane trigonometry, by Sir Samuel Morland. 1663.

* One object of the mission of Professor Bolzani was, to take back with him to Russia such an account of the Mechanical Notation as might facilitate its teaching in the Russian Universities. I regret that it was entirely out of my power to assist him.

3. An original set of Napier's bones.
4. A small Arithmetical Machine, by Viscount Mahon, afterwards Earl Stanhope.   Without'date.
5. A larger Machine, to add, subtract, multiply, and divide, by Viscount Mahon.   1775.
6. Another similar Machine, of a somewhat different construction, for the same operations, by Viscount Mahon.   1777.
7. A small Difference Engine, made in London, in consequence of its author having read Dr. Lardner's article in the "Edinburgh Review" of July, 1834, No. CXX.

List of Mechanical Notations proposed to be Lent for the Exhibition.

1. All the drawings explaining the principles of the Mechanical Notation.
2. The complete Mechanical Notations of the Swedish Calculating Engine of M. Scheutz.

    These latter drawings had been made and used by my youngest son, Major Henry P. Babbage, now resident in India, in explaining the principles of the Mechanical Notation at the meeting of the British Association at Glasgow, and afterwards in London, at a meeting of the Institution of Civil Engineers.*
3. The Mechanical Notations of the Difference Engine No. 1

* See Proceedings of British Association at Glasgow, 1855, p. 203; also Minutes of Proceedings of the Institution of Civil Engineers, vol. xv., 1856. Reprinted in the present volume.

These had been made at my own expense, and
were finished by myself and my eldest son,
Mr. B. Herschel Babbage, now resident in
South Australia.

4. A complete set of the drawings of the Difference
Engine No. 2, for calculating and printing tables,
with seven orders of differences, and thirty places of
figures. Finished in 1849.

5. A complete set of the Notations necessary for the
explanation and demonstration of Difference Engine
No. 2, finished in 1849.

These drawings and notations would have required for their
exhibition about seven or eight hundred square feet of wall.
My letter to Mr. Gravatt was forwarded to the Commissioners
with his own application for space to exhibit them. The
Commissioners declined this offer; yet during the first six
weeks of the Exhibition there was at a short distance from
the Difference Engine an empty space of wall large enough for
the greater part of these instructive diagrams. This portion
of wall was afterwards filled up by a vast oil-cloth. Other
large portions of wall, to the amount of thousands of square
feet, were given up to other oil-cloths, and to numberless
carpets. It is evident the Royal Commissioners were much
better qualified to judge of furniture for the feet than of furniture for the head.

I was myself frequently asked why I did not employ a
person to explain the Difference Engine. In reply to some
of my friends, I inquired whether, when they purchased a
carriage, they expected the builder to pay the wages of their
coachman.

But my greatest difficulty was with foreigners; no explanation I could devise, and I tried many, appeared at all

to satisfy their minds. The thing seemed to them entirely incomprehensible.

That the nation possessing the greatest military and commercial marine in the world—the nation which had spent so much in endeavouring to render perfect the means of finding the longitude—which had recently caused to be computed and published at considerable expense an entirely new set of lunar Tables should not have availed itself *at any cost* of mechanical means of computing and stereotyping such Tables, seemed entirely beyond their comprehension.

At last they asked me whether the Commissioners were *bêtes*. I assured them that the only *one* with whom I was personally acquainted certainly was not.

When hard pressed by difficult questions, I thought it my duty as an Englishman to save my country's character, even at the expense of my own. So on one occasion I suggested to my unsatisfied friends that Commissioners were usually selected from the highest class of society, and that possibly four out of five had never heard of my name.

But here again my generous efforts to save the character of my country and its Commissioners entirely failed. Several of my foreign friends had known me in their own homes, and had seen the estimation in which I was held by their own countrymen and by their own sovereign. These were still more astonished.

On another occasion an anecdote was quoted against me to prove that my name was well known even in China. It may, perhaps, amuse the reader. A short time after the arrival of Count Strzelecki in England, I had the pleasure of meeting him at the table of a common friend. Many inquiries were made relative to his residence in China. Much interest was expressed by several of the party to learn on what subject the Chinese were most anxious to have information. Count Strzelecki told them that the subject of most frequent inquiry was Babbage's Calculating Machine. On being further asked as to the nature of the inquiries, he said they were most anxious to know whether it would go into the pocket. Our host now introduced me to Count Strzelecki, opposite to whom I was then sitting. After expressing my pleasure at the introduction, I told the Count that he might safely assure his friends in the Celestial Empire that it was in every sense of the word an *out-of-pocket* machine.

At last the Commissioners were moved, not to supply the deficiency themselves, but to address the Government, to whom the Difference Engine belonged, to send somebody to explain it. I received a communication from the Board of Works, inquiring whether I could make any suggestions for getting over this difficulty. I immediately made inquiries, and found a person who formerly had been my amanuensis, and had, under my direction, worked out many most intricate problems. He possessed very considerable knowledge of mathematics, and was willing, for the moderate remuneration of six shillings a day, to be present daily during nine hours to explain the Difference Engine.

I immediately sent this information to the Board of Works, with the name and address of the person I recommended. This, I have little doubt, was directly communicated to the Commissioners; but they did not avail themselves of his services.

It is difficult, upon any principle, to explain the conduct of the Royal Commissioners of the Exhibition of 1862. They were appointed by the Government, yet when the Government itself became an exhibitor, and sent for exhibition a Difference Engine, the property of the nation, these Commissioners placed it in a *small hole* in a *dark corner*, where it could, with some difficulty, be seen by six people at the same time.

No remonstrance was of the slightest avail; it was " Hobson's choice," that or none.   It was represented that all other space was occupied.

A trophy of children's toys, whose merits, it is true, the Commissioners were somewhat more competent to appreciate, filled one of the most prominent positions in the building.   On the other hand, a trophy of the workmanship of English engineers, executed by *machine tools* thirty years before, and admitted by the best judges to be unsurpassed by any rival, was placed in a position not very inappropriate for the authorities themselves who condemned it to that locality.

But no hired aristocratic * agent was employed to excite the slumbering perceptions of the Commissioners, who might have secured a favourable position for the Difference Engine, by practising on their good nature, or by imposing upon their imbecility.

It has been urged, in extenuation of the conduct of these Commissioners, that their duty as guardians of the funds intrusted to them, and of the interests of the Guarantors, compelled them to practise a rigid economy.

Rigid economy is to be respected only when it is under the control of judgment, not of favouritism.   If the machinery for making arithmetical calculations which was placed at the disposal of the Commissioners had been properly arranged, it might have been made at once a source of high gratification to the public and even of *profit* to the Exhibition.

Such a group of Calculating Machines might have been

* See "The Times," 19 Jan., 1863, and elsewhere.

placed by themselves in a small court capable of holding a limited number of persons. Round the walls of this court might have been hung the drawings I had offered to lend, containing the whole of those necessary for the Difference Engine No. 2, as well as a large number of illustrations for the explanation of the Mechanical Notation. The Swedish Difference Engine and my own might have been slowly making calculations during the whole day.

This court should have been open to the public generally, except at two or three periods of half an hour each, during which it should have been accessible only to those who had previously secured tickets at a shilling apiece.

During each half-hour the person whom I had recommended to the Commissioners might have given a short popular explanation of the subject.

This attraction might have been still further increased, and additional profit made, if a single sheet of paper had been printed containing a woodcut of the Swedish Machine, an impression from a page of the Tables computed and stereotyped by it at Somerset House, and also an impression from a stereotype plate of the Difference Engine exhibited by the Government.

A plate of the Swedish Machine is in existence in London. I am confident that, for such a purpose, I could have procured the loan of it for the Commissioners, and I would willingly have supplied them with the stereotype plate from which the frontispage of the present volume * was printed, together with from ten to twenty lines of necessary explanation.

These illustrations of machinery used for computing and printing Tables might have been put up into packets of dozens and half-dozens, and also have been sold in single

* " Passages from the Life of a Philosopher."

sheets at the rate of one penny each copy. There can be no doubt the sale of them would have been very considerable. As it was, I found the woodcut representing the Difference Engine No. 1 in great request, and during the exhibition I had numberless applications for it; having given away my whole stock of about 800 copies.

The calculating court might have held comfortably from sixty to eighty seats. Each lecture would have produced say 3*l*. This being repeated three times each day, together with the sale of the woodcuts, would have produced about 10*l*. per day, out of which the Commissioners would have had six shillings per day to pay the assistant who gave the required explanations.

If the dignity of the Commissioners would not permit them to make money by such means, they might have announced that the proceeds of the tickets would be given to the distressed population of the Manchester district, and there would then have been crowds of visitors.

But the rigid economy of the Commissioners, who refused to expend six shillings a day for an attendant, although it would most probably have produced a return of several hundred pounds, was entirely laid aside when their patronage was to be extended to a brother official.

Captain Fowke, an officer of engineers, whose high order of architectural talent became afterwards so well known to the public, and whose whole time and services were retained and paid for by the country, was employed to make a design for the Exhibition Building.

The Commissioners approved of this design, which comprised two lofty domes, uniting in themselves the threefold disadvantage of being ugly, useless, and expensive. They then proceeded to pay him five thousand pounds for the job.

2 E

This system of awarding large sums of money to certain favoured public officers who are already paid for their services by liberal salaries seems to be a growing evil. At the period of the Irish famine the under-secretary of the Treasury condescended to accept 2,500*l.* out of the fund raised to save a famished nation. Some inquiries, even recently, were occasionally made whether any similar deduction will be allowed from the liberal contributions to the sufferers by the cotton famine.

The question was raised and the practice reprobated in the House of Commons by men of opposite party politics. Mr. Gladstone remarked :—

" If there was one rule connected with the public service
" which more than any other ought to be scrupulously ob-
" served, it was this, that the salary of a public officer, more
" especially if he were of high rank, ought to cover all the
" services he might be called upon to render. Any departure
" from this rule must be dangerous." Hansard, vol. 101,
p. 138, 1848. Supply, 14 Aug. 1848. See also " The
" Exposition of 1851," 8vo., p. 217.

The following paragraph appeared in " The Times "* a short time since, under the head Naval Intelligence :—

" A reply has been received to the memorial transmitted
" to the Admiralty some few days since from the inspectors
" employed on the iron frigate ' Achilles,' building at Chatham
" dockyard, requesting that they may be placed on the same
" footing as regards increased pay as the junior officers and
" mechanics working on the iron frigate for the additional
" number of hours they are employed in the dockyard. The
" Lords of the Admiralty intimate that they cannot accede to
" the wishes of the memorialists, who are reminded that, as

* About the 20th of May, 1863.

" salaried officers of the establishment, the whole of their time
" is at the disposal of the Admiralty. This decision has caused
" considerable dissatisfaction."

It appears that the Admiralty wisely adopted the principle
enunciated by Mr. Gladstone.

It may, however, not unreasonably have caused dissatisfaction to those who had no interest to back them on finding
that such large sums are pocketed by those who are blessed
with influential friends in high quarters.

If the Commissioners had really wished to have obtained a
suitable building at a fair price their course was simple and
obvious. They need only have stated the nature and amount
of accommodation required, and then have selected half a
dozen of the most eminent firms amongst our great con-
tractors, who would each have given them an estimate of the
plans they respectively suggested.

The Commissioners might have made it one of the conditions that they should not be absolutely bound to give the
contract to the author of the plan accepted. But in case of
not employing him a sum previously stipulated should have
been assigned for the use of the design.

By such means they would have had a choice of various
plans, and if those plans had, previously to the decision of
the Commissioners, been publicly exhibited for a few weeks,
they might have been enlightened by public criticism. Such
a course would have prevented the gigantic job they afterwards perpetrated. It could therefore find no support from
the Commissioners.

The present Commissioners, however, are fit successors to
those who in 1851 ignored the existence of the author of the
" Economy of Manufactures " and his inventions. They seem
to have been deluded into the belief that they possessed

2 E 2

the strength, as well as the desire, quietly to strangle the Difference Engine.

It would be idle to break such butterflies upon its matchless wheels, or to give permanence to such names by reflecting them from its diamond-graven plates.*   Though the steam-hammer can crack the coating without injuring the kernel of the filbert it drops upon—the admirable precision of its gigantic power could never be demonstrated by exhausting its energy upon an empty nut-shell.

Peace, then, to their memory, aptly enshrined in unknown characters within the penetralia of the temple of oblivion.

These celebrities may there at last console themselves in the enjoyment of one enviable privilege denied to them during their earthly career—exemption from the daily consciousness of being "*found out.*"

It is, however, not quite impossible, although deciphering is a brilliant art, that one or other of them may have heard of the dread power of the decipherer.   Having myself had some slight acquaintance with that fascinating pursuit, it gives me real pleasure to relieve them from this very natural fear by assuring them that not even the most juvenile decipherer could be so stupid as to apply himself to the interpretation of—characters known to be meaningless.

Yet there is one name amongst, but not of them—a fellow-worshipper with myself at far other fanes, whose hands, like mine, have wielded the hammer, and whose pen, like mine, has endeavoured to communicate faithfully to his fellow-men

* For the purpose of testing the steadiness and truth of the tools employed in forming the gun-metal plates, I had some dozen of them turned with a diamond point.   The perfect equality of its cut caused the reflected light to be resolved into those beautiful images pointed out by Frauenhofer, and also so much admired in the celebrated gold buttons produced by the late Mr. Barton, the Comptroller of the Mint.

the measure of those truths he has himself laboriously extracted from the material world. With such endowments, it is impossible that *he* could have had any cognizance of this part of the proceedings of his colleagues.*

At the commencement of the Exhibition, Mr. Gravatt was constantly present, and was so kind as to explain to many anxious inquirers the nature and uses of the Difference Engine. This, however, interfered so much with his professional engagements as a Civil Engineer, that it would have been unreasonable to have expected its continuance. In fact, as not above half a dozen spectators could see the machine at once, it was a great sacrifice of valuable time for a very small result.

During the early part of my own examination of the Exhibition I had many opportunities of conversing with experienced workmen, well qualified to appreciate the workmanship of the Difference Engine; these I frequently accompanied to its narrow cell, and pointed out to them its use, as well as the means by which its various parts had received their destined form.

Occasionally also I explained it to some few of my personal friends. When Mr. Gravatt or myself were thus engaged, a considerable crowd was often collected, who were anxious to hear about, although they could not see, the Engine itself.

Upon one of these occasions I was insulted by impertinent questions conveyed in a loud voice from a person at a distance in the crowd. My taste for music, and especially for organs, was questioned. I was charitable enough to suppose that this was an exceptional case; but in less than a week another instance

* I have since learnt, with real satisfaction, that my friend, Mr. Fairbairn, was *not* a member of that incompetent Commission.

occurred. After this experience, of course, I seldom went near the Difference Engine. Mr. Gravatt who had generously sacrificed a considerable portion of his valuable time for the information and instruction of the public was now imperatively called away by professional engagements, and the public had no information whatever upon a subject on which it was really very anxious to be instructed.

Fortunately, however, the Exhibition took place during the long vacation; and a friend of mine, Mr. Wilmot Buxton, of the Chancery Bar, very frequently accompanied me in my visits. Possessing a profound knowledge of the mathematical principles embodied in the mechanism, I had frequently pointed out to him its nature and relations. These I soon found he so well apprehended that I felt justified in intrusting him with one of my keys of the machine, in order that he might have access to it without the necessity of my presence.

Whenever he opened it for his own satisfaction or for the instruction of his friends, he was speedily surrounded by a far larger portion of the public than could possibly see it, but who were still attracted by his lucid oral explanation.

It was fortunate for many of the visitors to the Exhibition that this occurred, for the demands on his time, when present, were incessant, and hundreds thus acquired from his explanations a popular view of the subject.

After the close of the Exhibition, Mr. Gravatt and myself attended to prepare the Difference Engine for its return to the Museum of King's College. To our great astonishment, we found that it had already been removed to the Museum at South Kensington. Not only the Difference Engine itself, but also the illustrations and all the unfinished portions of exquisite workmanship which I had lent to the Exhibition for its explanation, were gone.

On Mr. Gravatt applying to the Board of Works, it was stated that the Difference Engine itself had been placed in the Kensington Museum because the authorities of King's College had declined receiving it, and immediate instructions were of course given for the restoration of my own property.

NOTE ON THE APPLICATION OF MACHINERY TO THE COMPUTATION OF ASTRONOMICAL AND MATHEMATICAL TABLES.

By Charles Babbage, Esq., F.R.S.

From the Memoirs of the Astronomical Society of London.

Read 14th June, 1822.

It is known to several of the members of this society that I have been engaged during the last few months in the contrivance of machinery which, by the application of a moving force, may calculate any tables that may be required. I am now able to acquaint the society with the successful results at which I have arrived; and although it might at the first view appear a bold undertaking to attempt the construction of an engine which should execute operations so various as those which contribute to the formation of the numerous tables that are constantly required for astronomical purposes, yet to those who are acquainted with the method of differences the difficulty will be in a considerable degree removed.

I have taken the method of differences as the principle on which my machinery is founded; and in the engine which is just finished I have limited myself to two orders of differences. With this machine I have repeatedly constructed tables of square and triangular numbers, as well as a table from the singular formula x^2+x+41, which comprises amongst its terms so many prime numbers.

These, as well as any others which the engine is competent to form, are produced almost as rapidly as an assistant can write them down. The machinery by which these calculations are effected is extremely simple in its kind, consisting of a small number of different parts frequently repeated.

In the prosecution of this plan, I have contrived methods by which type shall be set up by the machine in the order determined by the calculation; and the arrangements are of such a nature that, if executed, there shall not exist the possibility of error in any printed copy of tables computed by this engine. Of several of these latter contrivances I have made models; and, from the experiments I have already made, I feel great confidence in the complete success of the plans I have proposed.

C. BABBAGE.

Devonshire Street, Portland Place,
June 2, 1822.

A LETTER TO SIR HUMPHRY DAVY, BART., PRESIDENT OF THE ROYAL SOCIETY, &c. &c., ON THE APPLICATION OF MACHINERY TO THE PURPOSE OF CALCULATING AND PRINTING MATHEMATICAL TABLES,

From CHARLES BABBAGE, ESQ., M.A., F.R.S., Lond. and Edin.,

Member of the Cambridge Philosophical Society, Secretary of the Astronomical Society of London, and Correspondent of the Philomathic Society of Paris. July 3, 1822.

MY DEAR SIR,

THE great interest you have expressed in the success of that system of contrivances which has lately occupied a considerable portion of my attention, induces me to adopt this channel for stating more generally the principles on which they proceed, and for pointing out the probable extent and important consequences to which they appear to lead. Acquainted as you were with this inquiry almost from its commencement, much of what I have now to say cannot fail to have occurred to your own mind: you will, however, permit me to restate it for the consideration of those with whom the principles and the machinery are less familiar.

The intolerable labour and fatiguing monotony of a continued repetition of similar arithmetical calculations, first excited the desire, and afterwards suggested the idea, of a machine, which, by the aid of gravity or any other moving power, should become a substitute for one of the lowest operations of human intellect. It is not my intention in the present Letter to trace the progress of this idea, or the means which I have adopted for its execution; but I propose stating some of their general applications, and shall commence with describing the powers of several engines which I have contrived: of that part which is already executed I shall speak more in the sequel.

The first engine of which drawings were made was one which is capable of computing any table by the aid of differences, whether they are positive or negative, or of both kinds. With respect to the number of the order of differences, the nature of the machinery did not in my own opinion, nor in that of a skilful mechanic whom I consulted, appear to be restricted to any very limited number; and I should venture to construct one with ten or a dozen orders with perfect confidence. One remarkable property of this machine is, that the greater the number of differences the more the engine will outstrip the most rapid calculator.

By the application of certain parts of no great degree of complexity, this may be converted into a machine for extracting the roots of equations, and consequently the roots of numbers: and the extent of the approximation depends on the magnitude of the machine.

Of a machine for multiplying any number of figures (m) by any other number (n), I have several sketches; but it is not yet brought to that degree of perfection which I should wish to give it before it is to be executed.

I have also certain principles by which, if it should be desirable, a table of prime numbers might be made, extending from 0 to ten millions.

Another machine, whose plans are much more advanced than several of those just named, is one for constructing tables which have no order of differences constant.

A vast variety of equations of finite differences may by its means be solved, and a variety of tables, which could be produced in successive parts by the first machine I have mentioned, could be calculated by the latter one with a still less exertion of human thought. Another and very remarkable point in the structure of this machine is, that it will calculate tables governed by laws which have not been hitherto shown to be explicitly determinable, or that it will solve equations for which analytical methods of solution have not yet been contrived.

Supposing these engines executed, there would yet be wanting other means to ensure the accuracy of the printed tables to be produced by them.

The errors of the persons employed to copy the figures presented by the engines would first interfere with their correctness. To remedy this evil, I have contrived means by which the machines themselves shall take from several boxes containing type, the numbers which they calculate, and place them side by side, thus becoming at the same time a substitute for the compositor and the computer: by which means all error in copying as well as in printing is removed.

There are, however, two sources of error which have not yet been guarded against. The ten boxes with which the engine is provided contain each about three thousand types; any box having, of course, only those of one number in it. It may happen that the person employed in filling these boxes shall accidentally place a wrong type in some of them; as for instance, the number

2 in the box which ought only to contain 7's. When these boxes are delivered to the superintendent of the engine, I have provided a simple and effectual means by which he shall, in less than half an hour, ascertain whether, amongst these 30,000 types, there be any individual misplaced, or even inverted. The other cause of error to which I have alluded, arises from the type falling out when the page has been set up: this I have rendered impossible by means of a similar kind.

The quantity of errors from carelessness in correcting the press, even in tables of the greatest credit, will scarcely be believed, except by those who have had constant occasion for their use. A friend of mine, whose skill in practical as well as theoretical astronomy is well known, produced to me a copy of the tables published by order of the French Board of Longitude, containing those of the Sun by Delambre, and of the Moon by Burg, in which he had corrected above *five hundred errors*: most of these appear to be errors of the press; and it is somewhat remarkable, that in turning over the leaves in the fourth page I opened we observed a new error before unnoticed. These errors are so much the more dangerous, because independent computers using the same tables will agree in the same errors.

To bring to perfection the various-machinery which I have contrived, would require an expense both of time and money, which can be known only to those who have themselves attempted to execute mechanical inventions. Of the greater part of that which has been mentioned, I have at present contented myself with sketches on paper, accompanied by short memorandums, by which I might at any time more fully develop the contrivances; and where any new principles are introduced, I have had models executed, in order to examine their actions. For the purpose of demonstrating the practicability of these views, I have chosen the engine for differences, and have constructed one of them, which will produce any tables whose second differences are constant. Its size is the same as that which I should propose for any more extensive one of the same kind: the chief difference would be, that in one intended for use there would be a greater repetition of the same parts in order to adapt it to the calculation of a larger number of figures. Of the action of this engine, you have yourself had opportunities of judging, and I will only at present mention a few trials which have since been made by some scientific gentlemen to whom it has been shown, in order to determine the rapidity with which it calculates. The computed table is presented to the eye at two opposite sides of the machine; and a friend having undertaken to write down the numbers as they appeared, it proceeded to make a table from the formula x^2+x+41. In the earlier numbers my friend, in writing quickly, rather more than kept pace with the engine; but as soon as four figures were required, the machine was at least equal in speed to the writer.

In another trial it was found that thirty numbers of the same table were calculated in two minutes and thirty seconds: as these contained eighty-two figures, the engine produced thirty-three every minute.

In another trial it produced figures at the rate of forty-four in a minute. As the machine may be made to move uniformly by a weight, this rate might be maintained for any length of time, and I believe few writers would be found to copy with equal speed for many hours together. Imperfect as a first machine generally is, and suffering as this particular one does from great defect in the workmanship, I have every reason to be satisfied with the accuracy of its computations; and by the few skilful mechanics to whom I have in confidence shown it, I am assured that its principles are such that it may be carried to any extent. In fact, the parts of which it consists are few but frequently repeated, resembling in this respect the arithmetic to which it is applied, which, by the aid of a few digits often repeated, produces all the wide variety of number. The wheels of which it consists are numerous, but few move at the same time; and I have employed a principle by which any small error that may arise from accident or bad workmanship is corrected as soon as it is produced, in such a manner as effectually to prevent any accumulation of small errors from producing a wrong figure in the calculation.

Of those contrivances by which the composition is to be effected, I have made many experiments and several models; the results of these leave me no reason to doubt of success, which is still further confirmed by a working model that is just finished.

As the engine for calculating tables by the method of differences is the only one yet completed, I shall, in my remarks on the utility of such instruments, confine myself to a statement of the powers which that method possesses.

I would however premise, that if any one shall be of opinion, notwithstanding all the precaution I have taken and means I have employed to guard against the occurrence of error, that it may still be possible for it to arise, the method of differences enables me to determine its existence. Thus, if proper numbers are placed at the outset in the engine, and if it has composed a page of any kind of table, then by comparing the last number it has set up with that number previously calculated, if they are found to agree, the whole page must be correct: should any disagreement occur, it would scarcely be worth the trouble of looking for its origin, as the shortest plan would be to make the engine recalculate the whole page, and nothing would be lost but a few hours' labour of the moving power.

Of the variety of tables which such an engine could calculate, I shall mention but a few. The tables of powers and products published at the expense of the Board of Longitude, and calculated by Dr. Hutton, were solely executed by the method of differences; and other tables of the roots of numbers have been calculated by the same gentleman on similar principles.

As it is not my intention in the present instance to enter into the theory of differences, a field far too wide for the limits of this letter, and which will probably be yet further extended in consequence of the machinery I have contrived, I shall content myself with describing the course pursued in one of the most stupendous monuments of arithmetical calculation which the world has yet produced, and shall point out the mode in which it was conducted, and what share of mental labour would have been saved by the employment of such an engine as I have contrived.

The tables to which I allude are those calculated under the direction of M. Prony, by order of the French Government,—a work which will ever reflect the highest credit on the nation which patronized and on the scientific men who executed it. The tables computed were the following.

1. The natural sines of each 10,000 of the quadrant calculated to twenty-five figures with seven or eight orders of differences.

2. The logarithmic sines of each 100,000 of the quadrant calculated to fourteen decimals with five orders of differences.

3. The logarithm of the ratios of the sines to their arcs of the first 5,000 of the 100,000ths of the quadrant calculated to fourteen decimals with three orders of differences.

4. The logarithmic tangents corresponding to the logarithmic sines calculated to the same extent.

5. The logarithms of the ratios of the tangents to their arcs calculated in the same manner as the logarithms of the ratios of the sines to their arcs.

6. The logarithms of numbers from 1 to 10,000 calculated to nineteen decimals.

7. The logarithms of all numbers from 10,000 to 200,000 calculated to fourteen figures with five orders of differences.

Such are the tables which have been calculated, occupying in their present state seventeen large folio volumes. It will be observed that the trigonometrical tables are adapted to the decimal system, which has not been generally adopted even by the French, and which has not been at all employed in this country. But, notwithstanding this objection, such was the opinion entertained of their value, that a distinguished member of the English Board of Longitude was not long since commissioned by our Government to make a proposal to the Board of Longitude of France to print an abridgment of these tables at the joint expense of the two countries; and five thousand pounds were named as the sum our Government was willing to advance for this purpose. It is gratifying to record this disinterested offer, so far above those little jealousies which frequently interfere between nations long rivals, and manifesting so sincere a desire to render useful to mankind the best materials of science in whatever country they might be produced. Of the reasons why this proposal was declined by our neighbours, I am at present uninformed: but, from a personal acquaintance with many of the distinguished foreigners to whom it was referred, I am convinced that it was received with the same good feelings as those which dictated it.

I will now endeavour shortly to state the manner in which this enormous mass of computation was executed; one table of which (that of the logarithms of numbers) must contain about eight millions of figures.

The calculators were divided into three sections. The first section comprised five or six mathematicians of the highest merit, amongst whom were M. Prony and M. Legendre. These were occupied entirely with the analytical part of the work; they investigated and determined on the formulæ to be employed.

The second section consisted of seven or eight skilful calculators habituated both to analytical and arithmetical computations. These received the formulæ from the first section, converted them into numbers, and furnished to the third section the proper differences at the stated intervals.

They also received from that section the calculated results, and compared the two sets, which were computed independently for the purpose of verification.

The third section, on whom the most laborious part of the operations devolved, consisted of from sixty to eighty persons, few of them possessing a knowledge of more than the first rules of arithmetic: these received from the second class certain numbers and differences, with which, by additions and subtractions in a prescribed order, they completed the whole of the tables above mentioned.

I will now examine what portion of this labour might be dispensed with, in case it should be deemed advisable to compute these or any similar tables of equal extent by the aid of the engine I have referred to.

In the first place, the labour of the first section would be considerably reduced, because the formulæ used in the great work I have been describing have already been investigated and published. One person, or at the utmost two, might therefore conduct it.

If the persons composing the second section, instead of delivering the numbers they calculate to the computers of the third section, were to deliver them to the engine, the whole of the remaining operations would be executed by machinery, and it would only be necessary to employ people to copy down as fast as they were able the figures presented to them by the engine. If, however, the contrivances for printing were brought to perfection and employed, even this labour would be unnecessary, and a few superintendents would manage the machine and receive the calculated pages set up in type. Thus the number of calculators employed, instead of amounting to ninety-six, would be reduced to twelve. This number might, however, be considerably diminished, because when an engine is used the intervals between the differences calculated by the second section may be greatly enlarged. In the tables of logarithms M. Prony caused the differences to be calculated at intervals of two hundred, in order to save the labour of the third section: but as that would now devolve on machinery, which would scarcely move the slower for its additional burthen, the intervals might properly be enlarged to three or four times that quantity. This would cause a considerable diminution in the labour of the second section. If to this diminution of mental labour we add that which arises from the whole work of the compositor being executed by the machine, and the total suppression of that most annoying of all literary labour, the correction of the errors of the press,* I think I am justified in presuming that if engines were made purposely for this object, and were afterwards useless, the tables could be produced at a much cheaper rate; and of their superior accuracy there could be no doubt. Such engines would, however, be far from useless: containing within themselves the power of generating, to an almost unlimited extent, tables whose accuracy would be unrivalled, at an expense comparatively moderate, they would become active agents in reducing the abstract inquiries of geometry to a form and an arrangement adapted to the ordinary purposes of human society.

I should be unwilling to terminate this Letter without noticing another class of tables of the greatest importance, almost the whole of which are capable of being calculated by the method of differences. I refer to all astronomical tables for determining the positions of the sun or planets: it is scarcely necessary to observe that the constituent parts of these are of the form $a \sin \theta$, where a is a constant quantity, and θ is what is usually called the argument. Viewed in this light they differ but little from a table of sines, and like it may be computed by the method of differences.

I am aware that the statements contained in this Letter may perhaps be viewed as something more than Utopian, and that the philosophers of Laputa may be called up to dispute my claim to originality. Should such be the case, I hope the resemblance will be found to adhere to the nature of the subject rather than to the manner in which it has been treated. Conscious, from my own experience, of the difficulty of convincing those who are but little skilled in mathematical knowledge, of the possibility of making a machine which shall perform calculations, I was naturally anxious, in introducing it to the public, to appeal to the testimony of one so distinguished in the records of British science. Of the extent to which the machinery whose nature I have described may be carried, opinions will necessarily fluctuate, until experiment shall have finally decided their relative value: but of that engine which already exists I think I shall be supported, both by yourself and by several scientific friends who have examined it, in stating that it performs with rapidity and precision all those calculations for which it was designed.

Whether I shall construct a larger engine of this kind, and bring to perfection the others I have described, will in a great measure depend on the nature of the encouragement I may receive.

Induced, by a conviction of the great utility of such engines, to withdraw for some time my attention from a subject on which it has been engaged during several years, and which possesses charms of a higher order, I have now arrived at a point where success is no longer doubtful. It must, however, be attained at a very considerable expense, which would not probably be replaced, by the works it might produce, for a long period of time, and which is an undertaking I should feel unwilling to commence, as altogether foreign to my habits and pursuits.

I remain, my dear Sir,

Faithfully yours,

C. BABBAGE.

Devonshire Street, Portland Place,
July 3rd, 1822.

---

* I have been informed that the publishers of a valuable collection of mathematical tables, now reprinting, pay to the gentleman employed in correcting the press, at the rate of three guineas a sheet, a sum by no means too large for the faithful execution of such a laborious duty.

ON THE THEORETICAL PRINCIPLES OF THE MACHINERY FOR CALCULATING TABLES.

In a Letter from Charles Babbage, Esq., F.R.S. Lond. and Edin. to Dr. Brewster.
6 November, 1822.

Reprinted from " Brewster's Journal."

My DEAR SIR,

    Having, during the last two or three months, laid aside the further construction of machinery for calculating tables, I have occasionally employed myself in examining the theoretical principles on which it is founded. Several singular results having presented themselves in these inquiries, I am induced to communicate some of them to you, less from the importance of the analytical difficulties they present, than from the curious fact which they offer in the history of invention.

    I had mentioned to you that, before I left London, I had completed a small engine which calculated tables by means of differences. On considering this machine, a new arrangement occurred to me, by which an engine might be constructed that should calculate tables of other species, whose analytical laws were unknown. On this suggestion, I proceeded to write down a table which might have been made, had such an engine existed; and finding that there were no known methods of expressing its nth term, I thought the analytical difficulty which was thus brought to light, was itself worthy of examination. The following are the first thirty terms of a series of this kind :

0... 2	11...222	22... 924
1... 2	264	1010
2... 4	310	1096
3... 10	356	25...1188
4... 16	15...408	1288
5... 28	468	1396
48	536	1510
76	610	1624
110	684	30...1742
144	20...762	1862
10...182	842	1984

    The law of formation of which is, that the first term is 2, its first difference 2, and its second difference equal to the units figure of the second term; and generally, the second difference corresponding to any term, is always equal to the units figure of the next succeeding term. This engine, when once set, would continue to produce term after term of this series without end, and without any alteration; but we are not in possession of methods of determining its nth term, without passing through all the previous ones. If u_n represent any term, then u_n must be determined from the equation

$$\Delta^2 u_n = \text{the units figure of } u_{n+1};$$

an equation of differences of a species which I have never met with in treatises on that subject.

If we push the inquiry one step further, it is possible to express the units figure of any number in an analytical form. Thus; let S_v represent the sum of the vth powers of the tenth roots of unity, then will

$$0\,S_v + 9\,S_{v+1} + 8\,S_{v+2} + \cdots \qquad 1\,S_{v+9}$$

represent the units figure of the number v. Now, if we put u_{n+1} instead of v in the above equation, we have

$$\Delta^2 u_n = 0\,S_{u_{n+1}} + 9\,S_{u_{n+1}+1} + 8\,S_{u_{n+1}+2} + \cdots \qquad 1\,S_{u_{n+1}+9}$$

an equation whose mode of solution is as yet quite unknown. Finding the difficulty of a direct attempt so considerable, I employed two other processes; one was a kind of induction, and the other was quite unexceptionable. From these I have deduced the following formula:

$$u_n + \overline{(a)} + 20b\,(10b + 2a - 1) + 2,$$

where a is the units figure of n; b is the number n, when its unit figure is cut off; and $\overline{(a)}$ represents whatever number is opposite to it in the subsidiary table below:

If $a = 0\ldots\quad 0$
$1\ldots\quad 0$
$2\ldots\quad 2$
$3\ldots\quad 8$
$4\ldots\ 14$ EXAMPLE: Required the 27th term of the series here, $a = 7$, and $b = 2$: hence,
$5\ldots\ 26$ $10b = 2a - 1 = 20 + 14 - 1 = 33.$
$6\ .\ .\ 46$
$7\ldots\ 74$
$8\ldots108$
$9\ldots142$

then

$\overline{(a)}$ or the number opposite 7, is   74
$20b\,(10b + 2a - 1) =$ 1320
$ 2 =$ 2
_____
$1396 = u_{27}$, or the 27th term.

similarly if $n = 1121, \quad a = 1, \quad b = 112,$
then $u_{1121} = 251106.$

Another series of a similar kind, but more simple in its form, is derived from the following equation:

$$\Delta u_z = \text{units figure of } u_z.$$

If the constant or first term is equal to 2, then we may express u_z thus,

$$u_z = 20b + 2^a,$$

where a is any of the numbers 1, 2, 3, 4, which, taken from z, leaves the remainder divisible by 4, and b is the quotient of that division: the series is,

$1\ldots\ 2$ 48   EXAMPLE: Let $z = 13$
$\ 4$ 56   1 being subtracted, 1
$\ 8$ 62   —
16 64   12 which,
$5\ldots22$ $15\ldots68$ divided by 4, gives   3, hence,
24 76 $a = 1, \quad b = 3, \quad u_{13} = 20.3 + 2 = 62.$
28 82
36 84
42 88
$10\ldots44$ $20\ldots96$

Innumerable other series might be formed by the same engine, the differences of any order depending on the value of the figure which might occur in the units, or the tens, or the hundreds place, or in any one or more determinate places of the same, or the next, or preceding terms.

Other laws might be observed by the same engine, of which the following is an example. A series of cube numbers might be formed, subject to this condition, that whenever the number 2 occurred in the tens' place, that and all the succeeding cubes should be increased by ten. In such a series. of course, the second figure would never be a 2, because the addition of ten would convert it into 3.

The Series of Cubes.	The Series Proposed.
1	1
8	8
27	* 37
64	74
125	135
216	* 236
343	363
512	532
729	749
1000	* 1030
1331	1361
1728	1758
2197	* 2237
2744	2784

the stars indicating the number at which the law takes effect. These, and other similar series, open a wide field of analytical inquiry—a subject which I shall take some other opportunity of resuming. I will, however, mention an unexpected circumstance, as it illustrates, in a striking manner, the connection between remote inquiries in mathematics, and as it may furnish a lesson to those who are rashly inclined to undervalue the more recondite speculations of pure analysis, from an erroneous idea of their inapplicability to practical matters. Amongst the singular and difficult equations of finite differences to which these series led, I recognised one which I had several- years since met with, in an analytical attempt to solve a problem considered by Euler and Vandermonde; it relates to the knight's move at chess. At that time, I had advanced several steps; but the equation in question proved an obstacle I was then unable to surmount. In its present shape, although I have not yet deduced the solution from the equation, yet, as I am in possession of the former, it is not too much to anticipate a general process applicable to this class of equations; and should that be the case, I shall be able to advance some steps further in a very curious and difficult inquiry, connected with the geometry of situation.

As an erroneous idea has been entertained relative to the nature of the machinery I have contrived, I will endeavour to state to you some of the mathematical principles on which it is founded. The contrivances of Pascal and others have, as far as I am aware, been directed to an entirely different object. Machinery which will perform the usual operation of common arithmetic will never, in my opinion, be of that essential utility which must arise from an engine that calculates tables; and although mine is not defective in these points, and will extract the roots of numbers, and approximate to the roots of equations, and even, as I believe, to their impossible roots, yet, had this been its only office, I should have esteemed it of comparatively but little value. As far as I have inquired, I believe the method of differences has now, for the first time, been embodied into machinery; and, in speaking of this method, I am far from meaning to confine myself to calculating tables by constant differences. The same mechanical principles which I have already proved, enable me to integrate innumerable equations of finite differences, if I may be allowed to use the term *integrate*, in a sense somewhat different from its usual acceptation. My meaning is, that the equation of differences being given, I can, by setting an engine, produce, at the end of a given time, any distant term which may be required; or, if a succession of terms are sought, commencing at a distant point, these shall be produced. Thus, although I do not determine the analytical law, I can produce the numerical result which it is the object of that law to give. Some kinds of equation of differences can be adapted to machinery with much greater facility than others; and hence it will become an object of inquiry how, when we wish to calculate that of any transcendant, we may deduce from some approximate equation the differences which may be suitable to our purpose. Thus, you see, one of the first effects of machinery adapted to numbers has been to lead us to surmount new difficulties in analysis; and should it be carried to perfection, some of the most abstract parts of mathematical science will be called into practical utility, to facilitate the formation of tables. The more I examine this theoretical part, the more I feel convinced that it will be long before the novel relations which it presents will be exhausted; and if the absence of all encouragement to proceed with the mechanism I have contrived, shall prove that I have anticipated too far the period at which it shall become necessary, I will yet venture to predict that a time will arrive when the accumulating labour which arises from the arithmetical applications of mathematical formulæ, acting as a constantly retarding force, shall ultimately impede the useful progress of the science, unless this or some equivalent method is devised for relieving it from the overwhelming incumbrance of numerical detail.

I remain,
My dear Sir,
Faithfully yours,
C. BABBAGE.

DEVONSHIRE STREET, PORTLAND PLACE,
6th Nov. 1822.

OBSERVATIONS ON THE APPLICATION OF MACHINERY TO THE COMPUTATION OF MATHEMATICAL TABLES.

By Charles Babbage, Esq., F.R.S., &c. &c.

From the Memoirs of the Astronomical Society.

Read 13th December, 1822.

Since I had the honour of communicating to the Astronomical Society a short account of an arithmetical engine for the calculation of tables, which has been examined by several of the members of this society, I have not added much to the practical part of the subject. I have however paid some attention to the improvements of which the machinery is susceptible, and which will, if another engine is made, be greatly improved.

The theoretical inquiries to which it has conducted me are however of a singular nature; and I shall take this opportunity of briefly explaining to the society some of the principles on which they depend, as far as the nature of the subject will permit me to do this without the introduction of too many algebraic operations, which are rarely intelligible when read to a large assembly.

Of the variety of tables which are required in the present state of science, by far the larger portion are intimately connected with that department of it which it is the peculiar object of this society to promote.

The importance of astronomical science, whether viewed as the proudest triumph of intellectual power, or considered as the most valuable present of abstract science to the comfort and happiness of mankind, equally claims for it the first assistance from any new method for condensing the processes of reasoning or abridging the labour of calculation. Astronomical tables were therefore the first objects on which I turned my attention, when attempting to improve the power of the engine, as they had formed the first motive for constructing it.

I have already stated to the society, in my former communication, that the first engine I had constructed was solely destined to compute tables having constant differences. From this circumstance it will be apparent that after a certain number of terms of a table are computed, unless, as rarely happens, it has a constant order of differences, we must stop the engine and place in it other numbers, in order to produce the next portion of the table. This operation must be repeated more or less frequently according to the nature of the table. The more numerous the order of differences, the less frequent will this operation become requisite. The chance of error in such computations arises from incorrect numbers being placed in the engine: it therefore becomes desirable to limit this chance as much as possible. In examining the analytical theory of the various differences of the sine of an arc, I noticed the property which it possesses of having any of its even orders of differences equal to the sine of the same arc increased by some multiple of its increment multiplied by a constant quantity. With the aid of this principle an engine might be formed which would require but little attendance, and I believe that it might in some cases compute a table of the form A sin θ from the 1st value of $\theta = 0$ up to $\theta = 90°$ with only one set of figures being placed in it.

It is scarcely necessary to observe what an immense number of astronomical tables are comprised under this form, nor the great accuracy which must result from having reduced to so few a number the preliminary computations which are requisite.

In pursuing into its detail the principle to which I have alluded, which lends itself so happily to numerical application, I have traced its application to other species of tables, and am enabled to point out a course of analytical investigation which will in all probability afford ready methods for constructing tables, even of the most complicated transcendent, in a manner equally easy.

I will now advert to another circumstance, which, although not immediately connected with astronomical tables, resulted from an examination of the engine by which they can be formed.

On considering the arrangement of its parts, I observed that a different mode of connecting them would produce tables of a new species altogether different from any with which I was acquainted. I therefore computed with my pen a small table such as would have been formed by the engine had it existed in this new shape, and I was much surprised at discovering that no analytical method was yet known for determining its nth term. The following is the first series I wrote down :—

	Series.	Diff.		Series.	Diff.		Series.	Diff.
0 ...	2	0	10 ...	222	42	20 ...	924	86
1 ...	2	2		264	46		1010	86
	4	6		310	46		1096	92
	10	6		356	52	25 ...	1188	100
	16	12	15 ...	408	60		1288	108
5 ...	28	20		468	68		1396	114
	48	28		536	74		1510	114
	76	34		610	74		1624	118
	110	34		684	78	30 ...	1742	120
	144	38	20 ...	762	80		1862	122
10 ..	182	40		842	82		1984	

The equation of finite differences from which it is produced is

$$\Delta^2 u_z = \text{units fig. of } u_{z+1}$$

which is one of a class of equations never hitherto integrated. I succeeded in transforming this equation into a more analytical form : but still it presented great difficulties; I therefore undertook the investigation in a different manner, and succeeded in discovering a formula which represented its nth term. It is the following :—

$$u_z = (\overline{a}) + 206 (106 + 2a - 1)$$

TABLE.	
0	2
1	2
2	4
3	10
4	16
5	28
6	48
7	76
8	110
9	144

where $(\overline{a})$ represents the number opposite a in the next subsidiary table, and a is the figure in the unit's place of z, and b is that number which arises from cutting off the last figure from z. Example: let the 17th term be required, then $z = 17$, and $a = 7$, $b = 1$; the number opposite 7 in the table is

$$(\overline{7}) = . \qquad . \qquad . \qquad 76$$
$$106 + 2a - 1 = 10 + 14 - 1 = 23$$
$$206 = 20 \qquad 206 (106 + 2a - 1) = 460$$
$$\overline{\phantom{536 = u_{17}}}$$
$$536 = u_{17}$$

I have formed other series of the same class, and have succeeded in expressing any term independent of all the rest by two distinct processes. Thus I have incidentally been able to integrate the equations I have mentioned. I will just state one other of a simple form; it is the equation

$$\Delta u_z = \text{units fig. } u_z$$

whose integral is

$$u_z = 20b + 2^a$$

2 G 2

where a is that one of the numbers 1, 2, 3, 4, which, taken from z, leaves the remainder divisible by 4, and b is the quotient of that division.

The table is as follows:—

$$
\begin{array}{ccc}
1 & - & 2 \\
  &   & 4 \\
  &   & 8 \\
4 & - & 16 \\
  &   & 22 \\
  &   & 24 \\
  &   & 28 \\
8 & - & 36 \\
  &   & 42 \\
\end{array}
$$

One of the general questions to which these researches give rise is, supposing the law of any series to be known, to find what figure will occur in the kth place of the nth term. That the mere consideration of a mechanical engine should have suggested these inquiries, is of itself sufficiently remarkable; but it is still more singular, that amongst researches of so very abstract a nature, I should have met with and overcome a difficulty which had presented itself in the form of an equation of differences, and which had impeded my progress several years since, in attempting the solution of a problem connected with the game of chess.

ADDRESS

OF

HENRY THOMAS COLEBROOKE, Esq., F.R.S.,

President of the Astronomical Society of London,

On Presenting the Gold Medal to Charles Babbage, Esq., F.R.S.

Awarded 13th July, 1823.

THIS country and the present age have been pre-eminently distinguished for ingenuity in the contrivance, or in the improvement, of machinery. In none has that been more singularly evinced, than in the instance to which I have the gratification of now calling the attention of the Society. The invention is as novel, as the ingenuity manifested by it is extraordinary.

In other cases, mechanical devices have substituted machines for simpler tools or for bodily labour. The artist has been furnished with command of power beyond human strength, joined with precision surpassing any ordinary attainment of dexterity. He is enabled to perform singly the work of a multitude, with the accuracy of a select few, by mechanism which takes the place of manual labour or assists its efforts. But the invention, to which I am adverting, comes in place of mental exertion: it substitutes mechanical performance for an intellectual process: and that performance is effected with celerity and exactness unattainable in ordinary methods, even by incessant practice and undiverted attention.

The invention is in scope, as in execution, unlike anything before accomplished to assist operose computations. I pass by, as what is obviously quite different, the Shwanpan, or Chinese abacus, the tangible arithmetic of FREND, NAPIER's rods, with the ruder devices of antiquity, the tallies, the cheque, and the counters. They are unconnected with it in purpose, as in form. Mechanical aid of calculation has in truth been before proposed by very eminent persons. PASCAL invented a very complicated instrument for the simplest arithmetical processes, addition and subtraction, and reaching by very tedious repetition to multiplication and division. LEIBNITZ proposed another, of which the power extends no further. DELEPINE's and BOITISSENDEAU's contrivances, which a century ago were applauded by the Academy of Sciences at Paris, are upon the model of PASCAL's, and may no doubt be improvements of it, but do not vary or enlarge its objects. MORELAND's instruments, described in an early volume of the *Philosophical Transactions* (the 8th), are confined, the one to addition and subtraction, the other to multiplication. The *Rotula Arithmetica* of BROWN, simpler in construction, reaches not beyond the four arithmetical operations.

The principle, which essentially distinguishes Mr. BABBAGE's invention from all these, is, that it proposes to calculate a series of numbers following any law by the aid of *differences;* and that, by setting a few figures at the outset, a long series of numbers is readily produced by a mechanical operation. The method of *differences,* in a very wide sense, is the mathematical principle of the contrivance. A machine to add a number of arbitrary figures together is no economy of time or trouble; since each individual figure must be placed in the machine. But it is otherwise when those figures follow some law. The insertion of a few at first determines the magnitude of the next; and these of the succeeding. It is this constant repetition of similar operations, which renders the computation of tables a fit subject for the application of machinery. Mr. BABBAGE's invention puts an engine in the place of the computer. The question is set to the instrument; or the instrument is set to the question: and, by simply giving it motion, the solution is wrought and a string of answers is exhibited.

Nor is this all; for the machine may be rendered capable of recording its answer, and even multiplying copies of it. The usefulness of the instrument is thus more than doubled: for it not only saves time and trouble in transcribing results into tabular form, and setting types for the printing of the table constructed with them, but it likewise accomplishes the yet more important object of ensuring accuracy, obviating numerous sources of error through the careless hands of transcribers and compositors.

On this part of the invention, which is yet a subject of experiment for the selection of the most eligible among divers modes of accomplishing it, I shall not dwell longer; as it is not for that superaddition, but for the machine in the finished form of a calculating instrument, that I am to make an acknowledgment, in the name of this Society, to Mr. BABBAGE for his very useful invention.

I speak of it as complete with reference to a model which satisfactorily exhibited the machine's performance, and am apprised that a more finished engine, which is in progress of preparation, may not yet for some time be in a forward state to be put in activity and receive its practical application.

In no department of science or of the arts does this discovery promise to be so eminently useful as in that of astronomy and its kindred sciences, with the various arts dependent on them. In none are computations more operose than those which astronomy in particular requires: in none are preparatory facilities more needful: in none is error more detrimental. The practical astronomer is interrupted in his pursuit, is diverted from his task of observation, by the irksome labour of computation; or his diligence in observing becomes ineffectual for want of yet greater industry of calculation. Let the aid, which tables previously computed afford, be furnished to the utmost extent which mechanism has made attainable through Mr. BABBAGE's invention, the most irksome portion of the astronomer's task is alleviated, and a fresh impulse is given to astronomical research.

Nor is it among the least curious results of the ingenious device, of which I am speaking, that it affords a new opening for discovery; since it is applicable, as has been shown by its inventor, to surmount novel difficulties of analysis. Not confined to *constant differences*, it is available in every case of differences that follow a definite law, reducible therefore to an equation. An engine, adjusted to the purpose, being set to work, will produce any distant term, or succession of terms, required: thus presenting the numerical solution of a problem, even though the analytical solution of it be yet undetermined.

It may not therefore be deemed too sanguine an anticipation, when I express the hope, that an instrument, which in its simpler form attains to the extraction of the roots of numbers and approximates to the roots of equations, may in a more advanced state of improvement rise to the approximate solution of algebraic equations of elevated degrees. I refer to solutions of such equations proposed by LA GRANGE, and more recently by other analysts, which involve operations too tedious and intricate for use, and which must remain without efficacy, unless some mode be devised of abridging the labour or facilitating the means of its performance. In any case this engine tends to lighten the excessive and accumulating burden of arithmetical application of mathematical formulæ, and to relieve the progress of science from what is justly termed by the author of this invention, the overwhelming incumbrance of numerical detail.

For this singular and pregnant discovery, I have the authority of the Astronomical Society of London to present to Mr. BABBAGE its Gold Medal, which accordingly I now do, as a token of the high estimation in which it holds his invention of an engine for calculating mathematical and astronomical tables.

ON MR. BABBAGE'S NEW MACHINE FOR CALCULATING AND PRINTING MATHEMATICAL AND ASTRONOMICAL TABLES.

By Francis Baily, Esq., F.R.S. & L.S.

From M. Schumacher's "Astronomische Nachrichten," No. 46.

Reprinted in the " Philosophical Magazine" for May, 1824.

This invention of Mr. Babbage's is one of the most curious and important in modern times: whether we regard the ingenuity and skill displayed in the arrangement of the parts, or the great utility and importance of the results. Its probable effect on those particular branches of science which it is most adapted to promote, can only be compared with those rapid improvements in the arts which have followed the introduction of the steam-engine; and which are too notorious to be here mentioned.

The object which Mr. Babbage has in view, in constructing his machine, is the formation and printing of mathematical tables of all kinds, totally free from error in each individual copy: and, from what I have seen of the mechanism of the instrument, I have not the least doubt that his efforts will be crowned with success. It would be impossible to give you a correct idea of the form and arrangement of this machine, or of its mode of operation, without the help of various plates, and a more minute description than is consistent with the nature of your journal. But it will be sufficient to say that it is extremely simple in its construction, and performs all its operations with the assistance of a very trifling mechanical power. Its plan may be divided into two parts, the mechanical and the mathematical.

The mechanical part has already been attained by the actual construction of a machine of this kind: a machine for computing numbers with two orders of differences only, but which I have seen perform all that it was intended to do, not only with perfect accuracy, but also with much greater expedition than I could myself perform the same operations with the pen. From the simplicity of the mechanism employed, the same principles may be applied in forming a much larger machine for computing tables depending on any order of differences, without any probability of failure from the multitude of wheels employed. The liberality of our Government (always disposed to encourage works of true science and real merit) has induced and enabled Mr. Babbage to construct a machine of this kind, capable of computing numbers with four orders of differences; and which will shortly be completed. To this machine will be attached an apparatus that shall receive, on a soft substance, the impression of the figures computed by the machine: which may be afterwards stereotyped or subjected to some other process, in order to ensure their permanency. By this means, each individual impression will be perfect.

The mathematical part depends on the *method of differences* to which I have above alluded: a principle well known to be, at once, simple and correct in its nature, and of very extensive use in the formation of tables, for the almost unlimited variety of its applications. It has been already successfully applied in the computation of the large tables of logarithms in France; and is equally applicable in the construction not only of astronomical tables of every kind, but likewise of most of the mathematical tables now in use.

But the full and complete application of this, and indeed of every other principle in the formation of tables, has been hitherto very much impeded by the impossibility of confining the attention of the computers to the dull and tedious repetition of many thousand consecutive additions and subtractions, or other adequate numerical operations. The substitution, however, of the unvarying action of machinery for this laborious yet uncertain operation of the mind, confers an extent of practical power and utility on the method of differences, unrivalled by anything which it has hitherto produced : and which will in various ways tend to the promotion of science.

The great object of all tables is to save time and labour, and to prevent the occurrence of error in various computations. The best proof of their utility and convenience is the immense variety that has been produced since the origin of printing; and the diversity of those which are annually issuing from the press.

The *general* tables, formed for the purpose of assisting us in our computations, may be divided into two classes : 1°, those consisting of natural numbers; 2°, those consisting of logarithms. Of the former kind are the tables of the products and powers of numbers, of the reciprocals of numbers, of the natural signs, cosines, &c. &c. Of the latter kind are not only the usual logarithmic tables, whose utility and importance are so well known and duly appreciated, but also various other tables for facilitating the several calculations which are constantly required in mathematical and physical investigations. I shall allude to each of these in their order.

1°. Tables of the products of numbers. The numerous tables of this species which have been published at various times and in different countries, sufficiently attest their utility and importance : and there can be no doubt that, if their accuracy were undeniable, their employment would be much more frequent. One of the first tables of this class was published in "Dodson's Calculator;" and contains a table of the first nine multiples of all numbers from 1 to 1000. In 1775 this table was much extended, and printed in an octavo size : it comprehended the first nine multiples of all numbers from 1 to 10,000. Notwithstanding these and other tables of the same kind, the Board of Longitude considered that still more extended tables might be useful to science, and employed the late Dr. Hutton to form a multiplication table of all numbers from 1 to 1000, multiplied by all numbers less than 100. These were printed by their directions; and it is to be presumed that no expense was spared to render them accurate : yet in one page only of those tables (page 20) no less than forty errors occur, not one of which is noticed in the printed list of the errata. The French Government, likewise, sensible of the utility of such tables, ordered the construction of a still more extensive set for the use of several of its departments. These are comprised in one volume quarto, and extend from the multiplication of 1 by 1 to 500 by 500 : and in the year 1812 they caused a second edition of those tables to be printed. But the most convenient tables of this kind which have yet appeared were recently published at Berlin by M. Crelle, and comprise, in one octavo volume, double the quantity of the French tables. Another volume, of the same size, which is announced by the same author, will render these by far the most valuable of their kind, provided their accuracy can be relied on. The quantity of mental labour saved, in the construction of such tables, by the help of the machine, is literally infinite : for, in fact, no previous calculation is at all requisite; and it will be necessary merely to put into the machine, at the end of every two pages, the number whose multiples are required. This number will be successively 1, 2, 3, &c. . . . . to 500.

2°. Tables of Square Numbers. The squares of all numbers, as far as 1000, were a long time ago published on the Continent by M. Lambert. These have been since extended as far as the square of 10,000 by Mr. Barlow of the Royal Military Academy at Woolwich. The Board of Longitude employed the late Dr. Hutton to calculate a similar table as far as the square of 25,400. In computing a table of this kind by the machine, even if extended to the most remote point that could be desired, the whole of the mental labour would be saved : and when the numbers 1, 1, 2 are once placed in it, it will continue to produce all the square numbers in succession without interruption. This is, in fact, one of those tables which the engine already made is capable of computing, as far as its limited number of wheels will admit.

3°. Tables of Cube Numbers. Tables of this kind have likewise been already computed by Mr. Lambert and Mr. Barlow; and also by the late Dr. Hutton, by order of the Board of Longitude. In computing such a table by the machine, the whole of the mental labour would be in this case also saved : since it would be merely necessary to place in the machine the numbers 1, 7, 6, 6 ; and it would then produce in succession all the cube numbers. ·

4°. Tables of the higher Powers of Numbers. The Board of Longitude employed Dr. Hutton also to construct a limited table of this kind, which should contain the first ten powers of all numbers from 1 to 100. And Mr. Barlow has published, in his collection, a table of the fourth and fifth powers of numbers between 100 and 1000. Should it be thought desirable to recompute or extend these tables, the whole labour may be performed by the help of the machine, except the few figures required to be first placed in it, and which might perhaps occupy the computer about ten minutes for each power. In fact, the computation of these few fundamental figures would not occupy so much time, nor be so liable to error, as the calculation of *one* of the tabular numbers, according to the usual method.

5°. Tables of the Square Roots and Cube Roots of Numbers. A table of the first kind has been given by Mr. Lambert, and a more extended one by Mr. Barlow, in his Collection. The latter writer constructed his table by means of differences; an advantage which may be applied with greater effect to the table of Cube Roots, on account of the greater convergency of this order of differences.

6°. Tables of the Reciprocals of Numbers. These are amongst the most simple but most useful of arithmetical tables, and are peculiarly valuable in converting various series into numbers,—thus faciliating the calculation of differences for the more ready construction of other tables. In order, however, to be employed in such operations, it is absolutely necessary that they should be infallible. Several tables of this kind have been printed, the most recent and extensive of which are those of Mr. Barlow and Mr. Goodwin.

7°. Tables of Natural Sines, Cosines, Tangents, &c. The utility of tables of this kind is evident from the variety of forms in which they have been, from time to time, printed : and it is needless to insist on their importance at the present day, since no seaman dare venture out of sight of land without a knowledge of their use. In order to be of any real service, however, they should be accurate, and diligently revised from time to time : otherwise they may be worse than useless. The labour of computing tables of this kind will vary according to the number of figures contained in the result. It appears desirable that the larger tables of this sort should be printed with their several orders of differences to a much greater extent than formerly, for the purpose of making other tables, and for executing several mathematical operations beneficial to science. It would be difficult to state precisely the quantity of mental labour saved by the machine in constructing tables of the kind; but I believe it may be fairly reduced to the two thousandth part of the whole.

8°. Tables of the Logarithms of Numbers. Tables of this kind are in the hands of every person engaged in numerical investigations: and it is needless to dwell on their utility and importance. The logarithms of number from 1 to 108,000 have been already computed, with a greater or less number of figures; but this has been the work of various authors, and of several successive years : the labour is so immense that no human being has ventured to undertake the whole. The tables which now exist are chiefly copies from those original and partial computations. By the help of the machine, however, this immense labour vanishes, and new tables may be readily computed and recomputed as often as may be required by the public. It is probable that the present tables, if extended from 108,000 to 1,000,000, would be of greater utility than an extension of the present tables to a larger number of figures. The quantity of mental labour saved by the machine may be estimated in the following manner :—Suppose a machine constructed capable of computing with five orders of differences ; it would be necessary to calculate those differences for every thousandth logarithm only : consequently, if the table extended from 10,000 to 100,000, there would be but ninety sets of differences to compute. Any one of these sets being placed in the machine, with its first five differences, it will deliver the 500 preceding logarithms, and also the 500 succeeding ones ; thus producing a thousand logarithms: at the end of which term another set of differences must be substituted. With five orders of differences, a table of logarithms may be computed to eight places of figures, which shall be true to the last figures, and it would not require more than half an hour to compute each set of differences : particularly as the higher numbers require very little labour, two or three terms of the series being quite sufficient.

9°. Tables of Logarithmic Sines, Cosines, Tangents and Cotangents. The remarks which have been made in the preceding article will apply with nearly equal propriety to the tables here alluded to. The mental labour required for their construction by the machine is reduced to a very insignificant quantity when compared with the prodigious labour employed in the usual way.

10°. Tables of Hyperbolic Logarithms. Some small tables of this kind have been printed in several works, and are useful in various integrations ; but the most comprehensive set was computed by Mr. Barlow, which contains the hyperbolic logarithms of all numbers from 1 to 10,000. The labour of computing them is very great, which is the cause of their not being more extended. From a slight examination of the subject, it would appear that the mental labour may, in this case, be reduced by the machine to about a two-hundredth part of what was formerly necessary.

11°. Tables for finding the Logarithms of the sum or difference of two quantities, each of whose logarithms is given. This table, which was first suggested by Mr. Gauss, has been printed in at least three different forms. It is extremely convenient when many similar operations are required: the whole of it was computed by the method of differences, and consequently nearly the whole of the labour may be saved by the help of the machine.

12°. Other general tables might also be here mentioned which have been of great service in various mathematical investigations, and have been computed and printed by different authors: such as tables of the powers of ·01, ·02, ·03, &c.; tables of the squares of the natural sines, cosines, tangents, &c.; tables of figurate numbers, and of polygonal numbers; tables of the length of circular arcs; tables for determining the irreducible case of cubic equations; tables of hyperbolic functions, viz., hyperbolic sines, cosines, &c., and logarithmic hyperbolic sines, cosines, &c. These, and various other tables, which it is needless here to mention, may be computed by the machine with very little mental labour, and with the greatest accuracy.

Besides the *general* tables above alluded to, there are many others which are applicable to *particular* subjects only: the most important of which are those connected with astronomy and navigation. When we contemplate the ease and expedition with which the seaman determines the position of his vessel, and with what confidence he directs it to the most distant quarter of the globe, we are not, perhaps, aware of the immense variety of tables which have been formed almost solely for his use, and without the aid of which he dare not venture on the boundless ocean. Not only must the general tables of the sun and moon be first computed, together with the various equations for determining their apparent places, but those places also for every day in the year are prepared solely for his use, and even for different hours in the same day. The places of certain stars must likewise be given; and, as these depend on precession, aberration, and nutation, tables of this kind also must be formed for each star. Then come the lunar distances, which are computed for every third hour in the day, and which depend likewise on a variety of other complicated tables. After these come the Requisite Tables, published by order of the Board of Longitude, and the usual Logarithmic Tables for facilitating the computations, both of which are dependent on other tables from which they have been deduced or copied. Now, when it is considered that an error in any one of these multifarious tables will affect the last result, and thereby render the navigator liable to be led into difficulties, if not danger, it must be acknowledged that it is of very essential importance that all such tables should be computed and printed in so perfect a manner that they may in all cases be depended upon. This, however, in the present mode of constructing them, is scarcely possible. I have myself discovered about five hundred errors in the work containing the Tables of the Sun and Moon, from which (till lately) the annual volumes of the Nautical Almanac have been computed; and a respectable author has asserted that, in the first edition of the Requisite Tables, published by order of the Board of Longitude, there were above a thousand errors. Many of the subsidiary tables, above alluded to, have not been computed since they were first printed: for the mental and even manual labour of calculating them has been so great that the world has been obliged to remain contented with those original computations; and they are consequently subject to all the errors arising from subsequent editions and copies.

In the calculations of astronomical tables the machine will be of very material assistance: not only because an immense variety of subsidiary tables are required to determine the place of the sun, moon, and planets, and even of the fixed stars, but likewise on account of the frequent change which it is found necessary to introduce in the elements from which those tables are deduced, and which vary from time to time according to the improvements in physical astronomy and the progress of discovery.

Within the last twenty years it has been found necessary to revise almost all the tables connected with the solar system; and already many of these have been found inefficient for the refined purposes of modern astronomy. But the great expense of time and labour and money has been the principal obstacle to the advancement of this part of the science: since each revision has been attended with the introduction of new equations, which consequently require new tables. And, to this day, we have not been furnished with any tables whatever of three (out of the four) new planets that have been discovered since the commencement of the present century; nor can the places of many thousands of the fixed stars be readily determined for want of the subsidiary tables necessary for that purpose.

It may perhaps be proper to state that *all* astronomical tables (with very few trifling exceptions) are deduced by the two following methods: 1°, by the addition of certain constant quantities, whereby the *mean* motions of the body are determined; 2°, by certain corrections (of those motions) which depend on the sine or cosine of a given arc, and which are called *equations* of the mean motions. The mean motions of any of the celestial bodies may be computed by means of the machine without any previous calculation; and those quantities depending on the sine and cosine may in *all* cases be computed by the machine, with the help of two previous calculations of no great length or labour, and in *most* cases with the help of one only.

In the year 1804 Baron de Zach published his Tables of the Sun; and within two years of that date, Mr. Delambre published similar tables. In 1810 Mr. Carlini published his Tables of the Sun, on a new construction, so that, within the space of six years, it was considered necessary by these distinguished astronomers to publish these three interesting and highly useful works.

In the year 1806 Mr. Bürg published his very valuable Tables of the Moon—a work which superseded the use of Mason's tables — and was rewarded with a *double* prize by the French Government. It was received with gratitude by the scientific in every nation, and opened a new era in the history of astronomy and navigation. These were followed by the tables of Burckhardt in 1812, which are still more accurate than those of Bürg; and at the present moment the elements of some new tables have been deduced by Mr. Damoiseau. But it is the elements only which have yet been deduced, since it is these alone which can be expected to engage the attention of the profound mathematician. Nevertheless the laborious, yet useful, operation of computation cannot safely be left to inferior hands. The merit of each is, however, very unequally estimated by the world. Euler had three *hundred* pounds granted him by the English Government *for furnishing the elements*, and Mayer three *thousand* for the *actual computation* of the tables of the moon, which were published by the Board of Longitude in the year 1770.

The elements of Mr. Damoiseau have been already two years before the public; but the time and labour necessary to compute the tables therefrom are so great that they have not yet appeared. In order to deduce the place of the moon from these elements, no less than 116 different equations are requisite, all depending on the sine or cosine of different arcs. The labour of computing each equation with the pen would be immense, and liable to innumerable errors; but, with the assistance of the machine, they are all deduced with *equal* facility and safety, and without much previous computation.

In the year 1808 Mr. Bouvard published his tables of Jupiter and Saturn; but in 1821, owing to the progress of discovery and the advancement of physical astronomy, it was found necessary to revise the elements; and an entire new set of tables was then published by this distinguished astronomer. . In order to deduce the geocentric places of these planets, it is requisite to compute no less than 116 tables depending on the sine or cosine of certain arcs.

I shall not intrude further on the time of your readers by alluding to the tables of the other planets, which are *all* liable to similar observations; but I shall take the liberty of calling their attention to those very useful tables which have from time to time appeared for determining the apparent places of the fixed stars, and which generally assume the title of "Tables of Aberration and Nutation." Tables of this kind are of vast importance to the practical astronomer, since they save a great deal of time and labour in the reduction of observations; and it is believed that many valuable observations remain unreduced for want of convenient tables of this sort.

The first general tables of this kind were published by Mr. Mezger at Mannheim in 1778, and contained the corrections of 352 stars. In 1807 Mr. Cagnoli extended these tables to the corrections of 501 stars; and in the same year Baron de Zach published at Gotha his "Tabulæ speciales Aberrationis et Nutationis," which contained the corrections for 494 zodiacal stars. But already these tables have nearly outlived their utility. Independent of their very limited extent, the elements from which they were deduced have been superseded by others more agreeable to actual observation; which, together with their exclusion of the solar nutation, and other minute quantities which cannot safely be neglected in the present state of astronomy, renders these tables of doubtful utility to the practical astronomer.

The number of zodiacal stars (without including the very minute ones) is considerably above a thousand; each of which may, in the course of a revolution of the nodes, suffer an occultation by the moon. These occultations are ascertained to be visible at sea, or even from the unsteady deck of a vessel under sail, and afford the surest means of determining the longitude provided the position of the star could be well ascertained. In order to furnish the corrections for *each* star, ten equations are requisite, depending on the sine and cosine of given arcs; so that it would require the computation of upwards of ten thousand subsidiary tables in order to produce the necessary corrections—a labour so gigantic as to preclude all hope of seeing it accomplished by the pen. By the help of the machine, however, the manual labour vanishes, and the mental labour is reduced to a very insignificant quantity. For, as I have already stated, astronomical tables of *every* kind are reducible to the same general mode of computation, viz., by the continual addition of certain constant quantities, whereby the mean motions of the body may be determined *ad infinitum*, and by the numerical computation of certain circular functions for the correction of the same. The quantities depending on these circular functions, let them arise from whatever source they may, or let them be dependent on any given law whatever, are deducible with *equal* ease, expedition, and accuracy by the help of the machine. So that, in fact, there is no limit to the application of it in the computation of astronomical tables of every kind.

I might now direct your attention to those other subjects of a particular nature to which the machine is applicable, such as tables of Interest, Annuities, &c. &c., all of which are reducible to the same general principles, and will be found to be capable of being computed by the machine with equal facility and safety. But I trust that enough has been said to show the utility and importance of the invention; an invention inferior to none of the present day, and which, when followed up by the construction of a machine of larger dimensions now in progress (by which alone its powers and merit can be duly appreciated), will tend considerably to the advancement of science, and add to the reputation of its distinguished inventor.

I have omitted to state that this machine computes, in all cases, to the nearest figure, whatever it may be. That is, after the required number of figures are computed, if the next following figure should be a 5 or upwards, the last figure is increased by unity without any attention on the part of the operator.

But it is not in these mechanical contrivances alone that the beauty and utility of the machine consist. Mr. Babbage, who stands deservedly high in the mathematical world, considers these but of a secondary kind, and has met with many curious and interesting results, which may ultimately lead to the advancement of the science. The machine which he is at present constructing will tabulate the equation $\Delta^4 u_z = c$; consequently there must be a means of representing the given constant c, and also the four arbitrary ones introduced in the integration. There are five axes in the machine, in each of which one of these may be placed. It is evident that the arbitrary constant must be given *numerically*, although the numbers may be any whatever. The multiplication is not like that of all other machines with which I am acquainted, viz., a repeated addition—but is an actual multiplication: and the multiplier as well as the multiplicand may be decimal. A machine possessing five axes (similar to the one now constructing) would tabulate, according to the peculiar arrangement, any of the following equations :—

$$\Delta^5 u_z = a u_{z+1} \qquad\qquad \Delta^5 u_z = a u_{z+2}$$
$$\Delta^5 u_{z+1} = a u_z + \Delta^4 u_z \qquad\qquad \Delta^5 u_{z+1} = a\Delta^2 u_{z+1} + \Delta^4 u_z.$$

If the machine possessed only three axes, the following series, amongst others, might be tabulated :—

$$\Delta^2 u_{z+1} = a\Delta u_z + \Delta^2 u_z \qquad \Delta^3 u_z = a u_z.$$

If there were but two axes, we might tabulate

$$\Delta^2 u_z = a u_{z+1}.$$

These equations appear to be restricted, and so they certainly are. But since they can be computed and printed by machinery of no very great complication, and since it is not necessary (after setting the machine at the beginning) to do anything more than turn the handle of the instrument, it becomes a matter of some consequence to reduce the mode of calculating our tables to such forms as those above alluded to.

A table of logarithms may be computed by the equation $\Delta^4 u_z = c$; but in this case the intervals must not be greater than a few hundred terms. Now, it may be possible to find some equation similar to those above mentioned which shall represent a much more extensive portion of such tables—possibly many thousand terms; and the importance that would result from such an equation renders it worthy the attention of mathematicians in general.

A table of sines may, for a small portion of its course, be represented by the equation $\Delta^2 u_z = c$, but it may be represented in its whole extent by the equation $\Delta^2 u_z = a u_{z+1}$. Now, this is precisely one of the equations above quoted; and if a proper machine were made (and it need not be a large one) it would tabulate the expression $A \sin \theta$ from one end of the quadrant to the other at any interval (whether minutes or seconds) by only once setting it. It would not be very complicated to place three such machines by the side of each other, and cause them to transfer their results to a common axis, with which the printing apparatus might be connected. Such a machine would, amongst other tables, compute one from the expression

$$A \sin \theta + B \sin 2\theta + C \sin 3\theta,$$

the utility of which, in astronomy, is well known. In fact, Mr. Babbage is of opinion that it would not be impossible to form a machine which should tabulate almost any individual equation of differences.

Amongst the singular and curious powers produced by small additions to the machinery, may be reckoned the possibility of tabulating series expressed by the following equations:—

$\Delta^2 u_z = $ the unit's figure of u_z,

$\Delta^3 u_z = 2 \times$ the figures found in the ten's place of u_{z+1},

$\Delta^3 u_z = 4 \times$ the figures found in the unit's and ten's place of u_{z+1},

and many others similar thereto.

Again: Let the machine be in the act of tabulating any series, a part may be attached by means of which, whenever any particular figure (a 6 for example) occurs in the unit's place, any other number (23 for instance) shall be added to that and all the succeeding terms; and when, in consequence of this, another figure 6 occurs in the unit's place, then 23 more will be added to that and all the succeeding terms. Or, if it be preferred, the number added shall be added to the term ending in 6 only, and not to all succeeding ones.

These views may appear to some persons more curious than useful. They lead, however, to speculations of a very singular and difficult nature in determining the laws which such series follow, and they are not altogether so remote from utility as may be imagined. I avoid alluding to many other curious properties which this machine is capable of exhibiting, as they will scarcely be intelligible till the machine itself is more known in the world. Indeed I fear I have already tired your patience with this long letter.

MINUTES OF THE COUNCIL OF THE ROYAL SOCIETY RELATING TO THE REPORT OF THE COMMITTEE ON MR. BABBAGE'S CALCULATING MACHINE.—FEBRUARY 12, 1829.

THE Report of the Committee appointed to consider the letter of Mr. Stewart relative to Mr. Babbage's Calculating Machine was received and adopted.

Resolved,—That the thanks of the Council be given to the Committee for the pains they have bestowed upon the subject referred to them, and for their able Report.

Resolved,—That the following answer be sent to Mr. Stewart, viz.

The Council of the Royal Society having taken into consideration Mr. Stewart's letter, dated December, 1828, requesting their opinion, " whether the progress made by Mr. Babbage in the construction of his Machine confirms them in their former opinion, that it will ultimately prove adequate to the important objects which it was intended to attain," appointed a Committee, consisting of the President and Secretaries, Mr. Herschel, Mr. Warburton, Mr. F. Baily, Mr. Barton, Comptroller of the Mint, Mr. Brunel, F.R.S., civil engineer, Mr. Donkin, civil engineer, Mr. Penn, and Mr. G. Rennie, F.R.S., civil engineer ; from the result of whose examination of the drawings, the tools employed, and the work already executed, as detailed in the annexed Report, they have not the slightest hesitation in pronouncing their decided opinion in the affirmative.

The Council of the Royal Society cannot conclude without stating their full concurrence in the Report of their Committee, comprising, as it does, among its members several of the first practical engineers and mechanicians in the country ; nor without expressing a hope that while Mr. Babbage's mind is intently occupied on an undertaking likely to do so much honour to his country, he may be relieved, as much as possible, from all other sources of anxiety.

REPORT OF THE COMMITTEE APPOINTED BY THE COUNCIL OF THE ROYAL SOCIETY TO CONSIDER THE SUBJECT REFERRED TO IN MR. STEWART'S LETTER RELATIVE TO MR. BABBAGE'S CALCULATING ENGINE, AND TO REPORT THEREON.—Feb. 1829.

Your Committee, in this their Report, have no intention of entering into any consideration of the abstract mathematical principle on which the practicability of such a Machine as Mr. Babbage's relies, nor of its public utility when completed. They consider the former as not only sufficiently clear in itself, but as already admitted and acted on by the Council in their former proceedings. The latter they regard as obvious to every one who considers the immense advantage of accurate numerical Tables in all matters of calculation, especially in those which relate to astronomy and navigation, and the great variety and extent of those which it is professedly the object and within the compass of Mr. Babbage's Engine to calculate, and print with perfect accuracy.

The original object of the present Machine was to compute any tables which could be calculated by six orders of differences, and twelve figures in each, and sixteen figures in the table itself, in such a form that, by bestowing a very moderate degree of attention to their publication, it would be impossible for a single figure to be erroneous; and supposing any person employing them to entertain a doubt whether that moderate degree of care had been bestowed, he might in a short time himself verify the tables. The Machine was intended to produce the work stamped on plates of copper, or other proper material.

Besides the cheapness and celerity of calculation to be expected from it, the absolute accuracy of the printed results being one of the prominent pretensions of Mr. Babbage's undertaking, the attention of your Committee has been especially directed, both by careful examination of the work already executed and of the drawings, and by repeated conferences with Mr. Babbage to this point. And the result of their inquiry is, that such precautions appear to have been taken in every part of the contrivance and work which they have examined, and so fully aware does the inventor appear to be of every circumstance which may by possibility introduce error, that they have no hesitation in saying they believe these precautions effectual, and that whatever the Engine does it will do truly.

In the actual execution of the work they find that Mr. Babbage has made a progress, which, considering the very great difficulties to be overcome in an undertaking so novel, they regard as fully equalling any expectations that could reasonably have been formed; and that although several years have now elapsed since the first commencement, yet that when the necessity of constructing plans, sections, elevations, and working drawings of every part; that of constructing, and in many cases inventing, tools and machinery of great expense and complexity (and in many instances of ingenious contrivances, and likely to prove useful for other purposes hereafter), for forming with the requisite precision parts of the apparatus dissimilar to any used in ordinary mechanical works; that of making many previous trials to ascertain the validity of proposed movements; and that of altering, improving, and simplifying those already contrived and reduced to drawings; your Committee are so far from being surprised at the time it has occupied to bring it to its present state, that they feel more disposed to wonder it has been possible to accomplish so much.

The drawings form a large and most essential part of the work; they are executed with extraordinary care and precision, and may be regarded as among the best that have ever been constructed. On these all the contrivance has been bestowed, and all the alterations made, so that scarcely any work excepting drawing has been thrown away. When it is mentioned that upwards of 400 square feet of surface are covered with drawings, many of them of the most elaborate description, it will easily be understood that a very great expense of time, thought, and capital must have been incurred in producing them, but without which your Committee consider that success would have been impossible.

Nearly the whole of this department of the work (according to Mr. Babbage's statements, probably nine-tenths) is completed, and what remains is of a nature to afford no difficulty on the score of contrivance; so that there is no reason why the execution of the work (hitherto necessarily retarded till the completion of the drawings) could not now proceed with rapidity; and according to what the Committee have been enabled to gather fro   Mr. Babbage's statements and their own observations, and supposing no unexpected cause of delay, they regard a further period of three years as probably sufficient for its completion.

In judging of the adequacy of Mr. Babbage's work to complete the objects for which it was intended, there are two distinct questions—the adequacy of the contrivance, and that of the execution. On the former point every explanation has been afforded by Mr. Babbage, and both the drawings and the work executed have been unreservedly subjected to their discussion, and have been such as to excite a well-grounded confidence. The movements are combined with all the skill and system which the most experienced workmanship could suggest.

But in so complex a work, in which interrupted motions are propagated simultaneously along a great variety of trains of mechanism, it might be apprehended that obstacles would occur, or even incompatibilities arise, from the impracticability of foreseeing all the possible combinations of the parts, and of which, in a mere inspection, your Committee could not be expected to form a judgment. But this doubt, should it arise, your Committee consider as fully and satisfactorily removed by the constant employment by Mr. Babbage of a system of mechanical notation, devised by himself, and explained in a paper in the Transactions of this Society; which places at every instant the progress of motion through all parts of this or any other mechanism distinctly in view, and, by an exact tabulation of the times required for all the movements, renders it easy to avoid all danger of two contradictory impulses arriving at the same instant at any part.

Of the adequacy of the machinery to work under all the friction and strain to which it can ever be fairly exposed, and of its durability, your Committee have not the least doubt. Great precautions are taken to prevent the wear of the parts by friction; and the strength, solidity, and equilibrium, in the whole apparatus, ensure it from all danger on the score of violence or constant wear.

It ought also constantly to be borne in mind, that in all those parts of the Machine where the nicest precision is required, the wheelwork only brings them by a first approximation (though a very nice one) to their destined places; they are then settled into accurate adjustment by peculiar contrivances, which admit of no shake or latitude of any kind.

The Machine consists essentially of two parts, a calculating part and a printing part. These are both equally essential to the fulfilment of Mr. Babbage's views; for the whole advantage would be lost, if the computations made by the Machine were copied off by human hands and transferred to type by the usual process. The actual work of the calculating part is in great measure constructed, although not put together, a portion only having been temporarily fitted up for the inspection of the Committee; and from its admirable workmanship they have been able to form a confident opinion that it will execute the work expected from it. At the same time, the Committee cannot but observe that, had inferior workmanship been resorted to, such is the number and complexity of the parts, and such the manner in which they are fitted together, the success of the undertaking would have been hazarded; and they regard as extremely judicious, although, of course, very expensive, Mr. Babbage's determination to admit of nothing but the very best and most finished work in every part; a contrary course would have been false economy, and might have led to the loss of the whole capital expended on it.

In the printing part less progress has been made in actual execution than in the calculating. The reason being the greater difficulty of its contrivance, not for transferring the computations from the calculating part to the copper, or other plate ultimately destined to receive them, but for giving to the plate itself the number and variety of movements which the forms adopted in printed tables may call for in practice. The movements necessary for effecting this, being entirely such as might at any time be decided on, were purposely allowed to stand over till the more difficult parts should be fully developed. Taking the calculating and the printing part together, and regarding the tools and machinery already erected as available for the performance of what remains, the Committee regard it as not improbable that three-fifths of the work may be already completed, but they cannot be expected to state this with any degree of certainty.

With regard to the expense incurred, and likely to be incurred, Mr. Babbage states the sum already expended by him at £6000,* £1000 of which he states to have been laid out in preliminary

* See pages 115 and 120.

trials, which have not formed an object of inquiry with the Committee. Taking into consideration the extent of the work and drawings, which they have examined, and judging entirely from the general knowledge of the cost of these and similar works, which the professional experience of several individuals of the Committee has enabled them to acquire, they are no way surprised at the outlay thus stated to have been incurred. With regard to the future cost, they have, of course, less means of judging than of the past—of which they see the results, and the tools with which they have been produced. A probable conjecture might be grounded on the proportion of three-fifths, assumed as the proportion of the work already done ; but this would require to be received with very great latitude.

Finally, taking into consideration all that has been already said, and relying not less on the talent and skill displayed by Mr. Babbage as a mechanician, in the prosecution of this arduous undertaking, for what remains,—than on the matured and digested plan and admirable execution of what is accomplished—your Committee have no hesitation in giving it as their opinion, that " in the present state of Mr. Babbage's engine, they do regard it as likely to fulfil the expectations entertained of it by its inventor."

(Signed)   J. F. W. HERSCHEL, *Chairman.*

February, 1829.

ON A METHOD OF EXPRESSING BY SIGNS THE ACTION OF MACHINERY.

By Charles Babbage, Esq., F.R.S.

From the Philosophical Transactions. 1826. Vol. 2, page 250.

In the construction of an engine, on which I have now been for some time occupied, for the purpose of calculating tables, and impressing the results on plates of copper, I experienced great delay and inconvenience from the difficulty of ascertaining from the drawings the state of motion or rest of any individual part at any given instant of time; and if it became necessary to inquire into the state of several parts at the same moment, the labour was much increased.

In the description of machinery by means of drawings, it is only possible to represent an engine in one particular state of its action. If indeed it is very simple in its operations, a succession of drawings may be made of it in each state of its progress, which will represent its whole course; but this rarely happens, and is attended with the inconvenience and expense of numerous drawings. The difficulty of retaining in the mind all the contemporaneous and successive movements of a complicated machine, and the still greater difficulty of properly timing movements which had already been provided for, induced me to seek for some method by which I might at a glance of the eye select any particular part, and find at any given time its state of motion or rest, its relation to the motions of any other part of the machine, and if necessary trace back the sources of its movement through all its successive stages to the original moving power. I soon felt that the forms of ordinary language were far too diffuse to admit of any expectation of removing the difficulty, and being convinced from experience of the vast power which analysis derives from the great condensation of meaning in the language it employs, I was not long in deciding that the most favourable path to pursue was to have recourse to the language of signs. It then became necessary to contrive a notation which ought, if possible, to be at once simple and expressive, easily understood at the commencement, and capable of being readily retained in the memory from the proper adaptation of the signs to the circumstances they were intended to represent. The first thing to be done was obviously to make an accurate enumeration of all the moving parts, and to appropriate a name to each; the multitude of different contrivances in various machinery, precluded all idea of substituting signs for these parts. They were therefore written down in succession, only observing to preserve such an order that those which jointly concur for accomplishing the effect of any separate part of the machine might be found situated near to each other: thus in a clock, those parts which belong to the striking part ought to be placed together, whilst those by which the repeating part operates ought, although kept distinct, yet to be as a whole, adjacent to the former part.

Each of these names is attached to a faint line which runs longitudinally down the page, and which may for the sake of reference be called its *indicating line*.

The next object was to connect the notation with the drawings of the machine, in order that the two might mutually illustrate and explain each other.

It is convenient in the three representations of a machine, to employ the same letters for each part; in order to connect these with the notations, the letters which in the several drawings refer to the same parts, are placed upon the indicating lines immediately under the names of the things. If circumstances should prevent us from adhering to this rule, it would be desirable to mark those things represented in the plan by the ordinary letters of the alphabet, those pointed out for one of the other projections by the letters of an accented alphabet, and the parts delineated on the third projection by a doubly accented alphabet. In engines of so complicated a nature as to require sections at various parts, as well as the three projections, this system is equally applicable, and its advantage consists in this—that the number of accents on the letter indicates at once the number of the drawing on which it appears, and when it is intended to refer to several at the same time, the requisite letters may be employed and placed in the order in which the drawings will best illustrate the part under examination.

The next circumstance which can be indicated by the system of mechanical notation which I propose, more readily than by drawings, is the number of teeth on each wheel or sector, or the number of pins or studs on any revolving barrel. A line immediately succeeding that which contains the references to the drawings is devoted to this purpose, and on each vertical line indicating any particular part of the machine, is written the number of teeth belonging to it. As there is generally a great variety of parts of machinery which do not consist of teeth; of course every vertical line will not have a number attached to it.

The three lines immediately succeeding this, are devoted to the indication of the velocities of the several parts of the machine. The first must have, on the indicating line of all those parts which have a rectilinear motion, numbers expressing the velocity with which those parts move, and if this velocity is variable, two numbers should be written, one expressing the greatest, the other the least velocity of the part. The second line must have numbers expressing the angular velocity of all those parts which revolve; the time of revolution of some one of them being taken as the unit of the measure of angular velocity.

It sometimes happens that two wheels have the same angular velocity when they move; but from the structure of the machine, one of them rests one-half of the time during which the other is in action. In this case, although their angular velocities are equal, their comparative velocities are as 1 to 2; for the second wheel makes two revolutions, whilst the other only makes one. A line is devoted to the numbers which thus arise, and it is entitled, Comparative Angular Velocity.

The next object to be considered is the course through which the moving power is transmitted, and the particular modes by which each part derives its movement from that immediately preceding it in the order of action. The sign which I have chosen to indicate this transmission of motion (an arrow), is one very generally employed to denote the direction of motion in mechanical drawings; it will therefore readily suggest the direction in which the movement is transmitted. There are however various ways by which motion is communicated; and it becomes a matter of some importance to consider whether, without interfering with the sign just selected, some modification might not be introduced into its minor parts, which, leaving it unaltered in the general form, should yet indicate the peculiar nature of the means by which the movement is accomplished.

On enumerating those modes in which motion is usually communicated, it appeared that they may be reduced to the following:

One piece may receive its motion from another by being permanently attached to it as a pin on a wheel, or a wheel and pinion on the same axis. | This may be indicated by an arrow with a bar at the end.

One piece may be driven by another in such a manner that when the driver moves, the other also always moves; as happens when a wheel is driven by a pinion. | An arrow without any bar.

One thing may be attached to another by stiff friction. | An arrow formed of a line interrupted by dots.

One piece may be driven by another, and yet not always move when the latter moves: as is the case when a stud lifts a bolt once in the course of its revolution. | By an arrow, the first half of which is a full line, and the second half a dotted one.

One wheel may be connected with another by a ratchet, as the great wheel of a clock is attached to the fusee. | By a dotted arrow with a ratchet tooth at its end.

Each of the vertical lines, representing any part of the machine, must now be connected with that representing the part from which it receives its movement, by an arrow of such a kind as the preceding table indicates; and if any part derives motion from two or more sources, it must be connected by the proper arrows with each origin of its movement. It will in some cases contribute to the better understanding of the machine, if those parts which derive movement from two or more sources have their names connected by a bracket, with two or more vertical lines, which may be employed to indicate the different motions separately. Thus if a shaft has a circular as well as a longitudinal motion, the two lines attached to its name should be characterised by a distinguishing mark, such as (vert. motion) and (circ. motion). Whenever any two or more motions take place at the same time, this is essential; and when they do not, it is convenient for the purpose of distinguishing them.

All machines require, after their parts are finished and put together, certain alterations which are called adjustments. Some of these are permanent, and, when once fixed by the maker, require no further care. Others depend on the nature of the work they are intended to perform, as in the instance of a corn mill; the distance between the stones is altered according to the fineness of the flour to be ground: these may be called usual adjustments; whilst there are others depending on the winding up of a weight, or spring, which may be called periodic adjustments. As it is very desirable to know all the adjustments of a machine, a space is reserved, below that in which the connections of the moving parts are exhibited, where these may be indicated; if there are many adjustments, this space may be subdivided into three, and appropriated to each of the three species just enumerated, the permanent, the usual, and the periodic: if their number is small, it is better merely to distinguish them by a modification of their signs. It is sometimes impossible to perform such adjustment, except in a particular succession; and it is always convenient to adhere to one particular order. Numbers attached to their respective lines denote the order in which the parts are to be adjusted; and as it will sometimes happen that two or more adjustments must be made at the same time, in that case the same numbers must be written on the lines belonging to all those parts which it is necessary to adjust simultaneously.

If it is convenient to distinguish between the species of adjustments without separating them by lines, this may be accomplished by putting a line above or below the figures, or inclosing them in a circle, or by some similar mode. I have attached the letter P to those which are periodic.

It would add to the knowledge thus conveyed, if the sign indicating adjustment also gave us some information respecting its nature; there are, however, so many different species, that it is perhaps better in the first instance to confine ourselves to a few of the most common, and to leave to those whom may have occasion to employ this kind of notation, the contrivance of signs, fitted for their more immediate purpose.

One of the most common adjustments is that of determining the distance between two parts, as between the point of suspension and the centre of oscillation of a pendulum. This might be

indicated by a small line crossing the vertical line attached to the part. If the distance

between two parts, which are represented by different lines, is to be altered, their lines may be connected by an horizontal line.

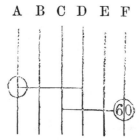

Thus the adjustment of A in the above figure depends on its distance from D.

This adjustment often determines the length of a stroke, and sometimes the eccentricity of an eccentric, which depends on the linear distance between its centre and its centre of motion.

The next most frequently occurring adjustment is that which is sometimes necessary in fixing two wheels, or a wheel and an arm on the same axis. This relation of angular position, may be indicated by a circle and two radii placed at the requisite angle, or that angle may be stated in figures and enclosed within the circle; this circle ought however to be connected by a line with the other part, with which the angle is to be formed, thus,

which means that an adjustment is to be made by fixing A on its axis, making a right angle with D and that F must also make with C an angle of sixty degrees. In speaking of these angles it should, always be observed that they refer to the angles made by the parts on one of the planes of projection.

When it is thought requisite to enter into this minute detail of adjustments, it will be necessary, in order to avoid confusion, to put the lines indicating the order of adjustment, above and distinct from these signs.

The last and most essential circumstance to be represented, is the succession of the movements which take place in the working of the machine. Almost all machinery, after a certain number of successive operations, recommences the same course which it had just completed, and the work which it performs usually consists of a multitude of repetitions of the same course of particular motions.

It is one of the great objects of the notation I am now explaining, to point out a method by which, at any instant of time in this course or cycle of operations of any machine, we may know the state of motion or rest of every particular part ; to present a picture by which we may, on inspection, see not only the motion at that moment of time, but the whole history of its movements, as well as that of all the contemporaneous changes from the beginning of the cycle.

In order to accomplish this, each of the vertical indicating lines representing any part of the machine, has, adjacent to it, other lines drawn in the same direction : these accompanying lines denote the state of motion or rest of the part to which they refer, according to the following rules :

1. Unbroken lines indicate motion.

2. Lines on the right side indicate that the motion is from right to left.

3. Lines on the left side indicate that the direction of the motion is from left to right.

4. If the movements are such as not to admit of this distinction, then, when lines are drawn adjacent to an indicating line, and on opposite sides of it, they signify motions in opposite directions.

5. Parallel straight lines denote uniform motion.

6. Curved lines denote a variable velocity. It is convenient as far as possible to make the ordinates of the curve proportional to the different velocities.

7. If the motion may be greater or less within certain limits ; then if the motion begin at a fixed moment of time, and it is uncertain when it will terminate, the line denoting motion must extend from one limit to the other, and must be connected by a small cross line at its commencement with the indicating line. If the beginning of its motion is uncertain, but its end determined, then the cross line must be at its termination. If the commencement and the termination of any motion are both uncertain, the line representing motion must be connected with the indicating line in the middle by a cross line.

8. Dotted lines imply rest. It is convenient sometimes to denote a state of rest by the absence of any line whatever.

9. If the thing indicated be a click, bolt, or valve, its dotted line should be on the right side if it is out of action, unbolted, or open, and on the left side if the reverse is the case.

10. If a bolt may rest in three positions : 1st, bolted on the right side ; 2nd, unbolted ; 3rd, bolted on the left side. When it is unbolted, and in the middle station, use two lines whilst in the act of unbolting, and two lines of dots, one on each side of, and close to the indicating line, whilst it rests in this position. When it is bolted on the right side, a line or a line of dots at a greater distance on the right hand from the indicating line will represent it. And if it is bolted on the left, a similar mode of denoting it must be used on that side. Any explanation may, if required, be put in words at the end of the notation, as will be observed in that of the hydraulic ram, Plate X.

I have now explained means of denoting by signs almost all those circumstances which usually occur in the motion of machinery : if other modifications of movement should present themselves, it will not be difficult for any one who has rendered himself familiar with the symbols employed in this paper, to contrive others adapted to the new combination which may present themselves.

The two machines which I have selected as illustrations of the application of this method, are, the common eight-day clock, and the hydraulic ram. The former was made choice of from its construction being very generally known, and I was induced to choose the latter from the apparent difficulty of applying this method to its operations.

The advantages which appear to result from the employment of this *mechanical notation,* are to render the description of machinery considerably shorter than it can be when expressed in words. The signs, if they have been properly chosen, and if they should be generally adopted, will form as it were an universal language; and to those who become skilful in their use, they will supply the means of writing down at sight even the most complicated machine, and of understanding the order and succession of the movements of any engine of which they possess the drawings and the mechanical notation. In contriving machinery, in which it is necessary that numerous wheels and levers, deriving their motion from distant parts of the engine, should concur at some instant of time, or in some precise order, for the proper performance of a particular operation, it furnishes most important assistance; and I have myself experienced the advantages of its application to my own calculating engine, when all other methods appeared nearly hopeless.

DESCRIPTION OF THE PLATES.*

Plates VII. and VIII. are different representations of an eight-day clock, for the purpose of comparing it with the notation.

Plate IX. represents the mechanical notation of the same clock.

Under the names of each part follow the letters which distinguish them in the plates.

The next line contains numbers which mark the number of teeth in each wheel, pinion, or sector.

The following line is intended to contain numbers expressing the linear velocity of the different parts: in the eight-day clock this line is vacant, because almost all the motions are circular.

The next line indicates the angular velocity of each part; and in order to render the velocity of the striking parts comparable with those of the time part, I have supposed one revolution of the striking fusee to be made in one minute: I have also taken one revolution of the scapement wheel, or one minute of time as the unit of angular velocity.

The space entitled comparative angular velocity, expresses the number of revolutions one wheel makes during one revolution of some other; it differs from the real angular velocity, because one wheel may be at rest during part of the time. Thus the clock strikes 78 strokes during twelve hours, or one revolution of the hour hand: if this be called unity, the pin wheel moves through 78 pins or $9\frac{3}{4}$ revolutions in the same time; its comparative angular velocity is therefore $9\frac{3}{4}$.

The space in which the origin of motion is given, will not require any explanation after reading the description of the signs employed in this paper.

The adjustments are numbered in the order in which they are to be made. No. 1, is attached to the crutch: the first adjustment is to set the clock in beat. No. 2, is to adjust the length of the pendulum to beat seconds. No. 3, occurs in three different places, at the hour and minute hands, and at the snail on the hour wheel. It is necessary that when the hour hand is at a given figure, three o'clock for instance, that the minute hand should be set to twelve o'clock; it is also necessary that the snail should be in such a position that the clock may strike three. These adjustments must be made at the same time. No. 4, is for the adjustment of the seconds hand to 60 seconds. No. 5, is double, and is for the adjustment of the minute and hour hand to the next whole minute to that which is indicated by the watch by which the clock is set. No. 6, is for the pendulum, which must be held aside at its extreme arc until the instant at which the watch reaches the time set on the face of the clock; it must then be set free.

The remaining part of the notation indicates the action of every part at all times; but as the whole cycle of twelve hours would occupy too much space, a portion only is given about the hour of four: from this the machine may be sufficiently understood. As an instance of its use, let us inquire what movements are taking place at seven seconds after four o'clock. On looking down on the lefthand side to the time just mentioned, we observe between the end of the sixth, and end of the seventh second, that the pendulum and crutch begin to move from the right to the left, increasing their velocity to a maximum, and then diminishing it; that the whole train of wheels of the time part are at rest during the greater part of that second, and all move simultaneously a little before its termination. The greater part of the train of the striking part is moving uniformly; but two parts, the cross piece, and the other moving the hammer, being at the commencement of this second in a state of motion from right to left, suddenly have that motion reversed for a short

---

* After this paper was printed, it was found that the copper plates by which it was illustrated had been destroyed. Impressions of them may, however, be found in the second volume of the Phil. Trans., 1826.

time : this is at the moment of striking : two other pieces, the hawk's bill, and the gathering pallet, appear to act at the same moment.

If the course of movement of any one part is required throughout the whole cycle of the machine's action, we have only to follow its indicating line. If it is required to find what motions take place at the same time, we have only to look along the horizontal line marked by the time specified.

Let us now inquire into the source of motion of the minute hand. On looking down to the space in which the origin of motion is given, we observe an arrow point, which conveys us to the

Cannon pinion, with which it is connected permanently.

The cannon pinion is driven by the centre or hour wheel, with which it is connected by stiff friction.

The hour wheel is driven by its pinion, to which it is permanently attached.

The hour wheel pinion is driven by the great wheel, into which it works.

The great wheel is driven by the fusee, with which it is connected by a ratchet.

The fusee is driven by the spring barrel or main spring, which is the origin of all the movements.

When that part of the notation which relates to the successive movements of the machine is of considerable extent, it is convenient to write on a separate piece of paper the names of every part, at the same distances from each other as the indicating lines, and exactly as they are placed at the top. By sliding this paper down the page to any part which is under consideration, the trouble of continual reference to the top of the drawing will be avoided.

Plate X. represents the hydraulic ram : its mechanical notation is added below it.

A, is the supplying pipe.

B, is the great valve.

C, the valve into the air vessel.

D, the air vessel.

E, the ascending water.

F, the small air valve : its office is to supply a small quantity of air at each stroke ; it opens when the valve C is just closed, and a regurgitation takes place in the supplying pipe just previous to the opening of the great valve. Without this contrivance, the pressure on the air in the air vessel would cause it to be soon absorbed by the water, and the engine would cease to act.

In this notation two indicating lines A, A, are allowed to the supplying water, because it takes three different courses during the action of the machine. The first of these marks the time of its motion when it enters the air vessel, and the second indicates its course when passing through the great valve, and also its course when, owing to the elasticity of the materials, its motion is for an instant reversed, at which moment air is taken in at the air valve F.

The action of the machine is as follows : the supplying water rushing along the great pipe passes out at the great valve ; it acquires velocity until the pressure of the effluent water against the under part of the great valve causes it to close suddenly. At this moment the whole momentum of the water is directed against the sides of the machine, and the air valve being the weakest part gives way, and admits a small quantity of water ; the air spring soon resists sufficiently to close the air valve : at this moment the elasticity of the apparatus reacting on the water in the great pipe, drives it back for an instant, during which the pressure of the atmosphere opens the air valve, and a small quantity of air enters ; this finds its way to the air chambers, which easily discharges it through the ascending pipe if too much air has entered.

LAWS OF MECHANICAL NOTATION.

(FOR CONSIDERATION.)

July, 1851.

This short paper, upon the principles of lettering mechanical drawings, was printed and given away by Mr. Babbage in considerable numbers, to foreigners as well as to his countrymen, during and after the Great Exhibition of 1851.

CHAPTER I.

ON LETTERING DRAWINGS.

ALL machinery consists of—

Framing.	Parts, or Pieces.
Fixed.　　　Moveable.	Moveable as axes, springs, &c.

Every *Piece* possesses one or more *Working Points*. These are divided into two classes, those by which the Piece acts on others, and those by which it receives action from them : these are called *Driving* and *Driven Points*. A *Working Point* may fulfil both these offices, as, for example, the same teeth which are driven by one wheel may in another part of their course drive other wheels.

The following alphabets of large letters are used in Drawings :—

Etruscan.	Roman.	Writing.
A *A*	A *A*	$\mathscr{A}$ $\mathscr{A}$
B *B*	B *B*	$\mathscr{B}$ $\mathscr{B}$
C *C*	C *C*	$\mathscr{C}$ $\mathscr{C}$
. . .	. . .	. . .

The following alphabets of small letters are used :—

$$a \quad b \quad c \quad d \quad . \quad . \quad . \quad .$$
$$a \quad b \quad c \quad d \quad . \quad . \quad . \quad .$$

It is most convenient, and generally sufficient, to use only the letters *a, c, e, i, m, n, o, r, s, u, v, w, x, z* of both these latter alphabets.

Rule 1.—Every separate portion of *Frame-work* must be indicated by a large *upright* letter.

Rule 2.—Every *Working Point* of Frame-work must be indicated by a *small* printed letter.

Rule 3.—*Frame-work* which is itself moveable must be represented by a *large upright* letter, with the sign of motion in its proper place below it (*see Signs of Motion*), as

$$\underline{G} \qquad \underset{\smile}{H}$$

Rule 4.—In lettering Drawings, commence with the axes. These must be lettered with *large inclined* letters of either of the three alphabets. Whenever the wheels or arms of any two or more adjacent axes cross each other on the plan, avoid denoting those axes by letters of the *same* alphabet.

Rule 5.—In lettering *Pieces*, as wheels, arms, &c., belonging to any axis, whether they are fixed to it or moveable upon it, always use *inclined capitals* of the *same* alphabet as that of the letter representing the Axis.

Rule 6.—Beginning with the lowest *Piece* upon an Axis, assign to it any *capital* letter of the *same* alphabet. To the Piece next above, assign any other *capital* letter which occurs *later* in the *same* alphabet. Continue this process for each *Piece*.

Thus, although the succession of the letters of the *same* alphabet need not be continuous, yet their occurrence in *alphabetic* order will never be violated.

Rule 7.—In lettering *Pieces* upon axes perpendicular to the elevation, or to the end views, looking from the left side, the earliest letters of the alphabet must be placed on the Pieces most remote from the eye.

Rule 8.—No axis which has a *Piece* crossing any other *Piece* belonging to an adjacent axis, must have the same identity as that axis.

If there are many *Pieces* on the same axis, it may be necessary to commence with one of the earlier letters of the alphabet.

Rule 9.—In placing letters representing any *Piece* on which portions of other *Pieces* are projected, it is always desirable to select such a situation that no doubt can be entertained as to which of those *Pieces* the letter is intended to indicate. This can often be accomplished by placing the letter upon some portion of its own *Piece* which extends beyond the projected parts of the other *Piece*.

Rule 10.—When *Pieces* are very small, or when they are crossed by many other lines, it is convenient to place the letter representing them outside the *Piece* itself, and to connect it with the *Piece* it indicates by an arrow. This arrow should be a short fine line terminated by a head, abutting on, or perhaps projecting into, the *Piece* represented by the letter.

Rule 11.—When upon any Drawing, a letter having a dot beneath it occurs, it marks the existence of a *Piece* below.

Rule 12.—In case another *Piece*, exactly similar to one already represented and lettered, exists below it, it cannot be expressed by any visible line. It may, however, be indicated by placing its proper letter outside, and connecting that letter with a *dotted* arrow abutting on the upper *Piece*.

Rule 13.—The permanent connexion of two pieces of matter, or the permanent gearing of two wheels, is indicated by a short line crossing, at right angles, the point of contact. The sign | indicates, in a certain sense, fixed connexion. This sign will be found very useful for indicating the boundaries of various pieces of framing.

It is to be observed that letters of the simplest and least ornamented style ought to be preferred : such are more quickly apprehended by the eye, and more easily recalled by the memory.

OF THE INDICES OF LETTERS.

Rule 14.—Various indices and signs may be affixed to letters. Their position and use are indicated in the subjoined letter :—

[Sign of Form.]

(Identity.) 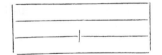 (Cir. Pos^{n}.)

(New Alph^{t}.) (Linear Pos^{n}.)

[Sign of Motion.]

Rule 15.—The index on the left-hand upper corner is used to mark the identity of two or more parts of a *Piece* which are permanently united; each being denoted by a letter with the *same* index.

Rule 16.—It is used also to connect any *Piece* itself with its various working points. Thus all the small letters which indicate the working points, must have the *same* index of identity as the letter expressing the *Piece* itself.

Rule 17.—Every *Working Point* must be marked by the *same small letter* as the *Working Poi* of the *Piece* upon which it acts.

Rule 18.—The bearings in which axes work, as well as the working surface of the axes themselves, and also the working surfaces of slides, are *Working Points*, and must be lettered as such.

OF THE INDEX OF LINEAR POSITION.

The successive order in which the various *Pieces* upon one axis succeed each other, is indicated by the alphabetic succession.

It may, however, in some cases be convenient to distinguish between the relative heights of the various arms or wheels which constitute one *Piece*.

This may be easily accomplished by means of the index of *linear position*.

Every Piece may be represented as a whole, by one letter, with its proper index of identity. If, however, it is necessary to distinguish the different arms or parts of which it is composed, so as to indicate their relative position above the plane of projection, this may be accomplished by means of the indices of linear position.

Rule 19.—If ^3P represent the whole of any *Piece*, $^3P_1$, $^3P_2$, $^3P_3$, $^3P_4$, &c., will represent in succession the several arms or parts of which ^3P is composed : $^3P_1$ indicating that which is most distant from the eye.

OF THE INDEX OF CIRCULAR POSITION.

It may occasionally be desirable to indicate the order of succession in angular position of the various arms belonging to the same *Piece*, when projected on a plane. The index on the right-hand upper corner is applied to this purpose.

Rule 20. ^6R representing any *Piece*,

$^6R^1$ will represent any arm as the origin,

$^6R^2$ the next arm in angular position in the direction " *screw,*"

that is, from left to right,

$^6R^3$ the next, &c.

Thus,

$$^6R_1^1, \, ^6R_2^2, \, ^6R_3^3, \, \ldots \, ^6R_n^n$$

would represent n arms placed spirally round an axis at various heights above it.

OF THE INDEX OF NEW ALPHABET.

In case the three alphabets given above are found insufficient, the index on the left lower side is reserved to mark new alphabets. In the most complicated drawing I have scarcely ever had occasion to use it. It might in some cases be desirable to have a fourth alphabet, differing in form from those already given.

The following lithographic Plate contains the signs of form and those of motion.

These signs of form have been the subject of much thought and discussion. A good test of their fitness arose under the following circumstances :—Three signs had been selected for the representation of various link motions, such as those of the parallel motions connected with the beam of a steam-engine.

Twelve of the motions of which links are susceptible are represented in the list; but, after a time, I observed that there were four other combinations which had not been represented, because they did not admit of motion.

On examining the combinations of these signs, it was found that, although not moveable, they represented real mechanical combinations.

The first twelve were formed according to the following laws :

1st. The circles at the ends of each line represent *axes* which are hollow if the axis is hollow, and are dark if the axis is solid.

2nd. If the axis is a *fixed* axis, then its circle has a vertical line passing through its centre.

It will be observed that links marked 23 are all moveable about their left-hand fixed centre, whilst those marked 25 are all moveable about their right-hand fixed centre.

Those links marked No. 24, which have no bar, are moveable centres like some of the rods of the parallel motion of a steam-engine.

There are, however, four other possible combinations of these signs.

It therefore becomes an interesting inquiry to ascertain whether these represent any known mechanical contrivances.

On interpreting them literally, it appears that the first is a bar having a solid stud fixed at each end, whilst the last is a bar having two holes in it, by which it may be screwed to any other piece of matter. The other pair represent a bar having a stud at one end, and a hole for a screw or bolt at the other.

There are two other chapters necessary to complete this subject.

CHAPTER II.—On the Notation of Periods.

The object of this is to give a minute account of the time at every motion and of every action throughout the cycle of the movement of the machine to be described.

CHAPTER III.—On the Trains.

The special object of this chapter is, to give an account of the directions of the various courses through which the active forces of the machine are developed. But the times of every action can be combined with it, and, to a certain extent, the forms of every moving part.

Some further notice of the mechanical notation will be found in the Introduction to this work

---

Mr. Babbage *will feel obliged by any criticisms, or additions to these Rules of Drawing, and to the Mechanical Alphabet, and requests they may be addressed to him by post, at No. 1, Dorset Street, Manchester Square.*

July, 1851.

ON MECHANICAL NOTATION, AS EXEMPLIFIED IN THE SWEDISH CALCULATING MACHINE OF MESSRS. SCHEUTZ.

By Major Henry P. Babbage.

[*Brit. Assoc. Sept.* 1855.]

Mr. H. P. Babbage said,—The system of describing machinery, of which I am about to give a brief outline, is not new. It was published by Mr. Babbage, in the 'Philosophical Transactions,' in the year 1826, where apparently it did not attract the notice of those most likely to find it practically useful. It had been used for some years before this in the construction, for the Government, of the Difference Engine, which is now in the Museum at King's College, London; and it was also used in the contrivance of the Analytical Engine, on which my father was engaged for many years. Indeed, without the aid of the mechanical notation, it would be beyond the power of the human mind to master and retain the details of the complicated machinery which such an engine necessarily requires. Its importance as a tool for the invention of machinery for any purpose is very great; since we can demonstrate the practicability of any contrivance, and the certainty of all its parts working in unison, before a single part of it is actually made. It is also important both as a means of understanding and of explaining to others existing machinery; for it is utterly impossible to make the notation of a machine without comprehending its action in every single part. There are also many other uses, which I shall not now stop to mention. The general principles of the notation are the same now as in 1826; but the practical experience of many years has, of course, suggested several alterations of detail, and led to the adoption of some important principles.

To understand the construction of a machine, we must know the size and form of all its parts, the time of action of each part, and the action of one part on another throughout the machine. The drawings give the size and form, but they give the action of the parts on each other very imperfectly, and scarcely anything of the time of action. The notation supplies these deficiencies, and gives at a glance the required information. When the drawings of a machine are made, it becomes necessary to assign letters to the different parts. Hitherto, I believe, this has been left much to chance; and each draughtsman has taken the letters of the alphabet, and used them with little or no system. With respect to lettering, the first rules are, that all *framework* shall be represented by *upright* letters. *Moveable pieces* shall be represented by *slanting* letters. Each piece has one or more working points; each of the *working points* must have its own *small* letter, the working points of *framework* having *small printed* letters, and the working points of the *moveable pieces* having *small written* letters.

Thus we have the machinery divided into Framing, indicated by large upright letters, as A, B, C, &c.; Moveable Pieces, indicated by large slanting letters, as *A, B, C,* &c.; Working points of Framing, indicated by small printed letters, as a, c, e, m, n, &c.; Working points of Moveable Pieces indicated by small written letters, as *a, c, e, m, n,* &c.

In lettering drawings the axes are to be lettered first. Three alphabets may be used—the Roman, Etruscan, and written, as—

A,   B,   C, &c.

A,   B,   C, &c.

𝒜,   ℬ,   𝒞, &c.

These should be selected as much as possible, so that no two axes which have arms or parts crossing each other should have letters of the same alphabet. Having lettered the axes, all the parts on them, whether loose or absolutely fixed to them, must be lettered with the same alphabet, care being taken that on each axis the parts most remote from the eye shall have letters earlier in the alphabet than those parts which are nearer. It is not necessary that the letters should follow each other continuously, as in the alphabet; for instance, *D, L, T,* may represent three wheels on the same axis: *D* must be the most remote, *L* the next, and *T* the nearest. The rule is, that on any axis, a part which is more remote from the eye than another, must invariably have a letter which occurs earlier in the alphabet. By these rules very considerable information is conveyed by the lettering on a drawing; but still more to distinguish parts and pieces, an index on the left-hand upper corner is given to each large letter; this is called the " index of identity," and all parts which are absolutely fixed to each other must have the same index of identity; no two parts which touch or interfere with or cross each other, on the drawings, must have the same index of identity. This may generally be done without taking higher numbers than 9. All pieces which are loose round an axis must have a letter of the same character, Roman, Etruscan, or writing; but a different index of identity will at once inform us that it is a separate piece, and not fixed on the axis. For example, ^6D, ^6L, ^6T, would indicate that the three wheels mentioned above were all *fixed* to the same axis; but ^6D, ^8L, ^6T, would at once show that D and T were fixed to the axis, and L *loose* upon it.

I shall now endeavour to explain how the transmission of motion and action of one piece on another is shown. Beginning from the source of motion, each part is written down with its working points; those of its points which are acted on are placed on the left-hand side; those points where it acts on other pieces are placed on the right hand: if there are several small letters, a bracket connects them with their own large letter—

$$\left.\begin{matrix} n \\ o \end{matrix}\right\} P \left\{\begin{matrix} u \\ a \\ e \end{matrix}\right.$$

The pieces being arranged, arrow-headed lines join each acting or driving point of one piece with the point of another piece, which it drives or acts on. When a machine is complicated, it is usually necessary to make two or three editions before all the parts can be arranged with simplicity; but, when done, the Trains, as they are called, indicate with the utmost precision the transmission of force or motion through the whole machine, from the first motive power to the final result. It is, however, one of the principles of the notation to give at one view the greatest possible amount of information, provided that no confusion is made; it has been found that, without in any way interfering with the simplicity of the Trains, a variety of information on other points may be conveyed. For instance, whilst looking at the Trains, it is often convenient or necessary to know something of the direction of motion of the piece under consideration, and, by the use of a few signs placed under the large letters, we can convey nearly all that is wanted in this respect. Again, though the drawings of a machine are specially intended to give the size and shape of each piece, yet, by the use of some signs of form which are placed above the letters, the form of each piece may be indicated. It is found that these signs do not confuse the Trains; but, on the contrary, extend their use, by making the information they convey more condensed, and more easily accessible.

I now pass on to the Cycles, as they are termed, or to that part of the notation which relates to the time of action of the different parts of a machine. The cycles give the action of every part during the performance of one complete operation of the machine, whatever that may be. Each piece has a column of its own, and the points by which it is acted on are placed on its left hand, and the points by which it acts on other parts are placed on its right; and each working point also has its own column. The whole length of the column indicates the time occupied in performing one operation, and we divide that time into divisions most suited to the particular machine. During each division of time that a piece is in motion, an arrow up or down its column indicates the fact; and during the time of action of each working point, an arrow in its own column shows the duration of its action. The times thus shown are, of course, only relative, and not absolute time: but it would be easy to show both, by making the divisions of the columns correspond with the number of seconds or minutes during which the machine performs one operation. The arrows which point upwards indicate circular motion in the direction, screw in, and the arrows which point downwards, screw out; where the motion is linear, the downward arrow indicates motion from right to left.

Mr. H. P. Babbage then illustrated this system of notation by directing attention to the notation of the Difference Engine of Messrs. Scheütz. This machine contains several hundred different pieces, yet the trains showed at one view how each piece was acted on, and how it acted on other pieces; the Cycles gave with equal clearness the time of action of each piece. In fact, the two pieces of paper before the Section gave a complete description of the machine, and, with the drawings, rendered further explanation unnecessary.

SCHEUTZ'S DIFFERENCE ENGINE, AND BABBAGE'S MECHANICAL NOTATION.

[*Minutes of Proceedings of Civil Engineers, May*, 1856.]

MR. HENRY P. BABBAGE exhibited some detailed diagrams, to illustrate the Difference Engine of Messrs. Scheütz, of Stockholm, as an example of the 'Mechanical Notation' of Mr. Babbage, showing, that even the most complicated machinery might be described by a method at once so clear, concise, and easy of reference, as to render any written description unnecessary.

Diagram No. 1, exhibited the principal groups of the machine, and their relative connection.

Diagram No. 2, gave these groups more in detail, and a very complete outline of the machine, without entering into minute mechanical details.

Diagram No. 3, contained the full details of the whole machine, of which it was, in fact, a complete description. Each single Piece of the machine was indicated by a capital letter: every working Point was indicated by a small letter. The motion of every Piece might be traced back, through any number of Pieces, to the first mover; or forward, to show how it contributed to the final result. To give a full description of this diagram, would be to give a full description of the machine itself, and would occupy a volume of some bulk, as in this Difference Engine there were about four hundred separate Pieces, and about two thousand working Points. Any one acquainted with the Mechanical Notation, would be able to refer, at once, to any part of the machine, and to understand its action quickly. Those conversant with descriptions of machinery, written in ordinary language, could say how laborious it was to wade through them, even when aided by a good index, and would, therefore, be able to appreciate this facility of reference. It had been observed by different persons who had examined these Notations, that the machine would be understood more quickly by studying the Notations, than by examining the machine itself; for in the Notations were seen, at once, everything that affected each Piece, and everything which it acted upon; whereas it was well known, that by looking at a machine at rest, this was almost impossible, and it was necessary to watch a machine in motion, for some time, in order to obtain the same information.

Diagram No. 4, contained that part of the Notation which was called the "Cycles," and related entirely to the time of motion of the Pieces (expressed by capital letters), and to the time of action of the working Points (expressed by small letters). Each Piece had a vertical column for itself; and an arrow in its column denoted the times and duration of motion of that Piece. On the left hand of each column for a Piece, there was a column for each of its driven Points; and on the right hand, a column for each of its driving Points. The arrows in these columns gave the times and duration of action of each working Point; sometimes the action of a working Point was simultaneous with the motion of its Piece; sometimes a Point was acting whilst its Piece was at rest. The Cycle for any machine was the time during which one complete operation was performed, after which all its motions recurred in the same order (in this instance, it was one calculation made, and punched on the lead), and the Cycle might be divided into any number. In this machine, the Cycle was divided into ninety-six parts, because one of the wheels having ninety-six teeth made one revolution during each operation, and it was convenient to count by it. By simply looking along the horizontal lines, it could be ascertained, at a glance, what Pieces were in motion, and what working Points were acting, at any instant.

The other illustrations consisted:—

1.—Of a plan and elevation of a small part of the "Analytical Engine," to give an idea of the system of lettering employed in the Notation.

2.—A copy of the trains of the printing part of the Analytical Engine. This was a description of machinery no part of which has yet been made.

3.—Signs of Motion employed in the Notation.

4.—Signs of Form employed in the Notation.

Mr. BABBAGE expressed his regret at the absence of the very beautiful calculating engine which had been invented and executed by Messrs. Scheutz, and which his son, Mr. H. P. Babbage, had described by the diagrams. He regretted it the more, because there was no Institution in the world in which mechanical ingenuity was better appreciated, or in which more sympathy would be felt with the genius of the inventor. He was, however, happy to add, that when the instrument was in this country, and also when in Paris, very important assistance had been rendered by Mr. Gravatt (M. Inst., C.E.), who took considerable pains to explain the structure of the machine, and to make its merits known to the scientific world.

Before explaining the Mechanical Notation, he would detain the meeting for a few moments to render that justice to Messrs. Scheutz which he did not think had yet been accorded them, at least to the extent which they really merited. It had been said, that their machine was, in a great measure, copied from or suggested by his own; but if the two engines were placed in juxtaposition, the difference between them would be immediately manifest. In the absence of the machines themselves, the diagrams would serve to illustrate the few remarks which he intended to offer.

He wished, in the first place, to call attention to a point which was the most important in machinery for making calculations of any kind: the method of adding number to number. It was not difficult to contrive machinery for that purpose: it had already been done in numerous instances. The simple addition of number to number was easy. The great difficulty was the carriage. In Messrs. Scheutz's machine there was a very beautiful contrivance to effect it, which, fortunately, admitted of description without drawings.

In that instrument, when fifteen figures were added to other fifteen, a carriage might occur at any one, or more of fourteen figures. There was, in this case, a certain analogy to a railway having fifteen stations. Messrs. Scheutz had contrived a travelling arm with a roller. When the pillar to which that arm was attached advanced, the roller ran along a railway from one end of the machine to the other. In the same way, a train travelling over fifteen stations, only stopped when the telegraph signal denoted that a passenger had to be taken up. In the machine, this travelling arm ran along the railway, until a signal that a carriage was wanted was made by a point being turned on the rail. The arm now went into a siding,—took up its passenger without stopping,—then re-entered the main line, and pursued its course until another point determined another carriage. The ingenuity of this contrivance would, he conceived, be universally admired.

He would mention only one other point of difference between the two machines, viz., the printing part. Mr. H. P. Babbage had given, on one of the diagrams, an analysis of the printing part of Messrs. Scheutz's machine; and Mr. Babbage had placed by its side another diagram, explanatory of the printing part of his own apparatus, by way of showing the striking difference between the respective works. He had mentioned a few of these decided contrasts, in order to do justice to the originality of an inventor who had, he believed, had as great difficulties to contend with as those Mr. Babbage had himself encountered.

In the remarks he had addressed to the President and Fellows of the Royal Society, on the occasion of the delivery of the Medals at the last Anniversary of that Society, he had made the following observations on this subject:—

"There is, however, an instrument to which we have given hospitality during many months in these apartments, which I think highly deserving of a Medal; and I had hoped, that on the present occasion, it might, at least, have been considered worthy of being placed among the list of candidates for that honour. I allude to the admirable Machine for Calculating and Printing Tables by Differences, and producing a mould for the stereotyped plates, to print the computed results,—an instrument we owe to the genius and persevering labour of Mr. Scheutz, of Stockholm. A Committee of the Royal Society has already reported upon the machine, and I can myself bear testimony to the care and attention which our Secretary bestowed upon that valuable Report. But as some misapprehension exists in the public mind, respecting the originality displayed in that invention, I trust, that having, as is well known, given much attention to the subject, I may be permitted briefly to explain some of its principles, and thus render justice to its Author.

"The principle of Calculation by Differences is common to Mr. Scheutz's engine and to my own, and is so obviously the only principle, at once extensive in its grasp and simple in its mechanical application, that I have little doubt it will be found to have been suggested by more than one antecedent writer.

"Mr. Scheutz's engine consists of two parts,—the Calculating and the Printing; the former oeing again divided into two,—the Adding and the Carrying parts.

2 L 2

" With respect to the Adding, its structure is entirely different from my own, nor does it even resemble any one of those in my drawings. The very ingenious mechanism for carrying the tens is also quite different from my own.

" The Printing part will, on inspection, be pronounced altogether unlike that represented in my drawings; which, it must also be remembered, were entirely unknown to Mr. Scheutz.

" The contrivance by which the computed results are conveyed to the printing apparatus, is the same in both our engines: and it is well known, in the striking part of the common eight-day clock, which is called the ' snail.'

" About 1834, or 1836, Mr. Scheutz, himself a Member of no Academy, a Professor at no University, but simply an eminent printer at Stockholm, first learnt, through the ' Edinburgh Review,' the existence of that Difference Engine, a small portion of which is now placed in one of the rooms of the adjoining building (the Museum of King's College.)

" Unfortunately for himself, Mr. Scheutz was fascinated by the subject, and was impelled by an irresistible desire to construct an engine for the same purposes. He has always avowed, in the most open and honourable manner, the origin of his idea; but his finished work contains undoubted proofs of great originality, and shows that little beyond the principle could have been borrowed from my previous work. Having formed the project, Mr. Scheutz immediately began to work upon it. After four years of labour and difficulties, which cost him a large portion of his fortune, he produced the first model. This, however, did not satisfy his wishes: but far from being disheartened, he immediately recommenced his experiments with renewed energy, expending on them all the remaining savings of an industrious life, as well as the whole of the time he could snatch from the labours on which the support of his family depended.

" His son, also, after completing his studies with great credit at the Technological School of Stockholm, was anxious to assist his father in this difficult task; and for that purpose, abandoned the career he had previously chosen.

" The father and son now worked together for several years, and at last produced a machine, in which were united all the requisite conditions of a Difference Engine. But the severe economy they had been compelled to use in the purchase of materials and tools, and probably the absence in Sweden of those precious but expensive machine-tools, which constitute the power of modern workshops, rendered this new model unsatisfactory in its operations, though perfectly correct in principle.

" Exhausted by the sacrifices thus made, yet convinced, that with better workmanship, a more perfect instrument was within their reach, Mr. Scheutz determined to apply for assistance to the Diet of Sweden.

" The Diet, with difficulty, consented to advance about 280l., on certain conditions and with a stipulated guarantee. This guarantee was as great a difficulty as the construction of the machine.

" Fortunately, however, amongst the Professors of the Academy of Stockholm, enlightened men were found, capable of sympathizing with moral and intellectual worth. To the enduring honour of the Swedish Academy, a numerous list was soon formed, in which each member became responsible for a part of the guarantee.

" Messrs. Scheutz, confident in ultimate success, further pledged their own credit, and after working night and day, with indefatigable industry, the last day of the allotted year saw the completion of their long-cherished hopes.

" Sweden has thus secured for herself the glory of having been the first nation, practically to produce a machine for Calculating Mathematical Tables by Differences, and printing the results. Wealthier and more powerful nations will regret that the country of Berzelius should thus have anticipated them, in giving effect to an invention which requires for its perfection the tools of nations more highly advanced in mechanical science. But there is still left to them the honour of acknowledging the services of a foreigner, from which the richest and most commercial countries will derive the greatest advantage.

" The machine was conveyed to Paris, and placed in the Great Exposition, and the Jury to which it was referred, after full examination, concurred with their distinguished colleague (M. Mathieu, Member of the Institute), in unanimously awarding to it the Gold Medal.

" The Emperor Napoleon, true to the inspirations of his own genius and to the policy of his dynasty, caused the Swedish engine to be deposited in the Imperial Observatory of Paris, and to be placed at the disposal of the Members of the Board of Longitude.

" Your Lordship is aware, that previously to awarding any of our medals, each Member of the Council may place one or more names on the list of candidates whose claims are to be discussed. I regret that (perhaps through inadvertence) the name of Mr. Scheutz was not placed upon that list, and I cannot, my Lord, sit down without expressing a hope that the Council of the ensuing year may more than repair the omission."

The Difference Engine of Messrs. Scheutz had supplied an excellent subject for illustrating the power of the language he had contrived for the purpose of describing machinery of all kinds. He had himself used the Mechanical Notation during the last thirty years. He was driven to contrive it, because he found himself completely stopped by the complexity of the movements necessary to be combined in the original Difference Engine. The language itself had been adapted to, and found sufficient for, all the most complicated requirements of the Analytical Engine, and it was, in his opinion, capable of being applied to the description of every kind of machinery. When it became more generally known, he had little doubt that it would be adopted as the ordinary language for all machinery, because by its means even the most complicated mechanism could be easily understood, without the aid of any written description. The signs themselves expressed everything without difficulty. As a proof of this, he might state that he had, at various periods, at least half-a-dozen draughtsmen working for him, who, after a short time, had each been able to use it with facility.

Machinery was usually represented by geometrical drawings. These gave the exact shape of every part, and the relative position of all the pieces. The art of mechanical drawing was of old date, but its principles had not yet been sufficiently discussed. By the use of a few simple rules to which the necessities of the Analytical Engine had given rise, its power of expressing machinery had been greatly augmented.

MECHANICAL NOTATION.

With respect to the Mechanical Notation there were three sections to which he wished particularly to direct attention. In order to explain the construction and action of machinery, they were all essentially necessary.

Section 1. The forms and exact magnitude of the whole machine and of all its separate parts. This was accomplished by the art of mechanical drawing.

Section 2. The connection between the first mover and the final results, including the courses through which each Piece received its various motions and transmitted them to others. The diagrams to be used for this purpose, he had called the Trains.

Section 3. The time of motion of every Piece and the time of action of every working Point: every successive time of motion, or rest of each Piece must be shown, and the contemporaneous action of every Piece at every instant of time. The diagrams which conveyed this information, he had called the Cycles.

The three great elements of the description of machinery were then:—Form, represented by the Drawings. Directive power, expressed by the Trains. Times of motion and of action, expressed by the Cycles.

Section I. Of Form.

The most important addition he had made to mechanical drawing consisted in having established rules for the use of letters. Hitherto, letters had only been employed for indicating certain points on the drawing. He had proposed, by a proper choice of various classes of letters, to convey other meanings, informing the observer by the nature of the letter itself, of certain characters belonging to the object it represented.

Thus the first great division of parts in every machine was into, 1st. Frame-work by which it was supported; 2nd. Movable Pieces which acted and were acted upon. He proposed to mark all frame-work by upright letters, and to indicate all moveable parts or Pieces by inclined letters. The two classes would at once be distinguished from each other upon all drawings, and the attention might be entirely confined to that class which the student wished to consider.

Every working part or single Piece of a machine might be considered either by itself as a whole, or as having Points and surfaces capable of acting and of being acted upon by those of other Pieces. Frame-work also might be viewed in the same light: it had its working Points, such as the supports of axes and slides: also Points against which Pieces were thrust. The rules he had proposed were first, that all working Points should be represented by small letters; and secondly, that all frame-work and every Piece should be represented by capital letters. The working Points of machines were very numerous, and a portion only of the small letters of the Roman and Italic alphabets were convenient for use. It was, therefore, necessary to distinguish the small letters from each other, and also to connect every small letter belonging to a Piece, with the capital letter which represented it as a whole. For this purpose he had established an index of identity, according to the following rules :—

1. A number placed on the upper left-hand corner of any letter, was called its index of identity.

2. The same index of identity which distinguished the letter denoting a Piece, must be applied to all small Italic letters denoting its working Points.

3. In Frame-work (indicated by upright letters), there were also working Points, such as the bearings of axes, &c. These might be marked by small printed letters, which were always upright letters.

4. Every working Point must be marked by the same small letter, as that of the working Point of the Piece on which it acted.

5. No two adjacent moveable Pieces could have the same index of identity, unless they were permanently fixed to each other.

6. No two adjacent parts of frame-work could have the same index of identity.

7. In lettering drawings, the axes were to be commenced with.

8. Axes must be marked by inclined capital letters of either of the six alphabets, hereafter mentioned.

9. Dead centres (being in fact frame-work), must be indicated by upright letters.

10. Whenever the wheels, arms, &c., of any two or more axes crossed each other, those axes must not be denoted by letters of the same alphabet.

11. No axis which had a Piece crossing any other Piece belonging to an adjacent axis, must have the same identity as that axis.

It was frequently necessary to refer from one drawing to another, in order to understand the relative position of the parts. Thus, in examining a plan, any axis having attached to it a considerable number of fixed or unfixed Pieces, it was very difficult, and in many cases impossible, to know the order of their superposition, without referring to the elevation or the end view. But this order might be made evident by a simple law of lettering. By first lettering the elevation, and marking the lowest Piece on the axis, by the letter A, the next above by the letter B, and so on ; then when these letters were respectively transferred to the plan, they would render the order of their superposition evident upon that plan. This system would, however, unnecessarily restrict the use of letters, and would cause an inconvenient frequency of occurrence of the earlier letters of the alphabet. The course proposed to be pursued was, therefore, to mark the lowest Piece upon an axis with any convenient letter. Then to place on the next Piece any subsequent letter of the alphabet, and so on, always observing that no letter earlier than the one last used could be applied to any Piece above it. This gave rise to the following rules :—

1. In lettering Pieces, to begin with the lowest, if on plan, or with the most remote, if in elevations or end view.

2. To place any letter on each Piece, provided the alphabetic order of the letters was never inverted.

On the drawings it often happened that several contiguous axes had upon them wheels, arms, &c., which spread over other wheels, arms, &c., belonging to other adjacent axes. In this case it was sometimes difficult to trace the pieces up to the axes with which they were connected. The index of identity would assist, in most cases, in tracing their origin, but it compelled an examination in detail of each Piece. The difficulty was much more readily solved by taking advantage of different forms of alphabets. He had found the following six alphabets to be quite sufficient for the purpose :—The Egyptian, the Roman, and the writing capital letters, inclined in the usual direction, namely, from left to right :—

$$A, B, \ldots A, B, \ldots \mathscr{A}, \mathscr{B},$$

and the same alphabets with the letters inclined in the opposite direction.

These were to be used according to the following rules :—

1. In assigning letters to the axes represented by circles in any drawing, never to allow any two adjacent axes, nor any two axes whose arms, levers, &c., interfered with each other, to have letters of the same alphabet.

2. The capital letters thus possessed peculiar characters, arising from their belonging to the Egyptian, Roman, or writing alphabets, inclined in the usual or in the opposite direction. The rule adopted for lettering the Pieces belonging to or connected with each axis was this :—Whatever

might be the character of letter which had been used for the axis or slide, letters of the same character or alphabet must be used for every Piece whether fixed or loose upon it.

It sometimes occurred that a boss or other piece of matter had several different arms forming part of it, which were placed at different heights upon it. It might be desirable to distinguish these projections from each other, and also from the Piece itself as a whole.

When this was necessary, it might be effected by the two following rules :—

1. Whenever it was necessary to distinguish the arms or parts of a Piece which projected from it, the same capital letter with the same index of identity must be used. But then, to that letter on the right-hand lower corner, an index of linear position must be added.

2. The lowest projection, or that which was most remote from the eye, must be marked by the index 1, the next lowest by 2, &c.

Under similar circumstances, if the various arms were placed in different angular positions, then the indices must be placed in the right-hand upper corner, which was reserved for the index of circular position.

By the use of both indices, it was possible to distinguish the circular as well as the linear position of every arm of any Piece. These two indices were indeed rarely wanted, but he had found them very useful in describing an axis with projecting arms placed spirally, which occurred in several of his drawings, as part of a means of carrying the tens.

In any drawing lettered according to this system, it was possible to distinguish at a glance,— Frame-work from moveable Pieces ;—the outlines of every Piece ;—all the Pieces connected with each axis ;—the order of superposition of each Piece upon its own axis ;—the order, both in elevation and angular position, of all the arms belonging to each Piece ;—the working points which acted and reacted on each other.

Section II. Of Trains.

In this section of the Mechanical Notation, he proposed to show almost at a glance, all the connections between the first mover and final result, and also the action which every working Point throughout the whole machine either gave or received. In very simple machines it was not difficult to effect this object ; but when it was attempted, by means of the drawings, for any machine of moderate complexity, the impossibility soon became apparent. He had given the general term of Trains to the system of signs he had employed for this purpose. Pieces and working Points had already been defined. The former, as represented by inclined capital letters ; the latter, as denoted by small inclined letters, if belonging to Pieces, and small printed letters, if part of frame-work. He would give an illustration. Supposing that ^2B represented a Piece, and a, e, n, t, v, were its working Points ; that of these a and n were Points by which ^2B was driven,—e and t Points by which it drove or acted on other Pieces ; and that the working Point v sometimes acted as a driving and at other times as a driven Point : then the Piece ^2B might be represented thus :—

$$
\left. \begin{matrix} a \\ n \\ v \end{matrix} \right\} \ ^2B \ \left\{ \begin{matrix} e \\ t \\ v \end{matrix} \right.
$$

He had proposed the following rules for the construction of the Trains :—

1. The object of Trains was to show the courses through which power was received and transmitted by every Piece.

2. Each axis must be taken in succession as it occurred in the machine, or as represented on the drawing, by its proper inclined capital letter and its index of identity. This letter should then be placed between two inverted parentheses, thus :—

$$
\left. \right\} \ ^3P \ \left\{ \right.
$$

3. Every Piece belonging to this axis which received motion, must have its capital letter placed next the left-hand side of the first bracket.

4. Every Piece which gave motion must have its letter with its identity placed on the right-hand of the second bracket.

5. When such Pieces were both acted and acted upon, their letter, with its index, must appear in both brackets, thus :—

$$\left.{}^3B\right\rbrace \ {}^3P \left\lbrace \begin{array}{l} {}^3D \\[4pt] {}^3B \end{array} \right.$$

^3B might represent a bevel wheel fixed to the axis ^3P, by which that axis was driven. This bevel wheel itself also drove some other wheel. ^3D might represent a spur wheel, higher up on the axis than ^3B, which gave motion to some other Piece. In this case the two wheels and their axis having the same index of identity, must act as one piece of matter, and the expression would be more conveniently expressed thus :—

$${}^3B \ {}^3P \left\lbrace \begin{array}{l} {}^3B \\[4pt] {}^3D \end{array} \right.$$

If the identity of the spur and bevel wheels were different from that of the axis, as ^2D and ^2B, they would then form one piece of matter loose upon their axis ^3P. If the wheels themselves had different identities, then each, separately, would be loose on the axis; the two cases would be thus represented :—

$${}^2B \ ({}^3P) \left\lbrace \begin{array}{l} {}^2B \\[4pt] {}^2D \end{array} \right. \qquad {}^2B \ ({}^3P) \left\lbrace \begin{array}{l} {}^2B \\[4pt] {}^5D \end{array} \right.$$

The parentheses ( ) in which the axis was included was so placed in consequence of the following rule :—Whenever all the Pieces belonging to an axis were loose upon it, the axis itself might be enclosed in parentheses.

The next step was to attach their working Points to these expressions. For this purpose the following rules must be observed :—

1. All working Points which were acted upon or driven by others must be written on the left hand of the letter representing the Piece to which they belonged.

2. If any working Point was acted upon by several different working Points, it must appear on its proper side as often as might be required.

3. All working Points which acted upon or drove others must be placed on the right hand of the Piece to which they belonged.

4. In Trains it was not necessary to put the index of identity to each working Point, because its juxtaposition to its own Piece sufficiently indicated its identity.

Resuming the first case of the three Pieces having the same identity. Supposing that the bevel wheel ^3B received motion from two different sources, and communicated motion to one Piece; and also, that the spur wheel ^3D communicated motion to two other Pieces : then the small letters indicating the working Points must be thus placed :—

$$\left.\begin{array}{l} a \\[10pt] a \end{array}\right\rbrace {}^3B \ ({}^3P) \left\lbrace \begin{array}{l} {}^3B \ a \\[8pt] {}^3D \left\lbrace \begin{array}{l} v \\[4pt] v \end{array}\right. \end{array}\right.$$

The other two cases would be thus represented :—

$$a \, {}^2B \ ({}^3P) \left\lbrace \begin{array}{l} {}^2B \ a \\[10pt] {}^2D \left\lbrace \begin{array}{l} v \\[4pt] v \end{array}\right. \end{array}\right. \qquad \left.\begin{array}{l} a \\[6pt] a \end{array}\right\rbrace {}^2B \left.\begin{array}{l} \\[10pt] v \, {}^5D \end{array}\right\rbrace ({}^3P) \left\lbrace \begin{array}{l} {}^2B \ a \\[10pt] {}^5D \left\lbrace \begin{array}{l} v \\[4pt] v \end{array}\right. \end{array}\right.$$

When every axis, slide, &c., throughout a machine had been thus represented, the whole of these expressions, each of which was called a Term, could be combined. This must be done according to the following rules :—

1. To write down the letter which represented the first mover, with its proper identity and all its driving Points. To connect each of its driving Points by an arrow-headed line, with the Point it drove, that being a small letter the same as itself.

2. These small letters would each be found on a Piece connected with some adjacent axis. The term expressing that Piece should be written down in such a manner that the driving Point of the first mover should be joined with the proper driven Point of the Term.

3. Each Term so written down would have its driving Points, and must be treated in the same manner, until every Term had been placed on the paper.

It would then probably be found that several of the driven Points remained without a driver. On further examination, their drivers would be discovered, and arrows must be placed to connect them. When it was found that every driving Point throughout the machine had been connected by an arrow with the driven Points on which it acted, the work of constructing the Trains was accomplished. The paper, it was true would probably present a mass of letters irregularly grouped and connected together by curved arrow-headed lines, crossing each other in every direction. This might be considered as the first edition of the Trains. The next step was by altering the position of a few of the Terms, and consequently of the connecting arrows, to bring nearer to each other those Terms which were most intimately connected, and thus to diminish the number of intersecting arrows. Without attempting to carry this process too far, it was best to make a rough copy, perhaps in ink, with these improvements. The same process must be repeated on each edition, until a tolerably clear and symmetrical representation had been obtained. It had not been an uncommon event, in respect to some of the Trains he had made, that they had passed through above a dozen editions. He believed the intimate knowledge gained by such a tedious process was, however, invaluable in enabling the inventor to effect improvements.

It would have been observed that mechanical drawings were the pictures of things, and that by the aid of lettering, those drawings had been rendered, as it were, transparent. It would also have been noticed, that through the same system of lettering, the constant reference from plan to elevation, section, or end view, had been very greatly abridged. In constructing the Trains, the letters of the drawings had been used as the signs, not as the pictures of things. The consequence of this was, that in preparing the Trains, an almost constant reference became necessary to the drawings themselves, and thus much time was consumed. Moreover, when the Trains themselves were studied, it was often difficult to find a letter on the drawing, because the letter itself conveyed no notion of the form or shape of the Piece it represented. In order to remedy this inconvenience, he devised a new system of signs, which he had called the Alphabet of Form. It consisted of a series of certain very simple signs to be placed above the letters, in order to express the species, but not the exact shape or magnitude of the Pieces they represented. That Alphabet would be difficult to explain in print, until special type had been cast for it. It might, however, be sufficient to state that he had endeavoured to make each sign, directly or constructively, resemble the Piece it represented. Thus :—a small hollow circle above a letter, represented a hollow axis; a circular dot of the same magnitude, denoted a solid axis; $^4\overset{\circ}{N}$ was a hollow axis; $^3\overset{\bullet}{P}$ was a solid axis. Another source of delay arose from the frequent necessity of referring from the Trains to the drawings, in order to ascertain the nature of the motions communicated to each Piece. To remove this difficulty, he adopted another Alphabet, that of Motion, which consisted of about ten signs. These were placed under the letter which received the motion. When a Piece had two or more motions, it must have as many signs of motion under the capital letter representing it. In such cases, the letter representing each working Point must have under it the sign of that motion which was communicated through it to its own Piece. In other cases the signs of motion were rarely required for the working Points.

The Trains were specially intended to represent the transmission of motion; but by placing the signs of Form above the letters, frequent references to the drawings would be saved. By placing signs of motion below these letters, a further economy of reference would be produced.

A still greater power of explanation might be given to the Trains, by adding to them information relative to the times of action of the working Points, which would afterwards be found in the Cycles. Under the arrow connecting each pair of working Points, the numbers in the Cycle at which the driving Point began its actions must be written. Above the arrow, and immediately over each time of commencing action, the number of units during which that action continued must be written. With these additions, the Trains alone contain every information relating to the working parts of a machine, except the exact form and position which drawings were alone competent to afford.

Section III. Of Cycles.

The first section had explained the external form of machinery, the second had explained the whole sequence of its actions. In this, the third section, he proposed to represent every circumstance relating to the time of action of all its parts.

In the motion of machinery, two cases must be distinguished. 1st. The time during which a Piece moved. 2nd. The time during which it acted upon another Piece. An axis having arms and wheels attached to it might move continuously or at intervals, but it need not necessarily be giving motion to other Pieces during the whole of that time. On the other hand, a Piece at rest might be the cause of relative motion to an arm pressed against it. The times of motion belonged naturally to the Pieces themselves, or to the capital letters which represented them. The times of action were determined by the position of the working Points, therefore the signs which indicated them must refer to those Points. In almost every machine there was a Cycle of actions which recommenced and repeated themselves at certain fixed intervals. In the machine of Messrs. Scheütz, after the calculation of the tabular number and its differences had been completed and stereotyped, a similar process recommenced. The length of this cycle depended on the nature of the machine; it was usually measured by the number of rotations of some axis, or by the number of teeth in some wheel. Having decided into how many units of time it was convenient to divide this period or cycle, a number of horizontal parallel lines must be ruled, at equal distances, on a large sheet of paper. The number of spaces between the lines must be a few more than the number of units in the intended cycle. These horizontal lines must be crossed by other lines at right angles to them, thus forming vertical columns, which must be equal in number to the sum of all the working Points added to the number of all the Pieces in the Trains. The horizontal divisions of units must be numbered from the top downwards.

With the Trains for a guide, the number of driven Points possessed by the first Term, and also the number of Points by which it drove, must be ascertained; then, as many of the headings of the first columns, as there were driven Points, must be filled up with the small letters by which they were indicated. In the heading of the next columns, the principal capital letter with which those working Points were connected must be written, and as many vertical columns must be used for the Piece as it had separate kinds of motion. In the heading of the columns which immediately followed this capital letter, all the small letters indicating its driving Points must be written. The vertical lines at the commencement and end of these headings must be made darker than the intermediate lines, or they might be drawn with red ink.

Between these dark vertical lines there was now the principal letter of the first term of the Trains, with all its driven Points in columns on its left side, and all its driving Points on the right side, each being indicated by its own small letter. Every succeeding capital letter occurring in the Trains, with each of its driven and driving Points must, in succession, be inserted in the remaining columns. When two or more connected Pieces had different letters, with the same index of identity, it was not necessary to assign a vertical column to more than one of those capital letters. Commencing on the left-hand vertical column, just above the letters thus placed, the natural numbers must be successively written: these were necessary for reference. The verticals which represented each Term of the Trains must now be connected together, just above the numbers, by a horizontal bracket. Above this bracket the Term itself must be written, just as it appeared in the Trains. Commencing with the first mover, the number of units of time each Piece moved, must now be ascertained from the structure of the machine itself. In the vertical column headed by the capital letter indicating each Piece, a thick arrow must be drawn vertically through those units during which it moved. If the Piece itself sometimes moved in one direction, and sometimes in the opposite, this must be indicated by placing the heads of the arrows in opposite directions. As soon as the time of working of each Piece had been entered in its proper column, the time of action or of receiving action, of each working Point must be ascertained, and must be entered with a fainter arrow in its proper vertical column.

When this process has been gone through for every working Point and Piece in the machine, the Cycles were complete. All the possible successive motions of each would have been registered in the vertical columns. All the contemporaneous motions at any part of the Cycle might be immediately known by inspecting the horizontal divisions indicating the point of time for which that information was required. Above the headings of the vertical columns extracts from the Trains were placed, in order to exhibit the relative position of the larger divisions of the machine, and also to assist in directing attention to any required Point. The number of the principal letter of each term in the Cycles might be placed in red ink just above the term itself in the Trains, thus affording easy reference from each illustration to the other.

One of the greatest advantages which arose from the Mechanical Notation, was the power it gave of looking at any machine in its most general conception, or in its most minute detail, and even at any intermediate degree of generality.

Three different degrees of generality of the Trains of the Swedish machine, prepared by Mr. H. P. Babbage, were exhibited in the diagrams. In the first illustration there was a handle which showed, by two arrows proceeding from it, that the machine consisted essentially of two parts. One of these, a mangle-wheel, governed the calculating portion, and conveyed the computed table to certain steel punches. The other, by means of cams, governed the counting apparatus, conveyed the counted number to other steel punches, and then pressed a leaden plate up to the punches, which thus received a stereotype impression. Such, probably, would have been the first conception of the Swedish engine in the inventor's mind, and such was its shortest and most popular description. The third illustration of the Trains descended to the minutest information as to the directive power. The immediate cause of every movement throughout the whole cycle of its operations appeared on that paper. Every channel by which power was conveyed might be immediately traced out, either back to its original source, or onward to the ultimate result. The second illustration of the Trains was made merely to show that any intermediate degrees of generality thought desirable might be attained. The largest illustration of the Cycles was nearly five yards long. It contained the answer to every question relating to the time of action, or of motion, of about six hundred Pieces, and nearly two thousand working Points.

The three sections of the Mechanical Notation which he had now described were, in his opinion, sufficient for all the present wants of machinery. For the contrivance of the Analytical Engine they were, however, found insufficient. They would, perhaps, also be insufficient for working out the construction of certain automatic machines, capable of playing games of skill and chance, which the structure of the Analytical Engine led him to imagine. These new difficulties had been surmounted by other sections of the Mechanical Notation, to which he had alluded in addressing another Institution.* But he abstained from further remark upon them, because they had not that practical and every-day utility which so eminently characterized the discussions of the Institution of Civil Engineers. During a long use of this language, he had had opportunities of observing the facility with which it could be acquired. In one instance, having himself made the Trains, and also a rough sketch of an imaginary machine, he placed the Trains alone in the hands of two of his draughtsmen, and desired each, without communicating with the other, to make a pencil sketch of the machine. On comparing the three sketches, they were found to be identical in principle. Of course as no drawing had been given, the magnitudes and relative positions of the wheels, cranks, &c., were not the same; but the ultimate effect was, in all three cases, arrived at through the same chain of intermediate means.

As an illustration of the practical effect on the mind of these illustrations of the Swedish engine, he might mention the fact, that two gentlemen, both highly conversant with mechanical art, had, at separate times, each observed that, although he had never seen the Swedish machine, he thought he understood more of it from those illustrations, than he should have done by inspecting the engine itself. He (Mr. H. P. Babbage) attributed this to several concurring causes. The relations and subordination of the great sections of the instrument, as exhibited on those illustrations, were distinctly visible at a distance, whilst all the minuter details were indistinct. The consciousness that, on a closer examination, even the minutest action of each Piece, throughout the whole machine, would become immediately apparent.

Again, in examining a machine, it was scarcely ever possible to see all the parts and their connections at one view. The time of action of the various parts of the machine itself could only be observed in succession and during their motion; but the Mechanical Notation revealed even its most fleeting movements. It had, as it were, photographed the footsteps of time, and with power more enduring than electric fire, it had conferred fixity and permanence on the swiftest motion. In thus making public those laws which he had framed for the description of machinery by a new and universal language, he might be permitted to observe that, however desirable it might be in his opinion that the whole should at once be introduced, yet it was quite possible to divide the subject into at least four distinct stages :—

1st. The art of lettering drawings.
2nd. The Trains.
3rd. The Alphabets of Form and Motion.
4th. The Cycles.
Whatever course might be chosen, he left it with the fullest confidence in the hands of a Society whose object was essentially progressive.

* *Vide* " Sur la Machine de M. Scheutz." Académie des Sciences : Comptes-rendus. Tome xli. October 1855.

INSTITUT IMPÉRIAL DE FRANCE.

ACADÉMIE DES SCIENCES.

Extrait des *Comptes rendus des séances de l'Académie des Sciences*, tome XLI.,
séance du 8 octobre 1855.

NOTE

Sur la Machine Suédoise de MM. SCHUTZ *pour calculer les Tables mathématiques par la Méthode
des Différences, et en imprimer les résultats sur des planches stéréotypes ;*

Par M. CHARLES BABBAGE.

" Pour faire connaître plus exactement le mode de construction ou la théorie de l'admirable
instrument exposé maintenant par MM. Schutz dans le Palais de l'Industrie, j'ai l'honneur de mettre
sous les yeux de l'Académie, d'après le désir qui m'en a été exprimé par quelques Académiciens, une
série de tableaux graphiques. Ces tableaux, exécutés par mon fils M. Henry Prevost Babbage,
offrent une application à la machine suédoise du système de notation employé par moi, depuis un
assez grand nombre d'années déjà, pour la description et l'explication des machines.

" La *Notation mécanique* comprend plusieurs sections dont voici les principales :

" 1°. Les principes de l'emploi des lettres sur les dessins des machines ;

" 2°. L'alphabet des formes ;

" 3°. L'alphabet des mouvements ;

" 4°. Les trains ;

" 5°. Les cycles ;

" 6°. Les notations d'opération ;

" 7°. Les notations analytiques.

" La forme exacte d'une machine se connaît en général par la vue des dessins. Dans un grand
nombre de circonstances néanmoins, ce moyen manque de la précision désirable.

" Jusqu'ici l'emploi des lettres marquant les différentes parties d'un dessin a été presque
entièrement arbitraire. Dans la première section de la *Notation mécanique* sont posées les règles
d'après lesquelles les lettres elles-mêmes deviennent directement des signes, et donnent des indications
importantes. Il est alors presque inutile de se reporter du plan à l'élévation. Voici quelques-unes
de ces règles :

" Les bâtis ou supports (frame-work) s'expriment par des capitales droites ou des minuscules
droites ; les pièces mobiles par des capitales penchées ou par des minuscules penchées.

Frame-work.	*Pièces mobiles.*
A, B, C, a, b, c,	*A, B, C. a, b, c.*

" L'avantage de cette règle, c'est que si vous examinez un dessin dans un but d'étude, pour
vous rendre compte de la destination ou du jeu de tel ou tel organe ; au lieu de fatiguer votre
attention, vous pouvez faire entièrement abstraction de toutes les parties qui se rapportent seulement
au *frame-work.*

" Pour comprendre parfaitement une machine, il est indispensable d'en connaître avec exactitude :

" 1°. La forme ;

" 2°. Les différents récepteurs par lesquels les mouvements sont transmis depuis le premier
moteur jusqu'au résultat final : c'est ce que j'appelle les *trains ;*

" 3°. Pour les diverses parties de la machine, le temps précis où commence le mouvement, et la durée de ce mouvement : c'est là le *cycle*.

" Une machine est composée de pièces fixes et de pièces mobiles, de pièces qui impriment le mouvement ou qui le reçoivent. Chacune de ces pièces peut être considérée isolément, c'est-à-dire en elle-même. Les points ou les surfaces par lesquels chaque pièce imprime le mouvement ou le reçoit, sont les parties principales de cette pièce. Ces points s'appellent *opérateurs* ou *working-points*.—De là une distinction importante à observer dans le choix des lettres.

" Pour marquer les rapports des *points opérateurs* de chaque pièce avec les grandes lettres qui les représentent, on a donné à chaque grande lettre un indice supérieur, du côté gauche. C'est l'*indice d'identité*.

" Les petites lettres représentant un des points opérateurs sont marquées respectivement du même signe d'identité ; ainsi *a*, *b*, *c*, etc., sur un dessin, appartiennent à une lettre capitale adjacente, ^3D par exemple, seront écrits : ^3a, ^3b, ^3c, etc.

" Les axes, comme les pièces glissantes, pourront être désignés par les grandes lettres de cinq ou six alphabets différents. Etrusque, **A, B, C** ; romain, *A, B, C* ; italique, $\mathscr{A}$, $\mathscr{B}$, $\mathscr{C}$, etc.

" Mais quand l'alphabet a été déterminé, les pièces quelconques, soit fixes, soit mobiles, se rapportant à cet axe, doivent être représentées par une lettre capitale de l'alphabet attribué à l'axe lui-même.

" A l'aide de ce moyen, on peut distinguer aisément, sur le dessin le plus chargé, l'axe auquel telle ou telle pièce appartient. En effet, ces caractères de l'alphabet affectés aux axes doivent être choisis de telle sorte que deux axes dont les pièces se croisent ne soient jamais marqués du même caractère.

" L'*exécution des dessins de la machine analytique* n'a présenté que très-peu de cas par lesquels il eût été désirable de se servir de quatre ou même d'un plus grand nombre des alphabets différents, ou même d'un plus grand nombre. En réalité, il n'en a pas été employé plus de trois.

" La position relative des pièces, soit fixes, soit mobiles, appartenant à des axes ou à des pièces glissantes, peut toujours se voir sur chaque projection si l'on adopte la règle ci-après.

" Commencer à noter les lettres sur la pièce la plus éloignée appartenant à l'axe, en prenant une lettre convenable ; ensuite, et par ordre de proximité, marquer chaque autre pièce des lettres qui, dans l'alphabet, viennent plus bas qu'aucune de celles dont on s'est servi pour les parties précédentes.

" Voici, en peu de mots, l'application de ce système de notation à la description de la machine suédoise.

" Dans la série des tableaux graphiques exposés maintenant sous les yeux de l'Académie, les seules sections du système de notation employées par M. Henry P. Babbage sont les *Lettres*, les *Trains*, les *Cycles*.

" Le principal tableau A, long de plus de 4 mètres, est relatif au cycle. Il contient environ 700 colonnes verticales. Le *temps du mouvement* de chaque pièce, et le *temps de l'action* de chacun de ses *points opérateurs*, peut immédiatement se voir avec la plus grande facilité.

" Les lettres rouges, au haut des colonnes, marquent la liaison des *cycles* avec les *trains*.

" Le *cycle*, ou la période de la machine, emploie 96 unités de temps. Pendant le cycle, un calcul se trouve complétement exécuté et imprimé.

" Sur le tableaux B sont distinctement indiquées, au nombre de plusieurs centaines, toutes les pièces mobiles de la machine.

" Chacune de ces piéces montre ces *points opérateurs*. On voit de quelle manière elle reçoit les mouvements, et comment elle les transmet à d'autres pièces.

" Le signe de la manivelle, à gauche des *trains*, est le point de départ du mouvement. Après s'être subdivisé, et recomposé sous une grande variété de formes, ce mouvement vient s'arrêter au côté droit des trains par le résultat collectif du jeu de toutes les parties de la machine.

" Une série de poinçons d'acier portant des caractères numériques disposés par le mécanisme d'après la loi de la table qu'il s'agit de calculer, se présente successivement, ligne par ligne, au-dessus d'une plaque de plomb. Cette plaque, après chaque opération, vient se presser contre les poinçons et reçoit ainsi, sans possibilité d'erreur, l'empreinte des chiffres qui sont le résultat définitif du calcul."

OBSERVATIONS ADDRESSED, AT THE LAST ANNIVERSARY, TO THE PRESIDENT AND FELLOWS OF THE ROYAL SOCIETY, AFTER THE DELIVERY OF THE MEDALS. By Charles Babbage, Esq., F.R.S.—*November* 30, 1855.

My Lord Wrottesley,—I beg leave to offer a few observations on the distribution of our Medals, but not with the intention of finding fault with their present allotment.

The distinguished foreigner whose valuable discoveries you have so ably explained to us is fully entitled to a Copley Medal. I join also most cordially in the justice of the award of the first Royal Medal to that eminent astronomer who has organized a system for the discovery of new planets, and who has himself already added ten to their number. With the researches rewarded by the second Royal Medal I am entirely unacquainted; but I am willing to assume that they have been duly considered and justly rewarded.

There is, however, an instrument to which we have given hospitality during many months in these apartments, which I think highly deserving of a medal; and I had hoped that on the present occasion it might at least have been considered worthy of being placed amongst the list of candidates for that honour. I allude to the admirable machine for Calculating and Printing Tables by Differences, and producing a mould for the stereotype plates to print the computed results—an instrument we owe to the genius and persevering labour of Mr. Scheutz, of Stockholm. A Committee of the Royal Society has already reported upon the machine, and I can myself bear testimony to the care and attention which our Secretary bestowed upon that valuable report. But as some misapprehension exists in the public mind respecting the originality displayed in that invention, I trust that having, as is well known, given much attention to the subject, I may be permitted briefly to explain some of its principles, and thus render justice to its author.

The principle of Calculation by Differences is common to Mr. Scheutz's engine and to my own, and is so obviously the only principle, at once extensive in its grasp and simple in its mechanical application, that I have little doubt it will be found to have been suggested by more than one antecedent writer.

Mr. Scheutz's engine consists of two parts—the Calculating and the Printing; the former being again divided into two—the Adding and the Carrying parts.

With respect to the Adding, its structure is entirely different from my own, nor does it even resemble any one of those in my drawings.

The very ingenious mechanism for carrying the tens is also quite different from my own.

The Printing part will, on inspection, be pronounced altogether unlike that represented in my drawings; which, it must also be remembered, were entirely unknown to Mr. Scheutz.

The contrivance by which the computed results are conveyed to the printing apparatus, is the same in both our engines: and it is well known, in the striking part of the common eight-day clock which is called "the snail."

About 1834 or 1836, Mr. Scheutz, himself a member of no Academy, a Professor at no University—but simply an eminent printer at Stockholm, first learnt, through the "Edinburgh Review," the existence of that Difference-Engine, a small portion of which is now placed in one of the rooms of the adjoining building.*

Unfortunately for himself, Mr. Scheutz was fascinated by the subject, and impelled by an irresistible desire to construct an engine for the same purposes. He has always avowed, in the most open and honourable manner, the origin of his idea. But his finished work contains undoubted proofs of great originality, and shows that little beyond the principle could have been borrowed from my previous work. Having formed the project, Mr. Scheutz immediately began to work upon it. After four years of labour and difficulties, which cost him a large portion of his fortune, he produced the first model. This, however, did not satisfy his wishes: but far from being disheartened, he immediately recommenced his experiments with renewed energy, expending on them all the remaining savings of an industrious life, as well as the whole of the time he could snatch from the labours on which the support of his family depended.

His son also, after completing his studies with great credit at the Technological School of

---

* In the Museum of King's College.

Stockholm, was anxious to assist his father in this difficult task; and for that purpose abandoned the career he had previously chosen.

The father and son now worked together for several years, and at last produced a machine, in which were united all the requisite conditions of a Difference-Engine. But the severe economy they had been compelled to use, in the purchase of materials and tools, and probably the absence in Sweden of those precious but expensive machine-tools, which constitute the power of modern workshops, rendered this new model unsatisfactory in its operations, although perfectly correct in principle.

Exhausted by the sacrifices thus made, yet convinced that with better workmanship a more perfect instrument was within their reach, Mr. Scheutz determined to apply for assistance to the Diet of Sweden.

The Diet with difficulty consented to advance 5,000 rix-dollars (about £280), on condition that the new machine should be completed within a year, and that the Messrs. Scheutz should give a guarantee to return that sum to the State if the machine did not fully attain the objects proposed.

To the already exhausted funds of Messrs. Scheutz, this *guarantee* became a greater difficulty than the construction of the machine: they therefore felt compelled to renounce the work. Thus would have ended, unknown and unappreciated, the vast exertions of two men of highly cultivated understanding, whose truth and simplicity of character had been amply tested by the severest labour, by the greatest sacrifices.

Fortunately, however, amongst the Professors of the Academy of Stockholm enlightened men were found, capable of sympathizing with moral and intellectual worth. To the enduring honour of the Swedish Academy a numerous list was soon formed, in which each name became responsible for that part of the amount annexed to it, and thus the State was secured from any possible loss.

Although the very limited amount thus raised was inadequate, the Messrs. Scheutz, confident in ultimate success, pledged their own credit for the further necessary advances, and after working night and day, with indefatigable industry, the last day of the allotted year saw the completion of their long-cherished hopes.

The Diet, though at first unfavourable to the invention, now granted a reward of 5,000 rix-dollars to the inventors; thus raising their total grant to 10,000 rix-dollars (about £560).

A glance at this machine will convince any competent judge that this sum must be very far from replacing the mere money expended, during a period of almost twenty years, in its contrivance and construction. But Sweden has thus secured for herself the glory of having been the first nation practically to produce a machine for Calculating Mathematical Tables by Differences, and Printing the Results. Wealthier and more powerful nations will regret that the country of Berzelius should thus have anticipated them, in giving effect to an invention which requires for its perfection the tools of nations more highly advanced in mechanical science. But there is still left to them the honour of acknowledging the services of a foreigner, from which the richest and most commercial countries will derive the greatest advantage.

The machine was conveyed to Paris, and placed in the Great Exposition. The jury to which it was referred contained many distinguished names, amongst them that of M. Mathieu, Member of the Institute, who having been for a long period entrusted by the Academy of Sciences with the arduous duty of reporting upon the numerous Calculating Machines submitted to that learned body, was already familiar with the history of the past. Availing himself of all the printed documents, relating to former Difference-Engines, and studying those latest illustrations of Mr. Scheutz's Machine, which had rendered visible to the eye, in one unbroken chain, the whole sequence of its minutest movements,* this eminent astronomer was in a position to pronounce with authority on the merit of the Swedish engine. That jury, after full examination, concurred with their distinguished colleague in *unanimously* awarding to it the Gold Medal.

The Emperor Napoleon, true to the inspirations of his own genius and to the policy of his dynasty, caused the Swedish engine to be deposited in the Imperial Observatory of Paris, and to be placed at the disposal of the Members of the Board of Longitude.†

Your Lordship is aware that previously to awarding any of our Medals, each Member of the Council may place one or more names on the list of candidates whose claims are to be discussed. I regret that (perhaps through inadvertence) the name of Mr. Scheutz was not placed upon that list, and I cannot, my Lord, sit down without expressing a hope that the Council of the ensuing year may more than repair the omission.

* These illustrations were made by my son, Mr. Henry Babbage, an officer of the Indian army, now on furlough in England. They consist of the complete "*Mechanical Notations*" of the Swedish machine, and were exhibited at the Meeting of the British Association at Glasgow, and afterwards sent to Paris for the use of the jury to whom that machine was referred.—See *Phil. Trans.* 1825, and *Comptes Rendus*, Oct. 8th, 1855, vol. xli.

† This fact was not stated at the Meeting, as it had not then reached the Author in an authentic form.

LETTER FROM M. MÉNABRÉA TO THE EDITOR OF "COSMOS."

T. vi., *Avril* 20, 1855.

VARIÉTIÉS.

" L'insertion dans le *Cosmos* du discours du comte Rosse, où le noble président de la Societé Royale plaidait avec tant de chaleur la cause de la machine à calcul de M. Babbage, nous a valu l'honneur d'une lettre de M. Ménabréa, savant italien très-distingué, qui s'est aussi occupé de cette question. Voici la lettre du mathématicien Turinais :

" En parcourant la livraison du 6 avril courant de votre intéressant journal le *Cosmos,* j'ai appris avec un vif plaisir que l'on s'occupait sérieusement en Angleterre de mettre à exécution la *Machine analytique* de M. Charles Babbage. Dans le discours où M. le comte Rosse rend compte des démarches qui ont déjà été faites dans ce but, j'ai surtout remarqué le passage suivant : 'Avant de prendre une détermination et de me mettre en campagne, j'écrivis à plusieurs hommes éminents par leur savoir, et leur demandai si, dans leur opinion, on aurait fait un grand pas au point de vue de la science théorique et pratique, si les vues de M. Babbage, telles qu'elles sont exposées dans le petit essai publié par lui, sous le titre de *Ménabréa,* étaient complétement réalisées. Leurs réponses furent unanimement et fortement affirmatives.'

" Comme je ne suis pas entièrement étranger à cette question, permettez-moi, Monsieur, de vous communiquer quelques détails qui ne seront pas sans intérêt pour les personnes qui s'occupent de science.—Je suis bien réellement l'auteur du petit écrit que M. le comte Rosse attribue à M. Babbage, sous le pseudonyme de Ménabréa. Il y a bien des années, M. Babbage, lors d'un voyage qu'il fit en Italie, s'arrêta quelque temps à Turin, où il eut la bonté de m'expliquer les dispositions principales de sa *machine analytique,* qui diffère essentiellement de la *machine aux différences* déjà connue du public. A dire vrai, le problème que s'était proposé M. Babbage était tellement singulier, que mon premier sentiment fut celui du doute. Mais en y réfléchissant, je parvins à me convaincre que l'idée de cet illustre savant était parfaitement rationnelle et, du consentement de l'auteur, je pris le parti de faire connaître les principes fondamentaux sur lesquels repose l'étonnant instrument dont il s'agit, dans un article qui fut publié dans la *Bibliothèque universelle de Genève,* n° 82, octobre 1842, page 352 et suivantes.

" Quelque temps après, parut une traduction anglaise de ce même écrit, intitulée : *Sketch of the Analytical Engine, invented by Charles Babbage, Esq., by L. F. Ménabréa of Turin, officer of the Military Engineers,* WITH NOTES BY THE TRANSLATOR. (*Extracted from the Scientific Memoirs, vol.* III. —London, *printed by Richard and John E. Taylor, Red Lion-court Fleet-street,* 1843.) Les notes qui accompagnaient la traduction de mon petit mémoire étaient extrêmement remarquables et annonçaient, dans leur auteur, une sagacité peu ordinaire.—J'ignorais le nom de cet auteur lorsque, à mon grand étonnement, j'appris de M. Babbage lui-même que la traduction et les notes étaient l'ouvrage de lady Ada Lovelace, de la fille de lord Byron, dame aussi distinguée par l'élévation de son esprit que remarquable par sa beauté, et que la mort a enlevée, il y a peu d'années, dans l'âge le plus brillant de la vie.

" Je croyais cet écrit oublié ; mais, puisqu'il vient de servir de point de départ aux démarches qui ont été faites auprès du gouvernement anglais pour réaliser l'invention de M. Babbage, j'ai cru devoir saisir cette occasion pour rendre hommage à la mémoire de la fille du grand poëte et attirer l'attention des savants, non pas sur mon ouvrage qui est bien modeste, mais sur les notes et commentaires qui en accompagnent la traduction et qui sont de nature à faire connaître le but et la puissance de la *Machine analytique.* Cette invention ne ressemble à rien de ce qui a été imaginé jusqu'à ce jour, et je ne pourrais, dans les limites d'une simple lettre, en faire connaître les principes ; mais qu'il suffise de dire que, par son moyen, on pourrait exécuter la série des opérations analytiques et numériques qu'exige la solution d'un problème déterminé, de la même manière que, dans le métier à la *Jacquard,* par l'emploi des *cartons,* on exécute les dessins d'une étoffe brochée.

"Agréez, Monsieur le directeur, l'expression de la considération très-distinguée avec laquelle j'ai l'honneur d'être.

"L. F. MÉNABRÉA,

"Membre de l'Académie R. des sciences et du Parlement sarde."

REPORT OF A COMMITTEE APPOINTED BY THE COUNCIL
TO EXAMINE THE CALCULATING MACHINE OF M. SCHEUTZ.

Inserted for the information of the Fellows by order of the President and Council.*

[*From the Proceedings of the Royal Society, January* 21, 1855.]

THE various applications of mathematics to physical questions, or to the transactions of common life, continually require the computation of numerical results. At one time isolated results have to be calculated from particular formulæ; at another it is required to calculate a series of values of the same analytical formula; in other words, to tabulate a function. It is only in the latter case that different instances have so much in common as to permit of the application of general methods irrespective of the particular function to be calculated. But even in the tabulating of functions one or other of two objects may be kept in view. At one time a result may be arrived at expressed in complicated, perhaps transcendental, formula, and the mathematician may desire to know merely the general progress of the function. In such a case it will be sufficient to calculate values at rather wide intervals, and the mode of calculation must depend upon the peculiar function. But at other times functions present themselves which are of such common occurrence, or of such practical importance, that it is desirable to tabulate them for the values of the variable, increasing by small steps. In these cases general methods of interpolation come into use; it is sufficient to perform the calculations directly for comparatively wide intervals of the variable, and the intervening values of the function can be supplied by the mere addition of differences.

It is well known that Mr. Babbage was the first person who conceived the idea of performing all these systems of additions mechanically, and thereby saving both the mental labour and the risk of error attending their calculation in the ordinary way. This idea was actually carried out, and resulted in the invention of his Difference Engine. The engine, so far as it has yet been executed, was constructed at the public expense, and is now deposited in the Museum of King's College, London. The part constructed contains 19 digits and 3 orders of differences; and as all the essential movements are comprised in this part, a more extended engine would consist merely of the same members oftener repeated, and would not involve any additional difficulty of construction. It was part of Mr. Babbage's original design that machinery for printing off the results calculated should be included in his engine, and some of the mechanism for this purpose was actually executed. The portion placed in King's College contains machinery for calculating only. It does not fall within the province of this report to do more than mention the Analytical Engine subsequently invented by Mr. Babbage, as the machine of M. Scheutz is a Difference Engine, and nothing more.

A full account of the principles and action of Mr. Babbage's Difference Engine, but without any details of its mechanism, was published in the "Edinburgh Review" for April to July, 1834. It was, as we are informed, the perusal of this paper which induced M. Scheutz to set about the invention of modes of mechanically executing the necessary changes. The result was the completion of the present engine, which has now for some time been in the apartments of the Royal Society. In this machine M. Scheutz has followed the general ideas of Mr. Babbage in the distribution of digits and differences, and in particular in throwing back the differences at every alternate order one stage, from whence results the possibility of acting simultaneously on all the odd and on all the even differences, and thereby making the machine advance one stage by two addition-motions only; whereas otherwise as many separate addition-motions would have been necessary as there were orders of differences retained. But the mechanism by which the additions and carriages are effected

---

* The Committee consisted of Prof. Stokes, Sec. R.S., Prof. W. H. Miller, Prof. Wheatstone, and the Rev. Prof. Willis.

in the machine of M. Scheutz is different from that of Mr. Babbage. The engine is also provided with mechanism for printing, or rather for furnishing stereotype plates of the calculated results.

As M. Scheutz has taken out a patent for his engine, it will be unnecessary to give a detailed description of the machinery, which may be obtained in the specification, a copy of which has been presented to the Royal Society. It will be sufficient to give an idea of its general construction and extent with a view of estimating its powers.

The machine takes in the function to be tabulated and the first four orders of differences, each to fifteen digits. Of these only the first eight (in the case of the function itself) are printed, the others being reserved to guard against errors arising from decimal places left out.

The places of the digits are represented by fifteen vertical spindles, around which, but not usually connected with which, are placed horizontal wheels in five separate tiers. Each wheel has its circumference divided into ten equal parts, and is marked with the digits 0, 1, 2, 3, 4, 5, 6, 7, 8, 9. In the normal state of the machine the numbers on the wheels of the highest tier represent the function (u_x) to be tabulated, and those on the tiers below represent respectively Δu_{x-1}, $\Delta^2 u_{x-1}$, $\Delta^3 u_{x-2}$, and $\Delta^4 u_{x-2}$. In each case the digits by which these numbers are represented run from left to right, as in print. The mechanism is such, that by turning a handle continuously in one direction an indefinite succession of movements is produced which are alternately backwards and forwards. The effect of the forward motion is, that the numbers on the third and fifth tiers (or as they may conveniently be called Δ^2 and Δ^4 tiers) add themselves respectively to those on the tier above, altering thereby the positions of the wheels of the Δ^1 and Δ^3 tiers, while the wheels of the Δ^2 and Δ^4 tiers remain at rest; and the backward motion does for the Δ^1 and Δ^3 tiers what the forward motion does for the Δ^2 and Δ^4 tiers. Thus the numbers on the several tiers will be as follows:—

At first	u_x	Δu_{x-1}	$\Delta^2 u_{x-1}$	$\Delta^3 u_{x-2}$	$\Delta^4 u_{x-2}$;
After the forward motion . .	u_x	Δu_x	$\Delta^2 u_{x-1}$	$\Delta^3 u_{x-1}$	$\Delta^4 u_{x-2}$;
After the complete motion . .	u_{x+1}	Δu_x	$\Delta^2 u_x$	$\Delta^3 u_{x-1}$	$\Delta^4 u_{x-1}$,

$\Delta^4 u_{x-1}$ in the last term being written instead of $\Delta^4 u_{x-2}$, which is allowable, since the fourth differences are supposed to be constant. Hence the effect of the complete motion, consisting of one forward and one backward motion, is to make all the numbers advance one stage; and therefore by continuing to turn the handle the numbers u_{x+1}, u_{x+2}, u_{x+3}, &c., will be calculated in succession. According as these numbers are calculated they are impressed, by the action of the machine itself, on a plate of lead, by means of steel punches, while a numerator at the same time impresses beside them the values of the argument x. These plates are afterwards taken out, and stamped on an easily fusible alloy just on the point of solidifying, and thus are obtained stereotype plates of the calculated results, fit for printing from.

In retaining a given number of decimals, it is usual to add one to the last figure if the first digit left out be 5 or a higher number. This is effected in the machine in the simplest possible manner, namely, by placing the cog which occasions the carriages from the ninth to the eighth place in the highest tier in such a position that the carriage takes place when the ninth wheel changes from 4 to 5, instead of from 9 to 0.

The principle of the machine is not of course dependent upon the circumstance that the radix of the scale of notation commonly employed has the particular value 10; and it would be as easy to construct a machine adapted to the senary or duodenary as to the denary scale. Not only so, but the machine actually constructed admits of being changed very readily from the denary to the senary scale, or rather to a mixture of the denary and senary scales, which is required in tabulating degrees, minutes, and seconds. For this purpose it is sufficient to take off the ordinary figure-wheels from those spindles which are to count by sixes, and put on spare wheels which are provided, adapted to the senary scale.

The machine works with the greatest freedom and smoothness. The parts move with the utmost facility, in fact, quite loosely. On this account no amount of dust which it would reasonably be expected to receive in any moderate time seems likely to interfere with its action. Besides, it can easily be taken to pieces and examined, if need be. Those motions which are not the direct consequences of the revolution of the handle acting through a train of rigid bodies are performed in consequence of gravity, no springs being employed in the whole construction except two, the office of which is quite subordinate. When the parts are moved, they remain in their new places either from their weight or from friction, there being nothing to disturb them. This circumstance, which renders a wilful derangement of the machine exceedingly easy, permits of great simplicity and consequent cheapness of construction; nor does the machine seem likely to get out of order if reasonable care be taken of it.

The machine is competent to tabulate to any extent a function whose fourth differences are constant, so long as the expression of the numerical value of the function does not involve more than eight digits. The most general form of such a function is of course

$$a + bx + cx^2 + dx^3 \qquad \cdots \qquad \cdots \qquad (1)$$

Were the machine restricted to such functions, its use would be limited indeed; its utility must of course depend upon its being applicable to functions in general, which, except in singular cases, may be expressed within a limited range of values of the variable x by a function of the above form. To estimate the capacities of the machine, or rather of a difference engine in general, whatever may be its particular construction, it will be necessary to investigate how soon the quantities neglected begin to tell in the result.

Now these quantities are of two kinds; first, the fifth and higher differences; secondly, the decimals of the fifteenth place. The effect of these may be examined separately. We may always suppose the first spindle to represent the first place of decimals, since it will only be necessary to multiply or divide by some power of 10 should that not be the case.

Suppose the machine set for u_x, and its first four differences (or to speak more exactly, the differences Δu_{x-1}, $\Delta^2 u_{x-1}$, $\Delta^3 u_{x-2}$, $\Delta^4 u_{x-2}$), and worked n periods, so as to give what ought to be u_{x+n}. We have

$$u_{x+n} = u_x + \frac{n}{1}\Delta u_x + \frac{n.\overline{n-1}}{1.2}\Delta^2 u_x + \qquad \cdots \qquad \cdots \qquad (2);$$

and since the machine would give u_{x+n} exactly if the fourth differences were constant, the error (E) will be

$$\frac{n.\overline{n-1}.\overline{n-2}.\overline{n-3}.\overline{n-4}}{1.2.3.4.5}\Delta^5 u_x + \frac{n.\overline{n-1}.\overline{n-2}.\overline{n-3}.\overline{n-4}.\overline{n-5}}{1.2.3.4.5.6}\Delta^6 u_x + \ldots *$$

The first term in this expression will usually be the most important; and for practical purposes the expression may be still further simplified. If n be tolerably large, the product $n.\overline{n-1}.\overline{n-2}.\overline{n-3}.\overline{n-4}$ may be replaced without material error by the fifth power of the arithmetic mean of the factors, or by $(n-2)^5$. Again if y be the variable of which u is a function, x being merely the numeral marking the number of increments of y, each equal to k, we shall have, near enough:

$$\Delta^5 u_x = k^5 \frac{d^5 u}{dy^5}$$

so that

$$E = \frac{1}{120}(n-2)^5 k^5 \frac{d^5 u}{dy^5}.$$

In expressing a number to eight decimal places, we are always liable to an error which may amount to 5 in the ninth place. Hence $10^{-9} \times 5$ may be regarded as the greatest allowable error, though in truth the error should not be allowed to amount to this, if we wish to have the last figure true to the nearest decimal. Equating then E to $10^{-9} \times 5$, we find

$$n = 2 + \left(\frac{\cdot 0000006}{\dfrac{d^5 u}{dx^5}} \right)^{\frac{1}{5}} \cdot \frac{1}{k}, \qquad \cdots \qquad \cdots \qquad (3)$$

which gives the greatest number n of times the machine may be worked without stopping and fresh setting, so far as the limitation depends on the cause of error now under consideration. The increment of y during the action of the machine, which is equal to nk, or to $(n-2)k$ nearly, n being large compared with 2, is therefore nearly independent of the closeness or wideness of the intervals for which the value of the function is required, a given range, so to speak, of the function being taken in. Hence, so far as this cause of limitation is concerned, the utility of the machine will be proportional to the closeness of the intervals for which it is desired to tabulate the function.

Let us now consider the effect of the decimals omitted, retaining only four orders of differences, since the effect of omitting the fifth and higher orders has been already investigated. Let E_1, E_2, E_3, E_4, be the errors left in the first, second third, and fourth differences in setting the machine. Then in the same manner as before these may without sensible error be regarded as the errors in

---

* This expression will not be absolutely exact, since it is Δu_{x-1}, $\Delta^2 u_{x-1}$, $\Delta^3 u_{x-2}$, $\Delta^4 u_{x-2}$, and not Δu_x, $\Delta^2 u_x$, $\Delta^3 u_x$, $\Delta^4 u_x$ that are given correctly; but the inaccuracy thus arising in the estimation of the error committed by leaving out the fifth, &c., differences will plainly be insignificant.

Δu_x, $\Delta^2 u_x$, $\Delta^3 u_x$ $\Delta^4 u_x$, although they are really the errors in Δu_{x-1}, &c., and we shall have for the error (E) in u_{x+n}

$$E = \frac{n}{1}\mathrm{E}_1 + \frac{n\cdot\overline{n-1}}{1.2}\mathrm{E}_2 + \frac{n\cdot\overline{n-1}\cdot\overline{n-2}}{1.2.3}\mathrm{E}_3 + \frac{n\cdot\overline{n-1}\cdot\overline{n-2}\cdot\overline{n-3}}{1.2.3.4}\mathrm{E}_4 ;$$

or, replacing the products as before,

$$E = n\mathrm{E}_1 + \frac{1}{2}\left(n-\frac{1}{2}\right)^2\mathrm{E}_2 + \frac{1}{6}(n-1)\,^3\mathrm{E}_3 + \frac{1}{24}\left(n-\frac{3}{2}\right)^4\mathrm{E}_4 \quad . \quad . \quad . \quad (4)$$

If each of the quantities E_1, E_2, E_3, E_4, be liable to be as great as $10^{-16}\times 5$, the last term in this expression will be the most important if n be considerably greater than 4. Equating this term to $10^{-9}\times 5$, the greatest allowable error in E, we find

$$n - \frac{3}{2} = (24 \times 10^7)^{\frac{1}{4}}, \qquad n = 126 \text{ nearly,}$$

so that the machine may be worked about 100 times without fresh setting.

In practice the limitation may be even less than this; for it may happen that $\Delta^4 u_x$ is smaller, perhaps much smaller, than $10^{-16}\times 5$, in which case the limitation will depend on the absolute value of $\Delta^4 u_x$ or the possible value $10^{-16}\times 5$ of E_3, as the case may be. Should the restriction arise from the latter cause, we get by equating the third term in the second member of (4) $10^{-9}\times 5$, $n = 392$ nearly.

To illustrate these limitations by an example, suppose that it was required to make a table of sines to every minute. In this case we have

$$u = \sin y, \quad k = \frac{\pi}{180\times 60} = \cdot0002909, \quad \frac{d^5 u}{dy^5} = \cos y.$$

Putting for this last differential coefficient its greatest value, unity, and substituting in (3), we get $n = 196$ nearly. The fourth difference is very nearly equal to $-k^4 \sin y$, which may contain figures in the fifteenth place, so that $n = 126$ is about the greatest allowable value of n in consequence of the restriction arising from decimals left out, which in this example is what limits the working. Should the intervals be a good deal wider than 1', as 5', it would then be the omission of fifth differences that would impose the limit, for the greatest allowable range on this account would be nearly the same as before, or about $3°$, which would contain only thirty-six values to be calculated. Should it happen that both causes of error were about equally restrictive, it must be remembered that the corresponding errors in u_x would be comparable with one another, and might be added together; and in this case it may easily be shown that $126 \times 2^{-\frac{1}{4}}$, or 106 nearly, is somewhat inferior to the greatest allowable value of n. Should eight figures not be required to be retained, but seven, six, or five be sufficient, the last one, two, or three of the first eight spindles might be used for calculating instead of printing: and since the greatest allowable value of n, so far as depends on omission of decimals, varies nearly as the fourth root of the greatest allowable error in u_x, that value would be increased in the ratio of 1 to the fourth root of 0, or 100, or 1000, and from 126 would become 224, or 398, or 708. The greatest allowable value of n as regards the omission of fifth differences would increase in a somewhat slower ratio, since it varies nearly as the fifth root of the greatest allowable error in u_x. If, for example, it were 196, it would become 311, or 492, or 780.

The above is a fair specimen of the application of the machine. The particular function chosen is, it is true, a familiar one, which has been long since tabulated, but it is not the worse fitted for an example on that account. It may be seen at once how much mental labour and risk of error is saved by the use of such a machine, when tables have to be calculated to close intervals. The whole exertion of mind is confined to calculating the function and its differences at wide intervals, say for every 100th or 60th number to be tabulated, and setting the machine. Even this exertion (except so far as relates to the setting, which is easy), might be reduced to one-half, if desired, by setting the machine to calculate backwards as well as forwards. In order to give in succession the numbers u_x, u_{x+1}, u_{x+2}, ... the machine has to be set to

$$u_x \quad \Delta u_{x-1} \quad \Delta^2 u_{x-1} \quad \Delta^3 u_{x-2} \quad \Delta^4 u_{x-2},$$

or to

$$u_x \quad \Delta\mathrm{D}^{-1}u_x \quad \Delta^2\mathrm{D}^{-1}u_x \quad \Delta^3\mathrm{D}^{-2}u_x \quad \Delta^4\mathrm{D}^{-2}u_x,$$

D denoting as usual the operation $1+\Delta$. In order to give in succession the numbers, u_x, u_{x-1}, u_{x-2}, .. the machine would simply have to be set to

$$u_x \quad \Delta'\mathrm{D}'^{-1}u_x \quad \Delta'^2\mathrm{D}'^{-1}u^x \quad \Delta'^3\mathrm{D}'^{-2}u_x \quad \Delta'^4\mathrm{D}'^{-2}u_x,$$

if $D'u_x$ be used to denote u_{x-1}, and Δ' to denote $D'-1$. But $D'=D^{-1}$, and $\Delta'=D'-1=D^{-1}-1=-D^{-1}\Delta$, so that the required numbers are

$$u_x \quad -\Delta u_x \quad \Delta^2 D^{-1}u_x \quad -\Delta^3 D^{-1}u_x \quad \Delta^4 D^{-2}u_x,$$

or

$$u_x \quad -\Delta u_x \quad \Delta^2 u_{x-1} \quad -\Delta^3 u_{x-1} \quad \Delta^4 u_{x-2}.$$

Hence the numbers on the top, Δ^2, and Δ^4 tiers are the same as for the forward calculation, while those on the Δ and Δ^3 tiers are the arithmetical complements of the numbers found on those tiers after the machine has made one complete movement in calculating forwards from u_x. The printing part, however, is not adapted to such a change: the numbers would be printed off correctly, but in a wrong order; so that unless some reversing movement were introduced into the printing part, the printed results would only serve to set types from.

In the example chosen above, and in similar cases, the differences required for setting the machine would be calculated from their mathematical expressions. It might, however, be required to tabulate for small intervals a function which had been given by observation for larger ones, or to tabulate a mathematical function of so complicated a form that the differences could not be got directly without great trouble. In such a case there would be no difficulty; the differences for the smaller intervals would first have to be calculated from those of the larger ones by formulæ in finite differences, and then the setting and working of the machine would proceed as before.

It must be confessed, however, that except in the case of mathematical tables like those of sines, cosines, logarithms, &c., it is not ordinarily required to tabulate functions to intervals at all approaching, in closeness, to those in the example selected. Hence it is mainly, as it seems to us, in the computation of mathematical tables, that the machine of M. Scheutz would come into use. The most important of such tables have long since been calculated; but various others could be suggested which it might be worth while to construct, could it be done with such ease and cheapness as would be afforded by the use of the machine. It has been suggested to us too, and we think with good reason, that the machine would be very useful even for the mere reprinting of old tables, because it could calculate and print more quickly than a good compositor could set the types, and that without risk of error.

G. G. STOKES.

21st Jan., 1855.

W. H. MILLER.
C. WHEATSTONE.
R. WILLIS.

P.S. Some time since, I received from Mr. Babbage, to whom I had written for information on one point connected with his machine, a letter, written subsequently to his first answer, in which he said he had forgotten to mention an addition to his machine which enabled it to calculate a function when the last differences, instead of being constant, were dependent on the functions then under calculation in the other parts of the machine, provided the coefficients of the variable parts were small enough to be expressed by a moderate number of digits. This was especially designed for the calculation of astronomical tables, where a difficulty occurs in the application of a machine with constant differences, arising from the circumstance that in the case of functions of short period the omitted differences soon become sensible, even though the coefficients be but small. Mr. Babbage did not then recollect that this contrivance was accessible to the public, but in a subsequent letter he pointed out that such was the case. The following is an extract from this letter :—

"1st. The portion at Somerset House contains axes specially prepared for what (at this instant) I recollect to have familiarly called 'eating its own tail.'

"2nd. The drawings contain the mode of governing those axes in the finished engine.

"These are public property, and open daily to public inspection, which I suppose must be considered as publication. On referring to the 9th Bridgewater Treatise, second edition, I find (p. 34) that I have used as an illustration a series computed by that very machine. * * * "

In the same letter Mr. Babbage refers to the following documents :—

Extract from a letter of Mr. Babbage to Sir H. Davy, 3 July, 1822, printed by order of the House of Commons. No. 370, 1823 :—

"Another machine, whose plans are more advanced than several of those just named, is one for constructing tables which have no order of differences constant (p. 2).

" I should be unwilling to terminate this letter without noticing another class of tables of the greatest importance, almost the whole of which are capable of being calculated by the method of differences. I refer to all astronomical tables for calculating the places of the sun and planets. It is scarcely necessary to observe that the constituent parts of these are of the form $a \sin \theta$." (p. 5.)

He refers also to an extract from the Address of H. T. Colbroke, Esq., President of the Astronomical Society, on presenting the first medal given to him by the Society, 1824 ; and to a description of his machine by the late Mr. Baily, published in Schumacher's ' Astronomische Nachrichten,' No. 46, and republished in the ' Philosophical Magazine ' for May 1824, p. 355. This last paper describes fully what could be done by the new contrivance.

I have ventured to insert this postscript without consulting my colleagues, as it is desirable not to delay the publication.

G. G. STOKES.

SCHEUTZ'S CALCULATING MACHINE.

[*Minutes of Proceedings of Civil Engineers, Jan.* 1857.]

AFTER the Meeting of January 27th, Messrs. Scheutz's Calculating Machine was exhibited in the Library, and was explained by Mr. Babbage and Mr. Gravatt, M. Inst. C.E. There was also shown a portion of a table of logarithms, which had been calculated, composed, and printed entirely by its aid, and without the use of types. It was estimated that these compound operations could be accomplished in less than half the time which a compositor would take to set the types; and at the same time all liability of error was avoided. The machine had been recently purchased by Mr. John F. Rathbone, of Albany, U.S., for presentation to the Dudley Observatory, U.S. America.

[*Minutes of Proceedings of Civil Engineers, April,* 1857.]

THE attention of the Members was especially directed to a small volume presented by Messrs. Scheutz, of Stockholm, through Mr. Gravatt (M. Inst. C.E.), entitled, "Specimens of Tables, calculated, stereo-moulded, and printed by Machinery." The book, which was with excellent feeling dedicated to Mr. Babbage, in recognition of the generous assistance he had afforded to the ingenious labourers in a similar field to that in which he had so long toiled, was preceded by a short memoir, describing the progress of the construction of the machine, under the most discouraging circumstances—the ultimate success obtained—the introduction of the machine in this country, through Count Sparre, to Messrs. Bryan, Donkin and Co., where Mr. Gravatt became interested in it, and placed it before the Royal Society, and the Institution of Civil Engineers—its success at the great Exhibition at Paris in 1855, where it obtained a gold medal—and finally its acquisition, through Professor B. A. Gould, for the Dudley Observatory, at Albany, U.S. America, as a gift to that establishment from Mr. John F. Rathbone, an enlightened and public-spirited merchant of that city.

The construction was briefly described, and it was shown that, at the average rate of working, one hundred and twenty lines per hour of arguments and results were calculated and stereotyped ready for the press. On trial, it was found that the machine would calculate and stereotype, without chance of error, two pages and a half of figures in the same time that a skilful compositor would take merely to set up the types for one single page.

There was also given an abstract of Mr. Gravatt's description of his manner of considering and working the machine, and then followed tables of logarithms of numbers from 1 to 10,000; and various examples of calculations performed with unerring accuracy. The remarkable and unique feature of the book itself was, that the tables and calculations were all printed from stereotyped plates produced directly from the machine, and without the use of any moveable type.

The special thanks of the Meeting were given to Messrs. Scheutz for the present, and to Mr. Gravatt for taking charge of this interesting addition to the Library.

ANALYTICAL ENGINE.

CATALOGUE OF 446 NOTATIONS.

WITH THEIR CLASSIFICATION.

No. of Notations.				From	1834
171	.	up to	.	31 AUG.,	1839.
257	.	up to	.	8 OCT.,	1841.
446	.	up to	.	8 SEPT.,	1847.

N.B.—The figures which are separated by a dot on the right hand of the numbers expressing the Notation, indicate the number of sheets of which it consists.

CATALOGUE OF NOTATIONS

OF THE

ANALYTICAL ENGINE.

END OF VOL I. OF NOTATIONS.

Vol. II.

74	Sheet 1. Multiplication, with Hoarded Carriages. Beginning with highest figure of Multiplier	4 Dec. 1837
	Sheet 2. Do. lowest do.	6 Dec.
75	Sheet 1. Division $p<r+1$	9 Dec.
	Sheet 2. Do. $p>r+1$	13 Dec.
76	2 Cycles of 15 and 20.	26 Dec.
77	Sheet 1. Mill Cycles of 15 and 20	8 Jan., 1838
	Sheet 2. Rack do. do.	8 Jan.
78	Verticals (superseded by Notation 93)	6 Feb.
79	General Notation, Addition and Subtraction, with Signs	28 Feb.
80	Sheet 1. Lockings	7 March
	Sheet 2. Do.	7 March
	Sheet 3. Do. Store	7 March
81	General Notation, Addition, and Subtraction	17 March
82	Sectors, Levers, &c., in Mill and Store	23 March
83	Approximate Multiplication	April
84	Addition and Subtraction, with Signs	16 April
85	Approximate Division	23 April
86	Directive of Multiplication (see 94)	26 June
87	Do. Division (superseded by 96)	26 June
88	Do. Addition and Subtraction	27 June
89	Do. Multiplication (superseded by 94)	30 June
90	Do. Addition	30 June
91	Do. do.	2 July
92	Do. do.	2 July
93	Sheet 1. Verticals. Multiplication	5 July
	Sheet 2. Do. Division	14 July
	Sheet 3. Do. Approximate Do.	18 July
94	Directive of Multiplication	16 July
94*	Do. Common and Approximate Multiplication	July
95		
96	Sheet 1. Do. Division	18 July
	Sheet 2. Do. introduced for Approximate Division	19 July
97	Cards, Multiplication	20 July
98	Do. Division and Approximate Division	20 July
99	Addition and Subtraction	24 July
100	Addition (superseded, no date)	
101 and 102	Do.	27 July
103	Do. (on small folio)	27 July
104	Algebraic Addition	27 July
105	Addition and Subtraction, with Signs—	
	Sheet 1. Give off arbitrarily	28 July
	Sheet 2. Do. after each addition	28 July
106	Directive of Division, No. 1	2 Aug.
107	Do. Multiplication, No. 1	2 Aug.
108	Multiplication, Sheet 1, Standard Case	Feb., 1839
	Sheet 2. Do.	Feb.
	Sheet 3. Standard Case of Approximate Multiplication	Feb.
	Sheet 4. Do. do.	Feb.
109	Sheets 1-4. Common Division	Feb.
	Sheets 5-12. Approximate do.	Feb.
110	Verticals for Multiplication, Notation 108	March
111	Sheet 1. Do. Division	March
	Sheets 2-4. Do. Approximate Division	March
112	Sheet 1. Cases of Common Multiplication. Sheet 2. Cases of Approximate Multiplication, with their corresponding variations	April
112*	Sheet 1. Do. do. Division. Sheet 2. Do. Division	April
112**	Cases of Multiplication without the Verticals	May
113	Sheets 1-3. Addition and Subtraction of 120 figures	1 July
114	Cards for Algebraic Addition (superseded by 123)	6 July
115	Do. do. (unfinished, to suit 114, superseded by 122)	8 July
116	Sheets 1-5. do. of more than 40 figures	9 July
117	Do. do.	12 July

The following pages contain the preceding catalogue classed alphabetically according to the various subjects those
Notations illustrate.

CLASSED CATALOGUE

OF THE NOTATIONS OF THE

ANALYTICAL ENGINE.

CATALOGUE OF THE DRAWINGS

OF THE

A N A L Y T I C A L   E N G I N E.

(*Contained in three Mahogany Cases, Numbered* 1, 2, *and* 3.)

Each Drawing measures 36 *by* 25 *inches.*

1		The first Drawing of Circular Arrangement of new Engine, Plan .	Sept. 1834
2		The first Drawing of do. do. Elevation	Sept.
5	Fig. 1.	Method of Addition by a Figure Wheel and Adding Wheel, lifted out of gear by Inclined Plane (all begin together) . . .	Sept.
	Fig. 2.	Method of Addition by Snail and Gathering-up Wheel . . .	Oct.
3	Fig. 1.	Unfinished Sketch of a Plan for Hoarding the Carriages . . .	9 Oct.
4	Fig. A.	Anticipating Carriage by Inclined Planes	16 Oct.
	Fig. B.	Improved do. do.	19 Oct.
	Fig. C.	Anticipating Carriage by Vertical Chain	25 Oct.
3	Fig. 2.	Improved do. do.	7 Nov.
7		Method of Addition, Driver lifted out of gear by Inclined Plane (all begin together)	12 Nov.
6		Improved do. do. and Sections of Parts, Snail, Wheels, Method of Bringing Figure Wheels to Zero, &c. .	27 Nov.
5	Figs. 3 & 4.	Sections of Parts and Gearings for Grouping of do., supposing every Adding Axis to have a Carriage of its own	29 Dec.
8	Fig. 1.	Method of giving Vertical Motion to Multiplying Pinions which work in Central Wheels. (*No date.*)	
	Fig. 2.	Method of Managing the Feelers for Comparing Divisor with Dividend, &c. (*No date.*)	
9		Group of Wheels, in which Carriage gears with central . . .	24 Jan. 1835.
15		Method of Grouping for large Machine. . . Between Jan. and April	
16		Do. do. . . . . . Do.	
17		Do. do.	April
11	Fig. 1.	Four Racks in one Cage: in Section and Detail . . .	7 May
10		Sketch for Examining System of two Racks	8 May
12		Sketch of a Method of Multiplying by Stepping the Directive Power (20 small Racks in each Cage)	16 May
13	Fig. 1.	First Method of Multiplying by Stepping the Numbers on Figure Axis	16 May
	Fig. 2.	Improved Arrangement of do.	25 May
11	Fig. 2.	Two Racks in one Cage: in Section and Detail . . .	30 May
	Fig. 3.	One do. do.	30 May
24		First Drawing of Selecting Apparatus	15 June
18		Sketch of Gearings for General Plan, No. 5. . . .	4 July
19		Do. do. No. 7.	4 July
20	Fig. 1.	Sketch of General Plan, No. 7.	27 July
	Fig. 2.	Do. do. No. 8.	31 July
21		Gearings, Lockings, &c., in detail for Sketch No. 8. . .	31 July
22		Do. do.	31 July
23		Do. do.	31 July

175* Printing Section . . . . 21 June, 1861
T △ 4 176 ₄ Detailed Plan of Adding and Carrying on Table Axis.
  △ 5 177 Bars and Levers for Lifting Axis.
178 Special Trains, 5 sheets.
179 Plan of Calculating Axis and Framing of Difference Engine.
180 Elevation of Travelling Platforms for governing universal and general
          Locking Plates (*see* 183).
180* Plan for raising Axes of Difference Engine.
180** Anticipating Carriage. Hoarding Carriage . . . . . 21 June, 1864
181 Plan and Elevation of Platform for giving Circular Motion to Axis
          (applied to Difference Engine, Drawing 163) . . . . 6 March, 1849
182 Plan of Locking Plates for Locking Axis circularly and vertically
          (applied to 163).
183 Plan of Universal and General Locking Plates. Plan and Elevation
          of Locked Arms, *see* 180.
      Total of Difference Engine No. 2. 20 Driving.
183* Hoarding and Anticipating Carriage . . . . . . 8 Feb., 1864
184 Anticipating Carriage. Half Zero Chain . . . . . . 28 May
185 Whole Zero Making and Breaking Chains . . . . . . 2 July,
186 Latest general view of Analytical Engine . . . . . . 14 Sept.
187 Various Drawings of Half Zero Apparatus . . . . . . Feb., 1865
188 Various Half Zeros (Models made) . . . . 17 March
189   Do.   Apparatus . . . . 13 April
190 Whole and Half Zero Carriages . . . . 30 May
191 Circular   do.   do. . . . . 22 June
192 Longitudinal Half Zero . . . . 30 June
193 Circular   do. . . . . July
194   Do.   do. . . . . Sept.

N.B.—A and Anal. mean that those Drawings belong to the Analytical Engine.
△ means   do,   do.   Difference Engine, No. 2.
T means   do.   do.   have had Tracing made of them.
The Small Numbers refer to the Tracings (12 in number) of Difference Engine, No. 2.
The Antique Numbers refer to the Complete Set of Drawings of Difference Engine, No. 2.
The total number of Drawings and Tracings of Difference Engine, No. 2, amounts to thirty-two sheets.

LIST OF OTHER DRAWINGS OF THE ANALYTICAL ENGINE.

(Contained in a Mahogany Case, Numbered 1.)

Numbers.		Dates.
1	Sections of Store Wheels and their Framing Selectors . . .	6 June, 1857
2	Digit Counting Apparatus, on the two upper Wheels Selectors .	2 Oct.
3	Carriages with Tens Warnings	2 Sept.
4	Digit Counting Apparatus	Nov.
5	Frame, Lifting Bar, and Selectors	14 Dec.
6	Sections of Framing and Racks	Jan., 1858
7	Digit Counting Apparatus, Spiral Axis	11 Jan.
8	First Draft of Selecting Apparatus	8 Dec., 1857
9	Lifting Apparatus by Screw	Dec.
10	Carriage and Racks in Sections	10 Mar., 1858
11	Plan of Mill with two Racks	Feb.
12	Selecting Apparatus	15 Jan.
13	Carriage Axis, Head and Tail of Product	13 Dec., 1857
14	Carriage	Mar., 1850
15	Selectors, Circular Framing for Mill	4 May, 1858
16	Circular Mill, various Plans	May
17	First Draft of Framing	19 Sept., 1857
18	Hoarding Carriage, Framing	19 Nov.
19	Side and End view of Framing: bearings for Rack.	
20	Elevation, Arrangements for Division	14 May, 1858
21	Digit Counting—Units and Tens	18 June
22	Revised Plan of Mill	26 June
23	Raising and Lowering Figure Wheel Axis	2 July
24	Plan of Mill, Table Wheels, Carriage	July
25	Platform Raising Apparatus.	1 July
26	Do. do.	June
27	Continuation of do.	21 July
28	General Plan	12 June, 1856
29	Draft of the Mill.	9 June, 1858
30	Plan of improved Mill	18 Sept.
31	Elevation of Mill.	25 Sept.
32	Long Stepping Mill driven by Pinions	10 July
33	Carriage Axis and Wheels: Sections shaded	9 Oct.
34	Plan of Mill and Store.	26 Nov.
35	Mill and Store: Plan and Elevation	14 Dec.
36	Plan of Mill	2 Nov.
37	Do	4 Nov.
38	Rising Platforms: Advancing Plate	8 Jan., 1859
39	Plan and Elevation of Bolts and of Advancing Plate . .	18 Jan.
40	Plan and Elevation of Bolts for Store	14 Feb.
41	Rising Platform: Plan and Elevation	7 March
42	Adding and Carriage	7 Aug.
43	Wanting Tooth for Long Stepping	2 June, 1858
44	Plan of Store with double Mill and Printing	14 Feb.
45	Printing and Card Punching	4 Dec.

There are also nineteen sheets of Drawings of Tools
The whole of the Drawings are upon paper about 38 inches wide by
    25 inches high.

LIST OF SCRIBBLING BOOKS.

The first rough notions of the Difference and of the Analytical Engines were developed by sketches and calculations in Nine Volumes. Some of these became the basis of the drawings in the preceding list.

VOL.	NO. OF PAGES.
I.—June, 1840–42. This was called the Tool Book. Discussion and Sketches of Tools.	
II.—Nov., 1835, to Feb., 1837	280
III.—Feb., 1837	550
IV.—July, 1838, to May, 1841	573
V.—May 29, 1841, to Dec. 14, 1844. Important matter of history in the Analytical Engine	562
VI.—Dec. 14, 1844, to March 25, 1859	558
VII.—March 4, 1858, to June, 1859	318
VIII.—June 16, 1859, to Dec. 2, 1864	430
IX.—Dec. 24, 1864, to Aug., 1865	194

NOTE.—*There are Four more Volumes of later date, H. P. B.*

Besides the above list of Scribbling Books, there were others in the following list which contained materials belonging to same subject. The dates are various and irregular.

VOL.	NO. OF PAGES.
X.—Labelled MSS. in 4to., vellum binding, 9 inches wide, 12 inches long, probably before 1825. Used in Devonshire Street	555
XI.—Sketch Book, vol. iii., 26 Oct., last date May, 1841.	
XII.—List of parts of the Engine, before Aug., 1821.	40
XIII.—Sketch of the History of the new Calculating Engine, dated 25 June, 1835	176
XIV.—Travelling Book, various dates	278
XV.—Folio, in Russia binding, pages 10½ by 16 inches; only 102 pages filled. Containing various developments of series for the Analytical Engine. Molecules elliptic, various developments	550
XVI.—Folio, in Russia binding, 9 by 14½ inches, various analytical developments adapted to Engine	800

ON THE ECONOMY OF MACHINERY AND MANUFACTURES.

(First Edition, Chapter XIX.)

CHAPTER XIX.

ON THE DIVISION OF MENTAL LABOUR.

(241.) WE have already mentioned what may, perhaps, appear paradoxical to some of our readers,—that the division of labour can be applied with equal success to mental as to mechanical operations, and that it ensures in both the same economy of time. A short account of its practical application, in the most extensive series of calculations ever executed, will offer an interesting illustration of this fact, whilst at the same time it will afford an occasion for shewing that the arrangements which ought to regulate the interior economy of a manufactory, are founded on principles of deeper root than may have been supposed, and are capable of being usefully employed in preparing the road to some of the sublimest investigations of the human mind.

(242.) In the midst of that excitement which accompanied the Revolution of France and the succeeding wars, the ambition of the nation, unexhausted by its fatal passion for military renown, was at the same time directed to some of the nobler and more permanent triumphs which mark the era of a people's greatness,—and which receive the applause of posterity long after their conquests have been wrested from them, or even when their existence as a nation may be told only by the page of history. Amongst their enterprises of science, the French government was desirous of producing a series of mathematical tables, to facilitate the application of the decimal system which they had so recently adopted. They directed, therefore, their mathematicians to construct such tables, on the most extensive scale. Their most distinguished philosophers, responding fully to the call of their country, invented new methods for this laborious task; and a work, completely answering the large demands of the government, was produced in a remarkably short period of time. M. Prony, to whom the superintendence of this great undertaking was confided, in speaking of its commencement, observes : *" Je m'y livrai avec toute l'ardeur dont j'étois capable, et je m'occupai " d'abord du plan général de l'exécution. Toutes les conditions que j'avois à remplir nécessitoient "l'emploi d'un grand nombre de calculateurs ; et il me vint bientôt à la pensée d'appliquer à la con-"fection de ces Tables la division du travail, dont les Arts de Commerce tirent un parti si avantageux "pour réunir à la perfection de main-d'œuvre l'économie de la dépense et du temps."* The circumstance which gave rise to this singular application of the principle of *the division of labour* is so interesting, that no apology is necessary for introducing it from a small pamphlet printed at Paris a few years since, when a proposition was made by the English to the French government, that the two countries should print these tables at their joint expense.

(243.) The origin of the idea is related in the following extract :—

"C'est à un chapitre d'un ouvrage Anglais,* justement célèbre, (I.) qu'est probablement due l'existence de l'ouvrage dont le gouvernement Britannique veut faire jouir le monde savant :—

"Voici l'anecdote : M. de Prony s'était engagé, avec les comités de gouvernement, à composer, pour *la division centésimale du cercle, des tables logarithmiques et trigonométriques, qui, non-seulement ne laissassent rien à désirer quant à l'exactitude, mais qui formassent le monument de calcul le plus vaste et le plus imposant qui eût jamais été exécuté, ou même conçu.* Les logarithmes des nombres de 1 à 200,000 formaient à ce travail un supplément nécessaire et exigé. Il fut aisé à M. de Prony de s'assurer que, même en s'associant trois ou quatre habiles co-opérateurs, la plus grande durée présumable de sa vie ne lui suffirait pas pour remplir ses engagements. Il était occupé de cette fâcheuse pensée lorsque, se trouvant devant la boutique d'un marchand de livres, il aperçut la belle édition Anglaise de Smith, donnée à Londres en 1776 ; il ouvrit le livre au hasard, et tomba sur le premier chapitre, qui traite de *la division du travail*, et où la fabrication des épingles est citée pour exemple. A peine avait-il parcouru les premières pages, que, par une espèce d'inspiration, il conçut l'expédient de mettre ses logarithmes en *manufacture* comme les épingles. Il faisait, en ce

---

* *An Enquiry into the Nature and Causes of the Wealth of Nations*, by Adam Smith.

2 R 2

moment, à l'école polytechnique, des leçons sur une partie d'analyse liée à ce genre de travail, *la méthode des différences*, et ses applications à *l'interpolation*. Il alla passer quelques jours à la campagne, et revint à Paris avec le plan de *fabrication*, qui a été suivi dans l'exécution. Il rassembla deux ateliers, qui faisaient séparément les mêmes calculs, et se servaient de vérification réciproque." *

(244.) The ancient methods of computing tables were altogether inapplicable to such a proceeding. M. Prony, therefore, wishing to avail himself of all the talent of his country in devising new methods, formed the first section of those who were to take part in this enterprise out of five or six of the most eminent mathematicians in France.

First Section.—The duty of this first section was to investigate, amongst the various analytical expressions which could be found for the same function, that which was most readily adapted to simple numerical calculation by many individuals employed at the same time. This section had little or nothing to do with the actual numerical work. When its labours were concluded, the formulæ on the use of which it had decided, were delivered to the second section.

Second Section.—This section consisted of seven or eight persons of considerable acquaintance with mathematics: and their duty was to convert into numbers the formulæ put into their hands by the first section,—an operation of great labour; and then to deliver out these formulæ to the members of the third section, and receive from them the finished calculations. The members of this second section had certain means of verifying the calculations without the necessity of repeating, or even of examining, the whole of the work done by the third section.

Third Section.—The members of this section, whose number varied from sixty to eighty, received certain numbers from the second section, and, using nothing more than simple addition and subtraction, they returned to that section the tables in a finished state. It is remarkable that nine-tenths of this class had no knowledge of arithmetic beyond the two first rules which they were thus called upon to exercise, and that these persons were usually found more correct in their calculations, than those who possessed a more extensive knowledge of the subject.

(245.) When it is stated that the tables thus computed occupy seventeen large folio volumes, some idea may perhaps be formed of the labour. From that part executed by the third class, which may almost be termed mechanical, requiring the least knowledge and by far the greatest exertions, the first class were entirely exempt. Such labour can always be purchased at an easy rate. The duties of the secon class, although requiring considerable skill in arithmetical operations, were yet in some measure relieved by the higher interest naturally felt in those more difficult operations. The exertions of the first class are not likely to require, upon another occasion, so much skill and labour as they did upon the first attempt to introduce such a method; but when the completion of a calculating-engine shall have produced a substitute for the whole of the third section of computers, the attention of analysts will naturally be directed to simplifying its application, by a new discussion of the methods of converting analytical formulæ into numbers.

(246.) The proceeding of M. Prony, in this celebrated system of calculation, much resembles that of a skilful person about to construct a cotton or silk-mill, or any similar establishment. Having, by his own genius, or through the aid of his friends, found that some improved machinery may be successfully applied to his pursuit, he makes drawings of his plans of the machinery, and may himself be considered as constituting the first section. He next requires the assistance of operative engineers capable of executing the machinery he has designed, some of whom should understand the nature of the processes to be carried on; and these constitute his second section. When a sufficient number of machines have been made, a multitude of other persons, possessed of a lower degree of skill, must be employed in using them; these form the third section: but their work, and the just performance of the machines, must be still superintended by the second class.

(247.) As the possibility of performing arithmetical calculations by machinery may appear to non-mathematical readers to be rather too large a postulate, and as it is connected with the subject of the *division of labour,* I shall here endeavour, in a few lines, to give some slight perception of the manner in which this can be done,—and thus to remove a small portion of the veil which covers that apparent mystery.

(248.) *That nearly all tables of numbers which follow any law, however complicated, may be formed, to a greater or less extent, solely by the proper arrangement of the successive addition and subtraction of numbers befitting each table,* is a general principle which can be demonstrated to those only who are well acquainted with mathematics; but the mind, even of the reader who is but very slightly acquainted with that science, will readily conceive that it is not impossible, by attending to the following example.

---

* Note sur la publication, proposée par le gouvernement Anglais des grandes tables logarithmiques et trigonométriques de M. de Prony.—De l'imprimerie de F. Didot, Dec. 1, 1820, p. 7.

The subjoined table is the beginning of one in very extensive use, which has been printed and reprinted very frequently in many countries, and is called *a Table of Square Numbers.*

Terms of the Table.	A. Table.	B. First Difference.	C. Second Difference.
1	1		
		3	
2	4		2
		5	
3	9		2
		7	
4	16		2
		9	
5	25		2
		11	
6	36		2
		13	
7	49		

Any number in the table, column A, may be obtained, by multiplying the number which expresses the distance of that term from the commencement of the table by itself; thus, 25 is the fifth term from the beginning of the table, and 5 multiplied by itself, or by 5, is equal to 25. Let us now subtract each term of this table from the next succeeding term, and place the results in another column (B), which may be called first-difference column. If we again subtract each term of this first difference from the succeeding term, we find the result is always the number 2, (column C;) and that the same number will always recur in that column, which may be called the second difference, will appear to any person who takes the trouble to carry on the table a few terms further. Now when once this is admitted, it is quite clear that, provided the first term (1) of the Table, the first term (3) of the first differences, and the first term (2) of the second or constant difference, are originally given, we can continue the table of square numbers to any extent, merely by addition :—for the series of first differences may be formed by repeatedly adding the constant difference (2) to (3) the first number in column B, and we then have the series of numbers, 3, 5, 7, &c. : and again, by successively adding each of these to the first number (1) of the table, we produce the square numbers.

(249.) Having thus, I hope, thrown some light upon the theoretical part of the question, I shall endeavour to shew that the mechanical execution of such an engine, as would produce this series of numbers, is not so far removed from that of ordinary machinery as might be conceived.* Let the reader imagine three clocks, placed on a table side by side, each having only one hand, and each having a thousand divisions instead of twelve hours marked on the face ; and every time a string is pulled, let them strike on a bell the numbers of the divisions to which their hands point. Let him further suppose that two of the clocks, for the sake of distinction called B and C, have some mechanism by which the clock C advances the hand of the clock B one division, for each stroke it makes upon its own bell : and let the clock B by a similar contrivance advance the hand of the clock A one division, for each stroke it makes on its own bell. With such an arrangement, having set the hand of the clock A to the division I., that of B to III., and that of C to II., let the reader imagine the repeating parts of the clocks to be set in motion continually in the following order : viz.—pull the string of clock A ; pull the string of clock B ; pull the string of clock C.

The table on the following page will then express the series of movements and their results.

* Since the publication of the Second Edition of this Work, one portion of the engine which I have been constructing for some years past has been put together. It calculates, in three columns, a table with its first and second differences. Each column can be expressed as far as five figures, so that these fifteen figures constitute about one-ninth part of the larger engine. The ease and precision with which it works, leave no room to doubt its success in the more extended form. Besides tables of squares, cubes, and portions of logarithmic tables, it possesses the power of calculating certain series whose differences are not constant ; and it has already tabulated parts of series formed from the following equations :

$$\Delta^3 u_x = \text{units figure of } \Delta u_x$$

$$\Delta^3 u_x = \text{nearest whole No. to } \left(\frac{1}{10,000} \Delta u_x \right)$$

The subjoined is one amongst the series which it has calculated :

0	3,486	42,972
0	4,991	50,532
1	6,907	58,813
14	9,295	67,826
70	12,236	77,602
230	15,741	88,202
495	19,861	99,627
916	24,597	111,928
1,504	30,010	125,116
2,340	36,131	139,272

The general term of this is :

$$u_x = \frac{x \cdot x-1 \cdot x-2}{1 \cdot 2 \cdot 3} + \text{the whole number in } \frac{x}{10} +$$

$$+ 10 \, \Sigma^3 \left(\text{units figure of } \frac{x \cdot x+1}{2} \right)$$

THE ECONOMY OF MANUFACTURES.

Repetitions of Process.	Movements.	Clock A. Hand set to I.	Clock B. Hand set to III.	Clock C. Hand set to II.
		TABLE.	First difference.	Second difference.
1	Pull A.	A. strikes . . . 1	. . . . . .	. . . .
	—— B.	The hand is advanced (by B.) 3 divisions .	B. strikes . . . 3	. . . .
	—— C.	. . . .	The hand is advanced (by C.) 2 divisions .	C. strikes 2
2	Pull A.	A. strikes . . . 4	. . . . . .	. . . .
	—— B.	The hand is advanced (by B.) 5 divisions .	B. strikes . . . 5	. . . .
	—— C.	. . . .	The hand is advanced (by C.) 2 divisions .	C. strikes 2
3	Pull A.	A. strikes . . . 9	. . . . . .	. . . .
	—— B.	The hand is advanced (by B.) 7 divisions .	B. strikes . . . 7	. . . .
	—— C.	. . . .	The hand is advanced (by C.) 2 divisions .	C. strikes 2
4	Pull A.	A. strikes . . . 16	. . . . . .	. . . .
	—— B.	The hand is advanced (by B.) 9 divisions .	B. strikes . . . 9	. . . .
	—— C.	. . . .	The hand is advanced (by C.) 2 divisions .	C. strikes 2
5	Pull A.	A. strikes . . . 25	. . . . . .	. . . .
	—— B.	The hand is advanced (by B.) 11 divisions	B. strikes . . . 11	. . . .
	—— C.	. . . .	The hand is advanced (by C.) 2 divisions .	C. strikes 2
6	Pull A.	A. strikes . . . 36	. . . . . .	. . . .
	—— B.	The hand is advanced (by B.) 13 divisions	B. strikes . . . 13	. . . .
	—— C.	. . . .	The hand is advanced (by C.) 2 divisions .	C. strikes 2

If now only those divisions struck or pointed at by the clock A be attended to and written down, it will be found that they produce the series of the squares of the natural numbers. Such a series could, of course, be carried by this mechanism only so far as the numbers which can be expressed by three figures; but this may be sufficient to give some idea of the construction,—and was, in fact, the point to which the first model of the calculating-engine, now in progress, extended.

NINTH BRIDGEWATER TREATISE.

PART OF CHAPTER II.

THE illustration which I shall here employ will be derived from the results afforded by the Calculating Engine;* and this I am the more disposed to use, because my own views respecting the extent of the laws of Nature were greatly enlarged by considering it, and also because it incidentally presents matter for reflection on the subject of inductive reasoning. Nor will any difficulty arise from the complexity of that engine; no knowledge of its mechanism, nor any acquaintance with mathematical science, are necessary for comprehending the illustration; it being sufficient merely to conceive that computations of great complexity *can* be effected by mechanical means.

Let the reader imagine that such an engine has been adjusted; that it is moved by a weight; and that he sits down before it, and observes a wheel, which moves through a small angle round its axis, at short intervals, presenting to his eye, successively, a series of numbers engraved on its divided circumference.

Let the figures thus seen be the series of natural numbers, 1, 2, 3, 4, 5, &c., each of which exceeds its immediate antecedent by unity.

Now, reader, let me ask how long you will have counted before you are firmly convinced that the engine, supposing its adjustments to remain unaltered, will continue whilst its motion is maintained, to produce the same series of natural numbers? Some minds perhaps are so constituted, that after passing the first hundred terms, they will be satisfied that they are acquainted with the law. After seeing five hundred terms, few will doubt; and after the fifty-thousandth term the propensity to believe that the succeeding term will be fifty thousand and one, will be almost irresistible. That term *will* be fifty thousand and one: the same regular succession will continue; the five-millionth and the fifty-millionth term will still appear in their expected order; and one unbroken chain of natural numbers will pass before your eyes, from *one* up to *one hundred million.*

True to the vast induction which has thus been made, the next succeeding term will be one hundred million and one; but after that the next number presented by the rim of the wheel, instead of being one hundred million and two, is one hundred million *ten thousand* and two. The whole series from the commencement being thus :—

$$
\begin{array}{r}
1\\
2\\
3\\
4\\
5\\
\cdot\ \ \cdot\ \ \cdot\\
\cdot\ \ \cdot\ \ \cdot\\
\cdot\ \ \cdot\ \ \cdot\ \ \cdot\\
\cdot\ \ \cdot\ \ \cdot\ \ \cdot\\
99{,}999{,}999\\
100{,}000{,}000\\
\text{regularly as far as }100{,}000{,}001\\
100{,}010{,}002\ :\text{—the law changes}\\
100{,}030{,}003\\
100{,}060{,}004\\
100{,}100{,}005\\
100{,}150{,}006\\
100{,}210{,}007\\
100{,}280{,}008\\
100{,}360{,}009\\
100{,}450{,}010\\
100{,}550{,}011\\
\cdot\cdot\cdot\ \ \cdot\cdot\ \ \cdot\cdot\cdot\\
\cdot\cdot\cdot\ \ \cdot\cdot\ \ \cdot\cdot\cdot
\end{array}
$$

* The reader will find a short account of this engine in the Appendix, Note B. (*See page* 304 *of this book.*)

The law which *seemed* at first to govern this series fails at the hundred million and second term. That term is larger than we expected, by 10,000. The next term is larger than was anticipated, by 30,000, and the excess of each term above what we had expected forms the following table :—

$$
\begin{array}{r}
10,000 \\
30,000 \\
60,000 \\
100,000 \\
150,000 \\
\cdots \cdots \\
\cdots \cdots
\end{array}
$$

being, in fact, the series of *triangular numbers*,* each multiplied by 10,000.

If we still continue to observe the numbers presented by the wheel, we shall find, that for a hundred, or even for a thousand terms, they continue to follow the new law relating to the triangular numbers ; but after watching them for 2761 terms, we find that *this* law fails in the case of the 2762d term.

If we continue to observe, we shall discover another law then coming into action, which also is dependent, but in a different manner, on triangular numbers. This will continue through about 1430 terms, when a new law is again introduced, which extends over about 950 terms ; and this too, like all its predecessors, fails, and gives place to other laws, which appear at different intervals.

Now it must be remarked, that the law *that each number presented by the Engine is greater by unity than the preceding number*, which law the observer had deduced from *an induction of a hundred million instances*, was not the true law that regulated its action ; and that the occurrence of the number 100,010,002 at the 100,000,002d term, was *as necessary a consequence* of the original adjustment, and might have been as fully foreknown at the commencement, as was the regular succession of any one of the intermediate numbers to its immediate antecedent. The same remark applies to the next *apparent* deviation from the new law, which was founded on an induction of 2761 terms, and to all the succeeding laws ; with this limitation only—that whilst their consecutive introduction at various definite intervals is a necessary consequence of the mechanical structure of the engine, our knowledge of analysis does not yet enable us to predict the periods at which the more distant laws will be introduced.

Such are some of the facts which, by a certain adjustment of the Calculating Engine, would be presented to the observer. Now, let him imagine another engine, offering to the eye precisely the same figures in the same order of succession ; but let it be necessary for the maker of that other engine, previously to each apparent change in the law, to make some new adjustment in the structure of the engine itself, in order to accomplish the ends proposed. The first engine must be susceptible of having embodied in its mechanical structure, that more general law of which all the observed laws were but isolated portions,—a law so complicated, that analysis itself, in its present state, can scarcely grasp the whole question. The second engine might be of far simpler contrivance ; it must be capable of receiving the laws impressed upon it from without, but is incapable, by its own intrinsic structure, of changing, at definite periods, and in unlimited succession, those laws by which it acts. Which of these two engines would, in the reader's opinion, give the higher proof of skill in the contriver? He cannot for a moment hesitate in pronouncing that that for which, after its original adjustment, no superintendence is required, displays far greater ingenuity than that which demands, at every change in its law, the direct intervention of its contriver.

The engine we have been considering is but a very small portion (about fifteen figures) of a much larger one, which was preparing, and is partly executed ; it was intended, when completed, that it should have presented at once to the eye about one hundred and thirty figures. In that

---

* The numbers 1, 3, 6, 10, 15, 21, 28, &c., are formed by adding the successive terms of the series of natural numbers thus ;

$$
\begin{array}{r}
1 = 1. \\
1 + 2 = 3. \\
1 + 2 + 3 = 6. \\
1 + 2 + 3 + 4 = 10, \&c.
\end{array}
$$

They are called triangular numbers, because a number of points corresponding to any term can always be placed in the form of a triangle, for instance :—

1   3   6   10

more extended form which recent simplifications have enabled me to give to machinery constructed for the purpose of making calculations, it will be possible, by certain adjustments, to set the engine so that it shall produce the series of natural numbers in regular order, from unity up to a number expressed by more than a thousand places of figures. At the end of that term, another and a different law shall regulate the succeeding terms; this law shall continue in operation perhaps for a number of terms, expressed perhaps by unity followed by a thousand zeros, or 10^{1000}; at which period a third law shall be introduced, and, like its predecessors, govern the figures produced by the engine during a third of those enormous periods. This change of laws might continue without limit; each individual law being destined to govern for millions of ages the calculations of the engine, and then give way to its successor to pursue a like career.*

Thus a series of laws, each simple in itself, successively spring into existence, at distances almost too great for human conception. The full expression of that wider law, which comprehends within it this unlimited sequence of minor consequences, may indeed be beyond the utmost reach of mathematical analysis: but of one remarkable fact, however, we are certain—that the mechanism brought into action for the purpose of changing the nature of the calculation from the production of the merest elementary operations into those highly complicated ones of which we speak, is itself of the simplest kind.

* It has been supposed that ten turns of the handle of the calculating engine might be made in a minute, or about five hundred and twenty six millions in a century. As in this case, each turn would make a calculation, after the lapse of a million of centuries, only the fifteenth place of figures would have been reached.

PART OF CHAPTER VIII.

Let the reader suppose himself placed before the Calculating Engine, and let him again observe and ascertain, by lengthened induction, the nature of the law it is computing. Let him imagine that he has seen the changes wrought on its face during the lapse of thousands of years, and that, without one solitary exception, he has found the engine register the series of square numbers. Suppose, now, the maker of that machine to say to the observer, "I will, by moving a certain " mechanism, which is invisible to you, cause the engine to make one cube number instead of a " square, and then to revert to its former course of square numbers;" the observer would be inclined to attribute to him a degree of power but little superior to that which was necessary to form the original engine.

But, let the same observer, after the same lapse of time—the same amount of uninterrupted experience of the uniformity of the law of square numbers, hear the maker of the engine say to him—"The next number which shall appear on those wheels, and which you expect to find a " square number, shall not be so. When the machine was originally ordered to make these cal- " culations, I impressed on it a law, which should coincide with that of square numbers in every " case, *except* the one which is now about to appear; after which no future exception can ever occur, " but the unvarying law of the squares shall be pursued until the machine itself perishes from " decay."

Undoubtedly the observer would ascribe a greater degree of power to the artist who had thus willed that event which he foretells at the distance of ages before its arrival.

If the contriver of the engine then explain to him, that, by the very structure of it, he has power to order *any* number of such apparent deviations from its laws to occur at any future periods, however remote, and that each of these may be of a different kind; and, if he also inform him, that he gave it that structure in order to meet events, which he foresaw must happen at those respective periods, there can be no doubt that the observer would ascribe to the inventor far higher knowledge than if, when those events severally occurred, he were to intervene, and temporarily to alter the calculations of the machine.

If, besides this, the contriver were so far to explain the structure of the engine that the observer could himself, by some simple process, such as the mere moving of a bolt, call into action those apparent deviations whenever certain combinations were presented to his eye; if he were thus to impart a power of predicting such excepted cases, dependent on the will, though in other respects beyond the limits of the observer's power and knowledge,—such a structure would be admitted as evidence of a still more skilful contrivance.

2 s

The engine which, in a former chapter, I introduced to the reader, possesses these powers. It may be set, so as to obey any given law; and, at any periods, however remote, to make one or more *seeming* exceptions to that law. It is, however, to be observed, that the *apparent* law which the spectator arrived at, by an almost unlimited induction, is not the full expression of the law by which the machine acts; and that the excepted case is as absolutely and irresistibly the necessary consequence of its primitive adjustment, as is any individual calculation amongst the countless multitude which it may previously have produced.

When the construction of that engine was first attempted, I did not seek to give to it the power of making calculations so far beyond the reach of mathematical analysis as these appear to be: nor can I now foresee a probable period at which they may become practically available for human purposes. I had determined to invest the invention with a degree of generality which should include a wide range of mathematical power; and I was well aware that the mechanical generalisations I had organised contained within them much more than I had leisure to study, and some things which will probably remain unproductive to a far distant day.

PART OF CHAPTER XIII.

In the present chapter it is proposed to prove, that—

It is more probable that any law, at the knowledge of which we have arrived by observation, shall be subject to one of those violations which, according to Hume's definition, constitutes a miracle, than that it should not be so subjected.

To show this, we may be allowed again to revert to the Calculating Engine: and to assume that it is possible to set the machine, so that it shall calculate *any algebraic law whatsoever*: and also possible so to arrange it, that at any periods, *however remote*, the first law shall be interrupted for one or more times, and be superseded by *any other law*; after which the original law shall again be produced, and no other deviation shall ever take place.

Now, as all laws, which appear to us regular and uniform in their course, and to be subject to no exception, can be calculated by the engine: and as each of these laws may also be calculated by the same machine, subject to any assigned interruption, at distinct and definite periods; each simple law may be interrupted at any point by the temporary action of a portion of any one of all the other simple laws: it follows, that *the class of laws subject to interruption is far more extensive than that of laws which are uninterrupted*. It is, in fact, infinitely more numerous. Therefore, the probability of any law with which we have become acquainted by observation being part of a much more extensive law, and of its having, to use mathematical language, singular points or discontinuous functions contained within it, is very large.

Perhaps it may be objected, that the laws calculated by such an engine as I have referred to are not laws of nature; and that any deviation from laws produced by human mechanism does not come within Hume's definition of miracles. To this it may be answered, that a law of nature has been defined by Hume to rest upon experience, or repeated observation, just as the truth of testimony does. Now, the law produced by the engine may be arrived at by precisely the same means—namely, repeated observation.

It may, however, be desirable to explain further the nature of the evidence, on which the fact, that the engine possesses those powers, rests.

When the Calculating Engine has been set to compute the successive terms of any given law, which the observer is told will have an apparent exception (at for example, the ten million and twenty-third term,) the observer is directed to note down the commencement of its computations; and, by comparing these results with his own independent calculations of the same law, he may verify the accuracy of the engine as far as he chooses. It may then be demonstrated to him, by the very structure of the machine, that if its motion were continued, it would, *necessarily*, at the end of a very long time, arrive at the tenth millionth term of the law assigned to it; and that, by an equal *necessity*, it would have passed through all the intermediate terms.* The inquirer is now desired to turn on the wheels with his own hand, until they are precisely in the same situation as they would have been had the engine itself gone on continuously, to the ten-millionth term. The

---

* This can be done in a few minutes.

machine is again put in motion, and the observer again finds that each successive term it calculates fulfils the original law. But, after passing twenty-two terms, he now observes *one* term which does not fulfil the original law, but which does coincide with the predicted exception.

The continued movement now again produces terms according with the first law, and the observer may continue to verify them as long as he wishes. It may then be demonstrated to him, by the very structure of the machine, that, if its motion were continued, it would be *impossible* that any other deviation from the apparent law could ever occur at any future time.

Such is the evidence to the observer; and, if the superintendent of the engine were, at his request, to make it calculate a great variety of different laws, each interrupted by special and remote exceptions, he would have ample ground to believe in the assertion of its director, that he could so arrange the engine that any law, however complicated, might be calculated to any assigned extent and then there should arise one apparent exception; after which the original law should continue uninterrupted for ever.

CHAPTER XV.

The great question of the incompatibility of one of the attributes of the Creator—that of fore-knowledge, with the existence of the free exercise of their will in the beings he has created,—has long baffled human comprehension; nor is it the object of this chapter to enter upon that difficult question.

As, however, some of the properties of the Calculating Engine seem, although but very remotely, to bear on a similar question, with respect to finite beings, it may, perhaps, not be entirely useless to state them.

It has already been observed, that it is possible so to adjust the engine, that it shall change the law it is calculating into another law, at any distant period which may be assigned.

Now, by a similar adjustment, this change may be made to take place at a time not foreseen by the person employing the engine. For example: when calculating a table of squares, it may be made to change into a table of cubes, the first time the square number ends in the figures—

269696;

an event which only occurs at the 99736th calculation; and whether that fact is known to the person who adjusts the machine or not, is immaterial to the result.

But the very condition on which the change depends, may be impossible. Thus, the change of the law from that of squares to that of cubes may be made to take place the first time the square number ends in 7. But it is known, that no square number can end in a seven; consequently the event, on the happening of which the change is determined, can itself never take place. Yet, the engine retains impressed on it a law, which would be called into action if the event on which it depends could occur in the course of the law it is calculating.

Nay, further, if the observer of the engine is informed, that at certain times he can move the last figure the engine has calculated, and change it into any other, in consequence of which it becomes possible that some future term may end in 7; then, after he has so changed the last figure, whenever that terminal figure arrives, all future numbers calculated by the machine will follow the law of the cubes.

These contingent changes may be limited to single exceptions, and the arrival of such an exception may be made contingent on a change which is only possible at certain rare periods. For example: the engine may be set to calculate square numbers, and after a certain number of calculations—ten million and fifty-three, for instance, it shall be possible to add unity to a wheel in another part of the engine, which in every other case is immovable. This fact being communicated to the observer, he may either make that addition or refrain from it: if he refrain, the law of the squares will continue for ever; if he make the addition, one single cube will be substituted for that square number, which ought to occur ten million and five terms beyond the point at which he made the addition; and after that no future addition will ever become possible, and no deviation from the law of the square ever can occur.

— — — — — —

— — — — — —

— — — — — —

— — — — — —

NOTE B. PAGE 33.

ON THE CALCULATING ENGINE.   (*See page* 299 *of this book.*)·

THE nature of the arguments advanced in this volume having obliged me to refer, more frequently than I should have chosen, to the Calculating Engine, it becomes necessary to give the reader some brief account of its progress and present state.

About the year 1821, I undertook to superintend for the Government, the construction of an engine for calculating and printing mathematical and astronomical tables. Early in the year 1833, a small portion of the machine was put together, and was found to perform its work with all the precision which had been anticipated. At that period, circumstances which I could not control, caused what I then considered a temporary suspension of its progress; and the Government, on whose decision the continuance or discontinuance of the work depended, have not yet communicated to me their wishes on the question. The first illustration (p. 33 to 43) I have employed is derived from the calculations made by this engine.

About October, 1834, I commenced the design of another, and far more powerful engine. Many of the contrivances necessary for its performance have since been discussed and drawn according to various principles; and all of them have been invented in more than one form. I consider them, even in their present state, as susceptible of practical execution; but time, thought, and expense, will probably improve them. As the remaining illustrations are all drawn from the powers of this new engine, it may be right to state, that it will calculate the numerical value of any algebraical function—that, at any period previously fixed upon, or contingent on certain events, it will cease to tabulate that algebraic function, and commence the calculation of a different one, and that these changes may be repeated to any extent.

The former engine could employ about 120 figures in its calculations; the present machine is intended to compute with about 4,000.

Here I should willingly have left the subject; but the public having erroneously imagined, that the sums of money paid to the workmen for the construction of the engine, were the remuneration of my own services, for inventing and directing its progress; and a Committee of the House of Commons having incidentally led the public to believe that a sum of money was voted to me for that purpose,—I think it right to give to that report the most direct and unequivocal contradiction.

THE EXPOSITION OF 1851.

PART OF CHAPTER XII.

INTRIGUES OF SCIENCE.

It now becomes necessary to take a very brief review of the conduct of Government with respect to the Difference Engine. Having contrived and executed a small model of a Difference Engine, I published a very short account of it in a letter to Sir Humphry Davy, in the year 1822. At the wish of the Government I undertook to construct for them an engine on a much larger scale, which should print its results. I continued to work at this Engine until 1834, refusing in the mean time other sources of profitable occupation, amongst which was an office of about 2,500*l.* a-year. Circumstances over which I had no control then caused the work to be suspended.

After eight years of repeated applications, and of the most harassing delay, at the end of 1842 the Government arrived at the resolution of giving up the completion of the Difference Engine, on the alleged ground of its expense.

In the mean time, new views had opened out to me the prospect of performing purely algebraic operations by means of mechanism. To arrive at so entirely unexpected a result I deemed worthy of any sacrifice, and accordingly spared no expense in procuring every subsidiary assistance which could enable me to attain it. Each successive difficulty was met by new contrivances, and at last I found that I had surmounted all the great difficulties of the question, and had made drawings of each distinct department of the Analytical Engine.

Having expended upwards of 20,000*l.* on the experiments and enquiries which had led me to these results, it would not have been prudent to attempt the *construction* of such an engine. I thought, however, that there were several offices in the appointment of Government for which I was qualified, and to which, under the circumstances, I had some claim. I hoped if I had obtained one of these, by fulfilling its laborious duties for a few years, and by allowing the whole salary to accumulate, that I might then have been able to retire, and adding the money thus earned to my own private resources, that I might yet have enough of life and energy left to *execute* the Analytical Engine, and thus complete one of the great objects of my ambition.

Having neither asked nor been offered any acknowledgment for all the sacrifices I had made, I felt that I had some just claims to one of these appointments. Every application was unsuccessful; whatever may have been the reasons, the conduct of Government has been exactly that which might have been expected had they been the *allies* or the *dupes* of the party which thought it necessary, from enmity to Sir James South, to " discredit " the author of the Analytical Engine.

One only of the many reports which were circulated, I thought it worth while to contradict, and that cost me more trouble, and wasted more of my time, than the refutation of the calumny was worth. It was boldly and perseveringly stated that I had received from the Government a large pecuniary reward for my services. The fact was, not merely that I never *did* receive any such reward, but that I was almost constantly *advancing money* to pay the engineer who was constructing the Engine for the Government, before I had myself received the amount of his bills from the Treasury.

On tracing up these rumours, they were usually found to arise from a species of dishonesty very difficult to convict. Thus one person circulated them widely; when asked for the grounds of the charge, he referred to certain Parliamentary Papers, and affected to believe that the sums paid *for the workmen* were paid to the *inventor*: of course *he* could no longer safely propagate the falsehood. Another then took up the tale, until he was met by the same question, when *he* not only expressed his delight at being informed of the truth, but half convinced his indignant, though credulous auditor, that *he* would assist in propagating the correction. Thus the assertion was continually repeated, until honourable and upright men, who had been deceived and discovered the deception, were so frequent in society, that it became dangerous to the character of the traducers to continue the circulation of the calumny.

Even since the first edition of this work has appeared, one of these calumnies has been again revived, in the statement that—

The reason why the Government gave up the construction of the original Difference Engine was, that Mr. Babbage refused to finish *it*, and wished them to take up the Difference Engine No. 2.

An attempt has been made to prove its truth by a quotation from this volume, in which the accuser, mistaking dates, assigns the drawings of the Difference Engine No. 2, which did not exist until 1847, as the causes of the discontinuance of No. 1, which was given up in 1843. This charge too is made in the face of a distinct denial by Mr. Babbage that the late Sir Robert Peel could have been influenced by any such *supposed* wish, because he had in his possession a written *dis-avowal* of it from Mr. B. himself; it is also made in the teeth of the very words used by the Chancellor of the Exchequer, who, in his letter to Mr. B. regretting the necessity of giving it up, assigns as its cause "*the expense*." Both these latter statements had been already published in 1848.

CHAPTER XIII.

CALCULATING ENGINES.

IT is not a bad definition of *man* to describe him as a *tool-making animal*. His earliest contrivances to support uncivilized life, were tools of the simplest and rudest construction. His latest achievements in the substitution of machinery, not merely for the skill of the human hand, but for the relief of the human intellect, are founded on the use of tools of a still higher order.

The successful construction of all machinery depends on the perfection of the tools employed, and whoever is a master in the art of tool-making possesses the key to the construction of all machines.

The Crystal Palace, and all its splendid contents, owe their existence to *tools* as the physical means :—to intellect as the guiding power, developed equally on works of industry or on objects of taste.

The contrivance and the construction of tools, must therefore ever stand at the head of the industrial arts.

The next stage in the advancement of those arts is equally necessary to the progress of each. It is the art of drawing. Here, however, a divergence commences : the drawings of the artist are entirely different from those of the mechanician. The drawings of the latter are Geometrical projections, and are of vast importance in all mechanism. The resources of mechanical drawing have not yet been sufficiently explored : with the great advance now making in machinery, it will become necessary to assist its powers by practical yet philosophical rules for expressing still more clearly by signs and by the letters themselves the mutual relations of the parts of a machine.

As we advance towards machinery for more complicated objects, other demands arise, without satisfying which our further course is absolutely stopped. It becomes necessary to see at a glance, not only every *successive* movement of each amongst thousands of different parts, but also to scrutinize all contemporaneous actions. This gave rise to the Mechanical Notation, a language of signs, which, although invented for one subject, is of so comprehensive a nature as to be applicable to many. If the whole of the facts relating to a naval or military battle were known, the mechanical notation would assist the description of it quite as much as it would that of any complicated engine.

This brief sketch has been given partly with the view of more distinctly directing attention to an important point in which England excels all other countries—the art of *contriving and making tools ;* an art which has been continually forced upon my own observation in the contrivance and construction of the Calculating Engines.

When the first idea of inventing mechanical means for the calculation of all classes of astronomical and arithmetical tables occurred to me, I contented myself with making simple drawings, and with forming a small model of a few parts. But when I understood it to be the wish of the Government that a large engine should be constructed, a very serious question presented itself for consideration :—

Is the present state of the art of making machinery sufficiently advanced to enable me to execute the multiplied and highly complicated movements required for the Difference Engine?

After examining all the resources of existing workshops, I came to the conclusion that, in order to succeed, it would become necessary to advance the art of construction itself. I trusted with some confidence that those studies which had enabled me to contrive mechanism for new wants, would be equally useful for the invention of new tools, or of other methods of employing the old.

During the many years the construction of the Difference Engine was carried on, the following course was adopted. After each drawing had been made, a new inquiry was instituted to determine the mechanical means by which the several parts were to be formed. Frequently sketches, or new drawings, were made, for the purpose of constructing the tools or mechanical arrangements thus contrived. This process often elicited some simpler mode of construction, and thus the original contrivances were improved. In the mean time, many workmen of the highest skill were constantly employed in making the tools, and afterwards in using them for the construction of parts of the engine. The knowledge thus acquired by the workmen, matured in many cases by their own experience, and often perhaps improved by their own sagacity, was thus in time disseminated widely throughout other workshops. Several of the most enlightened employers and constructors of machinery, who have themselves contributed to its advance, have expressed to me their opinion that if the Calculating Engine itself had entirely failed, the money expended by Government in the attempt to make it, would be well repaid by the advancement it had caused in the art of mechanical construction.

It is somewhat singular that whilst I had anticipated the difficulties of construction, I had not foreseen a far greater difficulty, which, however, was surmounted by the invention of the Mechanical Notation.

The state of the *Difference Engine* at the time it was abandoned by the Government, was as follows: A considerable portion of it had been made; a part (about sixteen figures) was put together; and the drawings, the whole of which are now in the Museum of King's College at Somerset House, were far advanced. Upon this engine the Government expended about £17,000.

The drawings of the *Analytical Engine* have been made entirely at *my own cost*: I instituted a long series of experiments for the purpose of reducing the expense of its construction to limits which might be within the means I could myself afford to supply. I am now resigned to the necessity of abstaining from its construction, and feel indisposed even to finish the drawings of one of its many general plans. As a slight idea of the state of the drawings may be interesting to some of my readers, I shall refer to a few of the great divisions of the subject.

ARITHMETICAL ADDITION.—About a dozen plans of different mechanical movements have been drawn. The last is of the very simplest order.

CARRIAGE OF TENS.—A large number of drawings have been made of modes of carrying tens. They form two classes, in one of which the carriage takes place successively: in the other it occurs simultaneously, as will be more fully explained at the end of this chapter.

MULTIPLYING BY TENS.—This is a very important process, though not difficult to contrive. Three modes are drawn; the difficulties are chiefly those of construction, and the most recent experiments now enable me to use the simplest form.

DIGIT COUNTING APPARATUS.—It is necessary that the machine should count the digits of the numbers it multiplies and divides, and that it should combine these properly with the number of decimals used. This is by no means so easy as the former operation: two or three systems of contrivances have been drawn.

COUNTING APPARATUS.—This is an apparatus of a much more general order, for treating the indices of functions and for the determination of the repetitions and movements of the Jacquard cards, on which the Algebraic developments of functions depend. Two or three such mechanisms have been drawn.

SELECTORS.—The object of the system of contrivances thus named, is to choose in the operation of Arithmetical division the proper multiple to be subtracted; this is one of the most difficult parts of the engine, and several different plans have been drawn. The one at last adopted is, considering the object, tolerably simple. Although division is an inverse operation, it is possible to perform it entirely by mechanism without any tentative process.

REGISTERING APPARATUS.—This is necessary in division to record the quotient as it arises. It is simple, and different plans have been drawn.

ALGEBRAIC SIGNS.—The means of combining these are very simple, and have been drawn.

PASSAGE THROUGH ZERO AND INFINITY.—This is one of the most important parts of the

Engine, since it may lead to a totally different action upon the formulæ employed. The mechanism is much simpler than might have been expected, and is drawn and fully explained by notations.

BARRELS AND DRUMS.—These are contrivances for grouping together certain mechanical actions often required; they are occasionally under the direction of the cards; sometimes they guide themselves, and sometimes their own guidance is interfered with by the Zero Apparatus.

GROUPINGS.—These are drawings of several of the contrivances before described, united together in various forms. Many drawings of them exist.

GENERAL PLANS.—Drawings of all the parts necessary for the Analytical Engine have been made in many forms. No less than thirty different general plans for connecting them together have been devised and partially drawn; one or two are far advanced. No. 25 was lithographed at Paris in 1840. These have been superseded by simpler or more powerful combinations, and the last and most simple has only been sketched.

A large number of Mechanical Notations exist, showing the movements of these several parts, and also explaining the processes of arithmetic and algebra to which they relate. One amongst them, for the process of division, covers nearly thirty large folio sheets.

About twenty years after I had commenced the first Difference Engine, and after the greater part of these drawings had been completed, I found that almost every contrivance in it had been superseded by new and more simple mechanism, which the construction of the Analytical Engine had rendered necessary. Under these circumstances I made drawings of an entirely new Difference Engine. The drawings, both for the calculating and the printing parts, amounting in number to twenty-four, are completed. They are accompanied by the necessary mechanical notations, and by an index of letters to the drawings; so that although there is as yet no description in words, there is effectively such a description by signs, that this new Difference Engine might be constructed from them.

Amongst the difficulties which surrounded the idea of the construction of an Engine for developing Analytical formulæ, there were some which seemed insuperable if not impossible, not merely to the common understandings of well-informed persons, but even to the more practised intellect of some of the greatest masters of that science which the machine was intended to control. It still seemed, after much discussion, at least highly doubtful whether such formulæ could ever be brought within the grasp of mechanism.

I have met in the course of my inquiries with four cases of obstacles presenting the appearance of impossibilities. As these form a very interesting chapter in the history of the human mind, and are on the one hand connected with some of the simplest elements of mechanism, and on the other with some of the highest principles of philosophy, I shall endeavour to explain them in a short, and, I hope, somewhat popular manner, to those who have a very moderate share of mathematical knowledge. Those of my readers to whom they may not be sufficiently interesting, will, I hope, excuse the interruption, and pass on to the succeeding chapters.

§ The first difficulty arose at an early stage of the Analytical Engine. The mechanism necessary to add one number to another, if the carriage of the tens be neglected, is very simple. Various modes had been devised and drawings of about a dozen contrivances for carrying the tens had been made. The same general principle pervaded all of them. Each figure wheel when receiving addition, in the act of passing from nine to ten caused a lever to be put aside. An axis with arms arranged spirally upon it then revolved, and commencing with the lowest figure replaced successively those levers which might have been put aside during the addition. This replacing action upon the levers caused unity to be added to the figure wheel next above. The numerical example below will illustrate the process.

597,999⎫ 201,001⎭	Numbers to be added.
798,990 1	Sum without any carriage. Puts aside lever acting on tens.
798,900 1	First spiral arm adds tens and puts aside the next lever.
798,000 1	Second spiral arm adds hundreds, and puts aside the next lever.
799,000	Third spiral arm adds thousands.

Now there is in this mechanism a certain analogy with the act of memory. The lever thrust aside by the passage of the tens, is the equivalent of the note of an event made in the memory, whilst the spiral arm, acting at an after time upon the lever put aside, in some measure resembles the endeavours made to recollect a fact.

It will be observed that in these modes of *carrying*, the action must be *successive*. Supposing a number to consist of thirty places of figures, each of which is a nine, then if any other number of thirty figures be added to it, since the addition of each figure to the corresponding one takes place at the same time, the whole addition will only occupy nine units of time. But since the number added may be unity, the carriages may possibly amount to twenty-nine, consequently the time of making the carriages may be more than three times as long as that required for addition.

The time thus occupied was, it is true, very considerably shortened in the Difference Engine : but when the Analytical Engine was to be contrived, it became essentially necessary to diminish it still further. After much time fruitlessly expended in many contrivances and drawings, a very different principle, which seemed indeed at first to be impossible, suggested itself.

It is evident that whenever a carriage is conveyed to the figure above, if that figure happen to be a nine, a new carriage must then take place, and so on as far as the nines extend. Now the principle sought to be expressed in mechanism amounted to this.

1st. That a lever should be put aside, as before, on the passage of a figure-wheel from nine to ten.

2d. That the engine should then ascertain the position of all those nines which by carriage would ultimately become zero, and give notice of new carriages; that, foreseeing those events, it should anticipate the result by making all the carriages simultaneously.

This was at last accomplished, and many different mechanical contrivances fulfilling these conditions were drawn. The former part of this mechanism bears an analogy to memory, the latter to foresight. The apparatus remembers as it were, one set of events, the transits from nine to ten : examines what nines are found in certain critical places : then, in consequence of the concurrence of these events, acts at once so as to anticipate other actions that would have happened at a more distant period, had less artificial means been used.

§ The second apparent impossibility seemed to present far greater difficulty. Fortunately it was not one of immediate *practical* importance, although as a question of philosophical inquiry it possessed the highest interest. I had frequently discussed with Mrs. Somerville and my highly-gifted friend the late Professor M'Cullagh of Dublin, the question whether it was possible that we should be able to treat algebraic formulæ by means of machinery. The result of many inquiries led to the conclusion that, if not really impossible, it was almost hopeless. The first difficulty was that of representing an indefinite number in a machine of finite size. It was readily admitted that if a machine afforded means of operating on *all* numbers under twenty places of figures, then that any number, or *an indefinite* number, of less than twenty places of figures might be represented by it. But such number will not be really indefinite. It would be possible to make a machine capable of operating upon numbers of forty, sixty, or one hundred places of figures : still, however, a limit must at last be reached, and the numbers represented would not be really *indefinite*. After lengthened consideration of this subject, the solution of the difficulty was discovered ; and it presented the appearance of reasoning in a circle.

Algebraical operations in their most general form cannot be carried on by machinery without the capability of expressing *indefinite* constants. On the other hand, the only way of arriving at the expression of an indefinite constant, was through the intervention of Algebra itself.

This is not a fit place to enter into the detail of the means employed, further than to observe, that it was found possible to evade the difficulty by connecting *indefinite* number with the *infinite in time* instead of with the *infinite in space*.

The solution of this difficulty being found, and the discovery of another principle having been made, namely—that *the nature of a function might be indicated by its position*—algebra, in all its most abstract forms, was placed completely within the reach of mechanism.

§ The third difficulty that presented itself was one which I had long before anticipated. It was proposed to me nearly at the same time by three of the most eminent cultivators of analysis then existing, M. Jacobi, M. Bessel, and Professor M Cullagh, who were examining the drawings of the Analytical Engine. The question they proposed was this :—How would the Analytical Engine be able to treat calculations in which the use of tables of logarithms, sines, &c., or any other tabular numbers should be required ?

My reply was, that as at the time logarithms were invented, it became necessary to remodel the whole of the formulæ of Trigonometry, in order to adapt it to the new instrument of calculation : so when the Analytical Engine is made, it will be desirable to transform all formulæ containing

2 T

tabular numbers into others better adapted to the use of such a machine. This, I replied, is the answer I give to you as mathematicians; but I added, that for others less skilled in our science, I had another answer: namely—

That the engine might be so arranged that wherever tabular numbers of any kind occurred in a formula given it to compute, it would on arriving at any required tabular number, as for instance, if it required the logarithm of 1207, stop itself, and ring a bell to call the attendant, who would find written at a certain part of the machine "Wanted log. of 1207." The attendant would then fetch from tables previously computed by the engine, the logarithm it required, and placing it in the proper place, would lift a detent, permitting the engine to continue its work.

The next step of the engine, on receiving the tabular number (in this case the logarithm of 1207) would be to *verify* the fact of its being really that logarithm. In case no mistake had been made by the attendant, the engine would use the given tabular number, and go on with its work until some other tabular number were required, when the same process would be repeated. If, however, any mistake had been made by the attendant, and a wrong logarithm had been accidentally given to the engine, it would have discovered the mistake, and have rung a louder bell to call the attention of its guide, who on looking at the proper place, would see a plate above the logarithm he had just put in with the word "*wrong*" engraven upon it.

By such means it would be perfectly possible to make all calculations requiring tabular numbers, without the chance of error.

Although such a plan does not seem absolutely impossible, it has always excited, in those informed of it for the first time, the greatest surprise. How, it has been often asked, does it happen if the engine knows when the *wrong* logarithm is offered to it, that it does not also know the right one; and if so, what is the necessity of having recourse to the attendant to supply it? The solution of this difficulty is accomplished by the very simplest means.

§ The fourth of the apparent impossibilities to which I have referred, involves a condition of so extraordinary a nature that even the most fastidious inquirer into the powers of the Analytical Engine could scarcely require it to fulfil.

Knowing the kind of objections that my countrymen make to this invention, I proposed to myself this inquiry :—

Is it possible so to construct the Analytical Engine, that after the cards representing the formulæ and numbers are put into it, and the handle is turned, the following condition shall be fulfilled?

The attendant shall stop the machine in the middle of its work, whenever he chooses, and as often as he pleases. At each stoppage he shall examine all the figure wheels, and if he can, without breaking the machine, move any of them to other figures, he shall be at liberty to do so. Thus he may from time to time, falsify as many numbers as he pleases. Yet notwithstanding this, the final calculation and all the intermediate steps shall be entirely free from error. I have succeeded in fulfilling this condition by means of a principle in itself very simple. It may add somewhat, though not very much, to the amount of mechanism required; in many parts of the engine the principle has been already carried out. I by no means think such a plan *necessary*, although wherever it can be accomplished without expense it ought to be adopted.

THE EXPOSITION OF 1851.

APPENDIX.

EXTRACT FROM WELD'S HISTORY OF THE ROYAL SOCIETY.

Reprinted with the permission of the Proprietor.

CHAPTER XI.

The Society receive a Letter from the Treasury respecting Mr. Babbage's Calculating Machine—Letter from Mr. Babbage to Sir H. Davy—A Committee appointed to consider Mr. Babbage's Plan—They Report in favour of it—Mr. Babbage has an interview with the Chancellor of the Exchequer—Government advance 1,500*l.*—Difference Engine commenced—Mr. Babbage gives all his labour gratuitously—Advice of the Society again requested—Mr. Babbage's Statement—Committee appointed to inspect the Engine—Their Report—Heavy Expenses not met by the Treasury—Meeting of Mr. Babbage's personal friends—Their Report—Duke of Wellington inspects the Works—His Grace recommends the Treasury to make further Payments—Letter from Mr. Babbage to the Treasury—Communication from the Treasury to the Council—Referred to a Committee—Report of Committee—They recommend the Works to be removed to the vicinity of Mr. Babbage's Residence—Government act on the Recommendation—Fire-proof Buildings erected—Misunderstanding with Mr. Clement—Works stopped—Mr. Babbage discovers new principles which supersede those connected with the Difference Engine—He requests an interview with Lord Melbourne—Letter to M. Quetelet explaining the principles of Analytical Engine—Mr. Babbage visits Turin—M. Menabrea's account of the Engine—Translated with Notes by Lady Lovelace—Mr. Babbage applies to Government for their Determination—Letter from the Chancellor of the Exchequer—Mr. Babbage's Answer—Government resolve not to proceed with the Engine—Mr. Babbage has an interview with Sir R. Peel—Difference Engine placed in the Museum of King's College—Present State of the Analytical Engine.

1820-25.

On the 1st April, 1823, a letter was received from the Treasury, requesting the Council to take into consideration a plan which had been submitted to Government by Mr. Babbage, for "applying machinery to the purposes of calculating and printing mathematical tables;" and the Lords of the Treasury further desired " to be favoured with the opinion of the Royal Society on the merits and utility of this invention." [1]

This is the earliest allusion to the celebrated Calculating Engine of Mr. Babbage, in the records of the Society.[2] But the invention had been brought before them in the previous year by a letter from Mr. Babbage to Sir H. Davy, dated July 3, 1822, in which he gives some account of a small model of his engine for calculating differences, which "produced figures at the rate of 44 a minute, and performed with rapidity and precision all those calculations for which it was designed."[3] He then proceeds to enumerate various tables which the machine was adapted to calculate, and concludes: "I am aware that these statements may perhaps be viewed as something more than Utopian, and that the philosophers of Laputa may be called up to dispute my claim to originality. Should such be the case, I hope the resemblance will be found to adhere to the nature of the

---

[1] In the following account of the Difference and Analytical Engines, besides the MS. documents in the Archives of the Royal Society, I have derived very valuable information from an unpublished statement drawn up by Mr. Babbage, which he has been so kind as to place in my hands. The original documents which are in Mr. Babbage's possession, and which are referred to, I have myself examined.

[2] The idea of a Calculating Engine is not new. The celebrated Pascal constructed a machine for executing the ordinary operations of arithmetic, a description of which will be found in the *Encycl. Méthod.*, and in the Works of Pascal, Tom. iv. p. 7, Paris, 1819. In his *Pensées* he says, alluding to this Engine: " *La machine arithmétique fait des effets qui approchent plus de la pensée que tout ce que font les animaux ; mais elle ne fait rien qui puisse faire dire qu'elle a de la volonté comme les animaux.*" Subsequently, Leibnitz invented a machine by which, says Mr. De Morgan, "arithmetic computations could be made." Polenus, a learned and ingenious Italian, invented a machine by which multiplication was performed—and mechanical contrivances for performing particular arithmetical processes were made about a century ago, but they were merely modifications of Pascal's. These Engines were very different to Mr. Babbage's Difference Engine.

[3] This letter was printed and published in July, 1822.

subject, rather than to the manner in which it has been treated. Conscious from my own experience of the difficulty of convincing those who are but little skilled in mathematical knowledge, of the possibility of making a machine which shall perform calculations, I was naturally anxious, in introducing it to the public, to appeal to the testimony of one so distinguished in the records of British science.[4] Induced by a conviction of the great utility of such engines, to withdraw for some time my attention from a subject on which it has been engaged during several years, and which possesses charms of a higher order, I have now arrived at a point where success is no longer doubtful. It must, however, be attained at a very considerable expense, which would not probably be replaced by the works it might produce for a long period of time, and which is an undertaking I should feel unwilling to commence, as altogether foreign to my habits and pursuits."

The Council appointed a Committee to take Mr. Babbage's plan into consideration, which was composed of the following gentlemen: Sir H. Davy, Mr. Brande, Mr. Combe, Mr. Baily, Mr. (now Sir Mark Isambard) Brunel, Major (now General) Colby, Mr. Davies Gilbert, Mr. (now Sir John) Herschel, Captain Kater, Mr. Pond (Astronomer-Royal), Dr. Wollaston, and Dr. Young. On the 1st May, 1823, the Committee reported: "That it appears that Mr. Babbage has displayed great talents and ingenuity in the construction of his machine for computation, which the Committee think fully adequate to the attainment of the objects proposed by the inventor, and that they consider Mr. Babbage as highly deserving of public encouragement in the prosecution of his arduous undertaking."[5]

This Report was transmitted to the Lords of the Treasury, by whom it was, with Mr. Babbage's letter to Sir H. Davy, printed and laid before Parliament.[6]

In July, 1823, Mr. Babbage had an interview with the Chancellor of the Exchequer, Mr. Robinson (now Earl of Ripon), to ascertain if it were the wish of Government that he should construct a large engine of the kind, which would also print the results it calculated. Unfortunately, no Minute of that conversation was made at the time, nor was any sufficiently distinct understanding arrived at, as it afterwards appeared that a contrary impression was left on the mind of either party.[7] Mr. Babbage's conviction was, that whatever might be the labour and difficulty of the undertaking, the engine itself would, of course, become the property of the Government, which had paid for its construction.

Soon after this interview with the Chancellor of the Exchequer, a letter was sent from the Treasury to the Royal Society, informing them that the Lords of the Treasury "had directed the issue of 1,500l. to Mr. Babbage, to enable him to bring his invention to perfection in the manner recommended."

These words "in the manner recommended," can refer only to the previous recommendation by the Royal Society; but it does not appear from their Report that any plan, terms, or conditions had been pointed out.

Towards the end of July, 1823, Mr. Babbage took measures for the construction of the present Difference Engine,[8] and it was regularly proceeded with for four years.

And here it is right to state that Mr. Babbage gave his mental labour gratuitously, and that from first to last he has not derived any emolument whatever from Government.[9] Sectional, and other drawings, of the most delicate nature had to be made; tools to be formed expressly to meet mechanical difficulties; and workmen to be educated in the practical knowledge necessary in the construction of the machine. The mechanical department was placed under the management of Mr. Clement, a draughtsman of great ability, and a practical mechanic of the highest order.[10] Money was advanced from time to time by the Treasury, the accounts furnished by the engineer

[4] Sir H. Davy had witnessed and expressed his admiration of the performances of the Engine.

[5] I am informed upon good authority, that Dr. Young differed in opinion from his colleagues. Without doubting that an engine could be made, he conceived that it would be far more useful to invest the probable cost of constructing such a calculating machine as was proposed, in the funds, and apply the dividends to paying calculators.

[6] Parliamentary Paper, No. 370, 1823.

[7] Mr. Babbage very justly observes, that had the mutual relations of the two parties, and the details of the plans then adopted, been clearly defined, there is little doubt but that the Difference Engine would long since have existed.

[8] It will be desirable to distinguish between,
  1. The small *Model* of the Original or Difference Engine.
  2. The Difference Engine itself, belonging to the Government, a part only of which has been put together.
  3. The designs for another Engine called the Analytical Engine.

[9] Sir R. Peel distinctly admitted this in the House of Commons in March, 1843.

[10] A curious anecdote is related illustrative of the great perfection to which Mr. Clement was in the habit of bringing machinery. He received an order from America to construct a large screw in the *best possible manner*, and he accordingly made one with the greatest mathematical accuracy. But his bill amounted to some hundreds of pounds, which completely staggered the American, who never calculated upon paying more than 20l. at the utmost for the screw. The matter was referred to arbitrators, who gave an opinion in favour of Mr. Clement.

undergoing the examination of auditors,[11] and passing through the hands of Mr. Babbage. Thus years elapsed, and public attention became at length directed to the fact that a large sum had been expended upon the construction of the engine, which was not completed. Again the advice of the Royal Society was solicited.

In December, 1828, Government begged the Council " to institute such enquiries as would enable them to report upon the state to which it (the machine) had then arrived ; and also whether the progress made in its construction confirmed them in the opinion which they had formerly expressed, that it would ultimately prove adequate to the important object which it was intended to attain."

Accompanying this communication was a statement from Mr. Babbage of the condition of the engine, in which he says :—

"The machine has required a longer time and greater expense than was anticipated, and Mr. Babbage has already expended about 6,000l. on this object. The work is now in a state of considerable forwardness, numerous and large drawings of it have been made, and much of the mechanism has been executed, and many workmen are occupied daily in its completion."

A Committee was appointed by the Council, consisting of Mr. Gilbert (President), Dr. Roget, Captain Sabine, Sir John Herschel, Mr. Baily, Mr. Brunel, Captain Kater, Mr. Donkin, Mr. Penn, Mr. Rennie, Mr. Barton, and Mr. Warburton.[12]

They minutely inspected the drawings, tools, and the parts of the engine then executed, and drew up a report, " declining to consider the principle on which the practicability of the machinery depends, and of the public utility of the object which it proposes to attain : because they considered the former fully admitted, and the latter obvious to all who consider the immense advantage of accurate numerical tables in all matters of calculation, which it is professedly the object of the engine to calculate and print with perfect accuracy."

They further stated, that " the progress made was as great as could be expected, considering the numerous difficulties to be overcome; and lastly, that they had no hesitation in giving it as their opinion that the engine was likely to fulfil the expectations entertained of it by its inventor."

The Council adopted the Report, expressing their trust that while Mr. Babbage's mind was intently occupied on an undertaking likely to do so much honour to his country, he might be relieved as much as possible from all other sources of anxiety.

It is clear that the Council of the Royal Society regarded Mr. Babbage's engine, as it then existed, in a favourable light, and were sanguine respecting its satisfactory completion.

Government acted on the foregoing Report; funds were advanced, the machinery was declared national property, and the works were continued. But there was evidently a misgiving on the part of the Lords of the Treasury, for the official payments soon failed to meet the heavy and increasing expenses incurred by Mr. Babbage.

Under these circumstances, by the advice of Mr. Wolryche Whitmore (Mr. Babbage's brother-in-law), a meeting of Mr. Babbage's personal friends was held on the 12th of May, 1829. It consisted of :—

> The Duke of Somerset, F.R.S.,
> Lord Ashley, M.P.,
> Sir John Franklin, Capt. R.N., F.R.S.,
> Mr. Wolryche Whitmore, M.P.,
> Dr. Fitton, F.R.S.,
> Mr. Francis Baily, F.R.S.,
> Sir John Herschel, F.R.S.,

They drew up the annexed Report :—

May 12, 1829.

" The attention of the undersigned personal friends of Mr. Babbage having been called by him to the actual state of his Machine for Calculating and Printing Mathematical Tables ; and to his relation to the Government on the one hand, and to the Engineers and workmen employed by him in its execution on the other, declare themselves satisfied, from his statements and from the documents they have perused, of the following facts.

" That Mr. Babbage was originally induced to take up the work on its present extensive scale, by an understanding on his part, that it was the wish of Government he should do so, and by an advance of 1,500l. in the outset, with a full impression on his mind that such further advances would be made as the progress of the work should require, and as should secure him from ultimate loss.

" That the public and scientific importance of the Engine has been acknowledged, in a Report of a Committee of the Royal Society, made at the time of its first receiving the sanction of His Majesty's Government, and that its actual state

---

[11] They were Messrs. Brunel, Donkin, and Field.
[12] Colonel Sabine informs me, that Dr. Whewell was afterwards added to the Committee.

of progress is such, as in the opinion of the most eminent Engineers and other Members of the Royal Society, as detailed in a further Report of a Committee of that body, to warrant their impression of the moral certainty of its success, should funds not be wanting for its completion.

"That it appears that Mr. Babbage's actual expenditure has amounted to nearly 7,000*l.* and that the whole sum advanced to him by the Government is 3,000*l.*

"That Mr. Babbage has devoted, from the commencement of his arduous undertaking, the most assiduous and anxious attention to the work in hand, to the injury of his health, and the neglect and refusal of other profitable occupations.

"That a very large expense still remains to be incurred, to the probable amount of at least 4,000*l.*, as far as he can foresee, before the Engine can be completed; but that Mr. Babbage's private fortune is not such as, in their opinion, to justify the sacrifices he must make in completing it without further and effectual assistance from Government; taking into consideration not only his own interest, but that of his family dependent on him.

"Under these circumstances, it is their opinion that a full and speedy representation of the case ought to be made to Government, and that in the most direct manner by a personal application to his Grace the Duke of Wellington.

"And that in case of such application proving unsuccessful in procuring effectual and adequate assistance, they must regard Mr. Babbage as no longer called on—considering the pecuniary and personal sacrifices he will then have made; considering the entire and *bonâ fide* expenditure of all that he will have received from the public purse on the object of its destination, and considering the moral certainty to which it is at length by his exertions reduced—as no longer called on to go on with an undertaking which may prove the destruction of his health, and the great injury, if not the ruin of his fortune.

"That it is their opinion that Mr. W. Whitmore and Mr. Herschel should request an interview with the Duke of Wellington for the purpose of making this representation.

(Signed,)   "SOMERSET.
" ASHLEY.
" JOHN FRANKLIN.
" W. W. WHITMORE.
" WM. HENRY FITTON.
" FRANCIS BAILY.
" J. F. W. HERSCHEL."

In consequence of what passed at this interview, which took place as suggested, the Duke of Wellington, accompanied by the Chancellor of the Exchequer (Mr. Goulburn) and Lord Ashley, inspected the *model* of the engine, the drawings, and parts in progress. The Duke recommended that a grant of 3,000*l.* should be made towards the completion of the machine, which was duly paid by the Treasury.

In the mean time, difficulties of another kind arose. The engineer who had constructed the Engine under Mr. Babbage's directions, had delivered his bills in such a state, that it was impossible to judge how far the charges were just and reasonable; and although Mr. Babbage had paid several thousand pounds, there yet remained a considerable balance, which could not be liquidated until the accounts had been examined, and the charges approved by professional engineers.

With a view of drawing attention to these charges, Mr. Babbage addressed the following letter to the Chancellor of the Exchequer :—

"MY LORD,   *Dorset Street, 21 December,* 1830.

"I beg to call your Lordship's attention to the enclosed account[13] of the expenses of the Machine for calculating and printing mathematical tables, by which it appears that a sum of 592*l.* 4*s.* 8*d.* remained due to myself upon the last account, and that a further sum of nearly 600*l.* has since become due to Mr. Clement.

"It is for the payment of this latter sum that I wish to call your Lordship's attention. Mr. Maudslay, one of the engineers appointed by the Government to examine the bills of Mr. Clement, having been unable from illness to attend, his report has been delayed, and Mr. Clement informs me that should the money remain unpaid much longer, he shall be obliged, from want of funds, to discharge some of the workmen; an event which I need not inform your Lordship would be very prejudicial to the progress of the machine.

"Another point which I wish to submit to your attention, when your Lordship shall have had leisure to examine personally the present state of the works, is, that since it is absolutely necessary to find additional room for the erection of the machine, it becomes a matter of serious consideration whether it would not contribute to the speedier completion of the machine, and also to economy in expenditure, to remove the works to the neighbourhood of my own residence.

"I have, &c.
"C. BABBAGE."

		£	s.	d.
[13] Expense to end of 1824		600	0	0
Ditto „ „ 1827		521	16	9
Mr. Clement's Bills to June, 1827		4,775	15	3
Ditto, 9th May, 1829		730	12	8
		6,628	4	8
Deduct old tools sold		36	0	0
		6,592	4	8
Mr. Clement's Bill to December 1830, *about.*		600	0	0
		7,192	4	8

The receipt of this letter caused the Treasury to make the following communication to the Secretary of the Royal Society :—

"Sir,   *Treasury, 24 December*, 1830.
    "The Lords Commissioners of H. M. Treasury, having had under their consideration a letter from Mr. Babbage, containing an account of the expense which has been incurred in the construction of the Machine for calculating and printing mathematical tables, amounting to the sum of 7,192*l.* 4*s.* 8*d.*, and requesting an advance of 600*l.* to defray a part of that expense; I am commanded by their Lordships to refer you to the Report of the Council of the Royal Society dated 16th February, 1829, which entirely satisfied their Lordships of the propriety of supporting Mr. Babbage in the construction of this machine, and to state that advances to the amount of 6,000*l.* have been made on this account, and that directions have been given for a further advance of 600*l.*

    "I am also to acquaint you that the Machine is the property of Government, and consequently my Lords propose to defray the further expense necessary for its completion. I am further to request you will move the Council of the Royal Society to cause the machine to be inspected, and to favour my Lords with their opinion whether the work is proceeding in a satisfactory manner, and without unnecessary expense, and what further sum may probably be necessary for completing it.

                                                        "I am, &c.,
"*The Secretary, Royal Society.*   "J. Stewart."

The consideration of this letter was referred to the same Committee which had previously been appointed for a similar purpose, with the addition of Sir John Lubbock and Mr. Troughton.

    Again the Committee met[14] Mr. Babbage, at No. 21, Prospect Place, Lambeth (where the construction of the engine was carried on), and minutely inspected the machinery and drawings.

    Their Report embodied the whole facts of the case :—the workmanship of the various parts of the machine was declared to have been executed with the greatest possible degree of perfection, and the pains taken to verify the charges on the part of the Government altogether satisfactory. It was recommended that the vacancy occasioned by the decease of Mr. Maudslay, who had been appointed to inspect the accounts, should be filled up by another engineer, conversant with the execution of machinery, and the value thereof. With respect to the suggested removal of the workshops nearer to Mr. Babbage's residence, the Committee gave their entire concurrence, on the ground that greater expedition would thereby be attained in carrying on the work, and that it was highly essential to secure all the machinery and drawings in fire-proof premises, without delay. A plot of ground held on lease by Mr. Babbage, adjacent to his garden at the back of his house in Dorset Street, was recommended as a desirable site for the contemplated erections, of which the plans and estimates had been submitted to the Committee. The framers of the Report stated in conclusion that :—

    "Such an arrangement would be eminently conducive to the speedy and economical completion of the Machine, as well as to the effectual working and employment of the same, after it shall have been completed.
    "That as to the sum which may be necessary for completing the Engine, they attach hereto the estimate of Mr. Brunel."[15]

    The Report, with Mr. Brunel's estimate, were sent to the Treasury on the 13th April, 1831: and having been approved by a Committee of practical engineers appointed by Government, the latter acted on the recommendations which it contained. The piece of ground adjoining Mr. Babbage's garden was taken, and a fire-proof building erected, designed to contain the plans and drawings, and also the engine when completed. But new and unforeseen difficulties arose. When about 17,000*l.* had been expended, further progress was arrested on account of a misunderstanding with Mr. Clement, who made the most extravagant demands as compensation for carrying on the construction of the engine in the new buildings. These demands could not be satisfied with proper regard to the justice due to Government. Mr. Clement accordingly withdrew from the undertaking, and carried with him all the valuable tools that had been used in the work; a proceeding the more

---

    [14] I have a letter of Sir J. Herschel's before me, expressing his regret at being unable to attend on this occasion, but that his faith in the engine and its inventor remained unshaken.
    [15] Mr. Brunel's estimate appears in the following letter to Mr. Warburton :—

    "Dear Sir,   *Feb.* 28, 1831.
    "Having taken in consideration the erection of the proposed shops, the removal of the machinery, the accommodation for it, and also for the maker : having also taken into consideration the further completion of the drawings, and the ultimate accomplishment of the Engine until it is capable of producing plates for printing; though I feel confident that the sum of 8,000*l.* will be ample to realize the objects that are contemplated, I should nevertheless recommend that the Government be advised to provide for the sum of 12,000*l.* by way of estimate, and that the yearly sum required, exclusive of the sum requisite for the buildings and removal (say 2,000*l.*), will not exceed from 2,000*l.* to 2,500*l.*
                                                        "I am, &c.,
                                                        "M. I. Brunel."
'*Henry Warburton, Esq.*

unfortunate, as many of them had been invented expressly to meet the unusual forms and combinations arising out of the novel construction.[16]

An offer was made to surrender the tools, for a given sum, which was declined, and the works came to a standstill.   But other circumstances interposed to prevent the completion of the original design.

During the suspension of the works, Mr. Babbage had been deprived of the use of his own drawings.   Having in the meanwhile naturally speculated upon the general principles on which machinery for calculation might be constructed, *a principle of an entirely new kind* occurred to him, the power of which over the most complicated arithmetical operations seemed nearly unbounded. This was the executing of analytical operations by means of an analytical engine.   On re-examining his drawings, when returned to him by the engineer, the new principle appeared to be limited only by the extent of the mechanism it might require.   The invention of simpler mechanical means for performing the elementary operations of the engine, now derived a far greater importance than it had hitherto possessed; and should such simplifications be discovered, it seemed difficult to anticipate, or even to over-estimate, the vast results which might be attained.

These new views acquired additional importance from their bearings upon the engine already partly executed for the Government; for, if such simplifications should be discovered, it might happen that the Analytical Engine would execute with greater rapidity the calculations for which the Difference Engine was intended; or that the Difference Engine would itself be superseded by a far simpler mode of construction.

Though these views might perhaps at that period have appeared visionary, they have subsequently been completely realized.   To have allowed the construction of the Difference-Engine to be resumed, while these new conceptions were withheld from the Government, would have been improper; yet the state of uncertainty in which those views were then necessarily involved, rendered any written communication respecting their probable bearing on that engine, a task of very great difficulty.   It therefore appeared to Mr. Babbage, that the most straightforward course was to ask for an interview with the head of the Government, and to communicate to him the exact state of the case.

On the 26th September, 1834, Mr. Babbage requested an audience of Lord Melbourne, for the purpose of placing these views before him; his Lordship acceded to the request, but from some cause the interview was postponed; and soon after, the ministry went out of office, without the desired conference having taken place.

The duration of the Duke of Wellington's administration was short; and no decision on the subject of the *Difference* Engine was obtained.

In May 1835, Mr. Babbage announced in a letter [17] to M. Quetelet, which was laid before the Academy of Sciences at Brussels, that he had " for six months been engaged in making the drawings of a new calculating engine of *far greater power than the first*."   " I am myself astonished," says Mr. Babbage, " at the power I have been enabled to give to this machine; a year ago I should not have believed this result possible.   This machine is intended to contain a hundred variables, or numbers susceptible of changing, and each of these numbers may consist of twenty-five figures. The greatest difficulties of the invention have already been surmounted, and the plans will be finished in a few months."

Subsequently to the date of this letter, Mr. Babbage visited Turin, where he explained to Baron Plana, M. Menabrea, and several other distinguished philosophers of that city, the mathematical principles of his Analytical Engine, and also the drawings and engravings of the more curious mechanical contrivances, by which those principles were to be carried into effect.   M. Menabrea, with Mr. Babbage's consent, published the information which he had received in the 41st volume of the *Bibliothèque Universelle de Génève*.   The article is remarkable as giving the first account of the Analytical Engine.[18]   An English translation, with copious original notes, made by a lady of distinguished rank and talent,[19] was published in the third volume of Taylor's *Scientific Memoirs*.

But it did not contain all the information respecting the Difference Engine that was desirable, and Mr. Babbage was consequently led to communicate a short article upon this subject to the

[16] This Mr. Clement had a legal right to do.   Startling as it may appear to the unprofessional reader, it is nevertheless the fact, that engineers and mechanics possess the right of property to all tools that they have constructed, although the cost of construction has been defrayed by their employers.

[17] Mr. Babbage informs me that this letter was intended only as a private communication.

[18] In the *Ninth Bridgewater Treatise*, Mr. Babbage has employed various arguments deduced from the Analytical Engine, which afford some idea of its powers.   See second edition.   In 1838, several copies of plans of this new engine, engraved on wood, were circulated amongst Mr. Babbage's friends at the Meeting of the British Association at Newcastle.

In 1840, Mr. Babbage had one of his general plans of the Analytical Engine lithographed at Paris.

[19] I am authorized by Lord Lovelace to say that the translator is Lady Lovelace.

Philosophical Magazine, which is inserted in the 23rd volume.[20] The more comprehensive statements and official documents which Mr. Babbage has placed at my disposal render it unnecessary to do more than allude to that article.

For nine years, that is, from the year 1833, when the construction of the Difference Engine was suspended, until 1842, no decision respecting the machine was arrived at, although Mr. Babbage made several applications to Government on the subject.

On the 21st October, 1838, he wrote to the Chancellor of the Exchequer, stating that the question he wished to have settled was:—" Whether the Government required him to superintend the completion of the Difference Engine, which had been suspended during the last five years, according to the original plan and principle, or whether they intended to discontinue it altogether." This letter produced no result. Time wore on, and Sir Robert Peel became Prime Minister. This was in 1841. Up to the termination of the Parliamentary Session in 1842, Mr. Babbage had received no other communication on the subject than a note from Sir George Clerk (Secretary to the Treasury), written in January of that year, stating that he feared the pressing official duties of Sir Robert Peel would prevent him turning his attention to the matter for some days.

Having availed himself of several private channels for recalling the question to Sir Robert Peel's attention without effect, Mr. Babbage, on the 8th of October, 1842, again wrote to him, requesting an early decision.

At last Mr. Babbage received the following letter:—

"MY DEAR SIR, *Downing Street, Nov. 3, 1842.*
" The Solicitor-General has informed me that you are most anxious to have an early and decided answer as to the determination of the Government with respect to the completion of your Calculating Engine. I accordingly took the earliest opportunity of communicating with Sir R. Peel on the subject.

" We both regret the necessity of abandoning the completion of a Machine on which so much scientific ingenuity and labour have been bestowed. But on the other hand, the expense which would be necessary in order to render it either satisfactory to yourself, or generally useful, appears on the lowest calculation so far to exceed what we should be justified in incurring, that we consider ourselves as having no other alternative.

" We trust that by withdrawing all claim on the part of the Government to the Machine as at present constructed, and by placing it at your entire disposal, we may, to a degree, assist your future exertions in the cause of science.
" I am, &c.,
" *Charles Babbage, Esq.* " HENRY GOULBURN.

" P.S.—Sir R. Peel begs me to add, that as I have undertaken to express to you our joint opinion on this matter, he trusts you will excuse his not separately replying to the letter which you addressed to him on the subject a short time since."

To this letter Mr. Babbage replied as follows :—

"MY DEAR SIR, *Dorset Street, Nov. 6, 1842.*
" I beg to acknowledge the receipt of your letter of the 3rd of Nov., containing your own and Sir Robert Peel's decision respecting the Engine for calculating and printing mathematical tables by means of Differences, the construction of which has been suspended about eight years.

" You inform me that both regret the necessity of abandoning the completion of the Engine, but that not feeling justified in incurring the large expense which it may probably require, you have no other alternative.

" You also offer, on the part of Government, to withdraw all claim in the Machine as at present constructed, and to place it at my entire disposal, with the view of assisting my future exertions in the cause of science.

" The drawings and the parts of the Machine already executed are, as you are aware, the absolute property of Government, and I have no claim whatever to them.

" Whilst I thank you for the feeling which that offer manifests, I must, under all the circumstances, decline accepting it.
" I am, &c.,
" C. BABBAGE."

Mr. Babbage had an interview with Sir R. Peel subsequently to the date of the foregoing letter: the result was, however, entirely unsatisfactory; and thus, with the communication from the then Chancellor of the Exchequer, terminated an engagement which had existed upwards of twenty years, during which period it is due to Mr. Babbage to state that he refused more than one highly desirable and profitable situation,[21] in order that he might give his whole time and thoughts

[20] " The Difference Engine could only tabulate, and was incapable by its nature of developing; the Analytical Engine was intended to either tabulate or develop. The Difference Engine is the embodying of one particular and very limited set of operations, the Analytical Engine, the embodying of the science of operations. The distinctive characteristic of the Analytical Engine, is the introduction into it of the principle which Jacquard devised for regulating by means of punched cards the most complicated patterns in the fabrication of brocaded stuffs. Nothing of the sort exists in the Difference Engine. We may say most aptly, that the Analytical Engine weaves *Algebraical patterns,* just as the Jacquard loom weaves flowers and leaves !"— Note to translation of Menabrea's Memoir. The 59th volume of the *Edinburgh Review* contains an able and elaborate article upon the Difference Engine, written by Dr. Lardner.
[21] Mr. Babbage has shown me letters by which it appears that he declined offices of great emolument, the acceptance of which would have interfered with his labours upon the Difference Engine.

2 U

to the fulfilment of the contract, which he considered himself to have entered into with the Government.

With respect to the Difference Engine little remains to be added. In 1843, an application was made to Government, by the Trustees of King's College, London, to allow the Engine, as it existed, to be removed to the museum of that institution. The request was complied with; and the Engine, enclosed within a glass case, now stands nearly in the centre of the Museum. It is capable of calculating to five figures, and two orders of differences, and performs the work with absolute precision; but no portion whatever of printing machinery, which was one of the great objects in the construction of the Engine, exists. All the drawings of the machinery and other contrivances are also in King's College.

Before closing this Chapter, it will not be out of place to put upon record the state of the Analytical Engine at this period (1848).

Mechanical Notations have been made, both of the actions of detached parts, and of the general action of the whole, which cover about four or five hundred large folio sheets of paper.

The original rough sketches are contained in about five volumes. There are upwards of one hundred large drawings. No part of the construction of the Analytical Engine has yet been commenced. A long series of experiments have, however, been made upon the art of shaping metals; and the tools to be employed for that purpose have been discussed, and many drawings of them prepared. The great object of these inquiries and experiments is, on the one hand, by simplifying the construction as much as possible, and on the other, by contriving new and cheaper means of execution, ultimately to reduce the expense within those limits which a private individual may command.

PROFESSOR DE MORGAN'S REVIEW OF WELD'S HISTORY OF THE ROYAL SOCIETY.

THE ATHENÆUM.

London, Saturday, October 14, 1848.

MR. BABBAGE'S CALCULATING MACHINE.

In our review of Mr. Weld's "History of the Royal Society," [*ante*, p. 621,] we noted that one chapter was devoted to the history of the celebrated undertaking above named. This chapter is taken from materials furnished by Mr. Babbage himself, all the documents having undergone the inspection of Mr. Weld. Of recent publications on the subject it may be well to note—1. A short account of the transactions with the Government, communicated by Mr. Babbage to the *Philosophical Magazine* for September, 1843. 2. A sketch of the *Analytical Engine* (on which Mr. Babbage is now at work, that commenced by the Government being the *Difference Engine*) written in Italian by Menabrea, and translated, with notes (and a list of all previous publications), by the Countess of Lovelace (August 1843). The statements put forward by Mr. Babbage have thus been in substance before the public for five years, without contradiction: for though the account (No. 1) was not signed, it was stated to be *from authority*, allowed to pass as such by the Editors of the magazine, and generally understood to emanate from Mr. Babbage. We are then bound to take this first statement as admitted by Government, more especially after the publication by Mr. Weld, avowedly made from the documents furnished by Mr. Babbage himself: and assuredly we understand Mr. Weld as conceiving himself to be distinctly informed by Mr. Babbage, that *all* documents of any importance had been communicated.

The heads of the public history of the *Difference Engine* are as follows:—In April, 1823, the Government requested the opinion of the Royal Society on Mr. Babbage's plan for "applying machinery to the purposes of calculating and printing mathematical tables." The Royal Society reported favourably, that the machine was "fully adequate to the objects proposed,"—and this report was laid before Parliament. In July, Mr. Babbage had an interview with the Chancellor of the Exchequer (Earl of Ripon) to ascertain if Government would wish him to construct for *printing* as well as *calculating*. There is no minute of this conversation, and the parties have different memories upon it. But soon after, the Treasury informs the Royal Society that 1,500*l.* was to be issued to Mr. Babbage "to enable him to bring his invention to perfection, in the manner recommended." Mr. Weld remarks that no plan had been pointed out; but it must be noticed that the original application was for an opinion upon *calculating and printing*, that the opinion spoke of the *full adequacy* of the plan for *the objects proposed*, and that the final determination of the Government was to proceed *as recommended*. Unless there were a previous understanding that all documents should either speak with the verbal completeness of an indictment or be wholly void, it is clear that the Government determined to assist Mr. Babbage in realizing the full invention, and told him so.[1]

The work went on for four years, under advances of money from time to time: the funds were applied by Mr. Babbage, and the accounts were audited by Messrs. Brunel, Donkin, and Field. We suppose that Government did not exceed the proposed advance of 1,500*l.*; but this is not expressly stated. In December, 1828, Government applied again to the Royal Society to report upon the state, progress, and prospects of the machine. Mr. Babbage at the same time stated that he had expended 6,000*l.*—meaning, we suppose, 4,500*l.* over and above the Government advance. A Committee, consisting of Messrs. Gilbert, Roget, Sabine, Herschel, Baily, Brunel (the elder), Kater, Donkin, Penn, Rennie, Barton, Warburton, declined to report on practicability or utility, considering both as fully established, and reported that, the difficulties considered, the progress was as great as could be expected, and that the engine was likely to fulfil the expectations of its inventor. On this report the Government made further advances, and the machine was declared national property. But the official payments soon failed: and Mr. Babbage called a meeting of private friends, in May 1829, who, on the representation that he had then advanced 4,000*l.* himself, in addition to the Government advance of 3,000*l.*, advised him strongly not to proceed without adequate help from the Government. On this representation, the Duke of Wellington, Mr. Goulburn, and Lord Ashley inspected what there was to show, and the Treasury advanced 3,000*l.* more. In December 1830, nearly 600*l.* was still due to Mr. Babbage, "upon the last account," and that sum to the superintendent, Mr. Clement. The Treasury gave directions for the advance of 600*l.* to pay Mr. Clement, and desired a fresh inspection and opinion from the Royal Society. The Committee above named (with the addition of Sir J. Lubbock and Mr. Troughton) reported (April 1831) as favourably as before on every point, and recommended attention to Mr. Babbage's suggestion that the workshops should be removed to the neighbourhood of his residence. With regard to probable expense, they subjoined Mr. Brunel's estimate that 8,000*l.* additional would be sufficient; but recommending that the Government be advised to provide for 12,000*l.* by way of estimate. A piece of ground adjoining Mr. Babbage's garden was taken, and a fire-proof building was erected. When about 17,000*l.* had been expended altogether, further progress was arrested by the extravagant demands made by Mr. Clement, as compensation for carrying on the construction in the new buildings. These were out of the question: and Mr. Clement withdrew, taking with him all the tools which had been used, many of which had been invented for the occasion. For it is the law that engineers and mechanics possess the right of property in all tools they have constructed, even though the cost of construction may have been defrayed by their employers. A special agreement ought, the reader will say, to have been made as to these tools; but whether the neglect is to be charged on Mr. Babbage, or on the Government, those must say who feel able. As it very seldom happens that the employer furnishes tools, it is easy to see how the necessity for a special agreement may have escaped the notice of all parties.

So far all is intelligible enough, and no blame attaches to either side, at least that we can venture to impute. But now the question divides in a curious way. While the works were suspended, Mr. Babbage reconsidered the whole question, and invented what he calls the *Analytical Engine*,—which we will take, on his word and Menabrea's publication, derived from his communications, to be immensely superior to the *Difference Engine*. To resume the latter, while Government was unacquainted with these new and more simple conceptions, would have been improper; to write on unfinished speculations would have been difficult. Mr. Babbage therefore (September 1834) requested a personal interview with Lord Melbourne; which was agreed to,—but

---

[1] By the words "*no plan*," the reviewer here evidently refers to the *mechanical and mathematical plan*, on the fitness of which the Royal Society had already, as he observes, made a report. Mr, Weld, on the other hand refers to the *mutual relations* of the two parties, Mr. Babbage and the Chancellor of the Exchequer, relative to the expenses and even to the ownership of the *Difference Engine*, as appears by the footnote (7) at page 256.—C. B. (*See page 312 of this book.*)

2 x

before it took place the ministry was dissolved. From this time until 1842 Mr. Babbage made applications to the various administrations, which remained unanswered; until at last, in November, 1842, a letter from Mr. Goulburn, in answer to a new application, informed Mr. Babbage that the Government intended to discontinue the project on the ground of expense.

In the meanwhile Mr. Babbage incurred severe censure in scientific circles, as being himself the cause of the delay. It was asserted that he had compromised the Royal Society, which had so strongly recommended his project to the Government. It was pretty generally believed that the delay arose from his determination that the Government should take up the new engine and abandon the old one.

But, until the statement made by him shall be proved either false or defective, it must stand that the Government never returned any answer to the question—Shall the new engine be constructed, or shall the old one be proceeded with? We are of opinion that they ought to have required him to proceed with the old one. They ought to have said—The public can only judge by results: how well satisfied soever men of science may be that the new machine is immeasurably superior to the old one, society at large will never comprehend the abandonment of a scheme on which so much has been expended; they will say—What if, in constructing No. 2, No. 3 should be discovered as much superior to No. 2 as No. 2 is to No. 1! And if Mr. Babbage had declined to proceed with his first project, when thus urged, it is our opinion that he would have richly deserved a very harsh censure. And of this we are sure, that if Government had allowed him to finish the first machine, and he had done so with success, the House of Commons would willingly have granted money for the second,—aye, and for the third and fourth, if he had invented them. But the Government itself prevented the matter from coming to any such issue. It is possible that Sir R. Peel and Mr. Goulburn allowed Mr. Babbage's well-known wish[2] to abandon the first plan in favour of the new one to influence their decision. It may be that they were startled at finding that 17,000l. expended upon one project was only the precursor of another. If so, we think they put themselves in the wrong by not fastening on Mr. Babbage the alternative of either proceeding with the existing construction, or taking the entire responsibility of refusal upon himself. As the matter now stands, and unless Mr. Babbage can be refuted, the answer to the question why he did not proceed is, that during the eight years in which he had to bear the blame of the delay he could not procure even the attention of the Government, much less any decision on the course to be taken.

It is generally understood that Mr. Babbage is determined to proceed with the *Analytical Engine*, gradually, and at his own expense; and that the drawings are in a state of great forwardness. According to Mr. Babbage himself, many experiments have been made with the object " on the one hand, by simplifying the construction as much as possible, and on the other, by contriving new and cheaper means of execution, ultimately to reduce the expense within those limits which a private individual may command."

In looking at all the circumstances of this statement, we regret its divided responsibility. Mr. Weld has seen Mr. Babbage's documents. Should he have made an insufficient selection, who is to blame? Mr. Weld says, " I have derived very valuable information from an unpublished statement drawn up by Mr. Babbage, which he has been so kind as to place in my hands. The original documents, which are in Mr. Babbage's possession, and which are referred to, I have myself examined." From all this we should conclude that if Mr. Weld had omitted anything material, or fallen into any misconception, Mr. Babbage would before this have set it right. But it would be more satisfactory if we had Mr. Babbage's own acceptance of the statement thus made, as being that on which he is content to rest his case; at least until some specific counter-statement should demand more detail of explanation. Continued silence will be tantamount to such acceptance.

There is also one piece of information which must be drawn out before the case can be finally adjudicated. We stand thus:—Scientific rumour states that Mr. Babbage compelled the Government to give him up by demanding permission to abandon the *Difference Engine* and substitute the *Analytical Engine*. To this, in the formal point of view, Mr. Babbage has fully answered, by showing that the Government never communicated to him that it was their pleasure he should proceed on the plan originally contemplated. The question now remains—Did Mr. Babbage, or did he not, in the several unanswered applications which he made to the Ministry, press the claims of the new machine and the abandonment of the old? If so, did he do it in such a manner as to give to understand, or make apparent, that he would not consent to recommence operations at the point of relinquishment? The "several applications" which were made from 1833 to 1838 are not particularized,

[2] It is scarcely possible that this *supposed* wish could have influenced Sir Robert Peel, because he had before him a written disavowal of it from Mr. Babbage himself.—C. B.

much less described as to contents.   But, in October 1838, Mr. Babbage wrote to the Chancellor of the Exchequer, stating, to use Mr. Weld's words, that "the question he wished to have settled" was, whether the Government required him to superintend the completion of the *Difference Engine* according to the original plan and principle, or whether they intended to discontinue it altogether." Now the words *quoted* are very like the idiom a person would employ who had in his mind that up to that time some other question had been among those proposed for discussion.   And it is worthy of note that all the communications are undescribed until we come to the one of October 1838; which shows that then at least, whether before or not, Mr. Babbage had put the question on the right issue.   Of what tenor, then, were the undescribed applications?[3] If of the same as that of October 1838, Mr. Babbage stands quite clear; but if they were such as fairly to give rise to the rumour above mentioned, then it must be said, that though *he* had every disposition to get wrong, Government always prevented him by blocking his path with an error of its own.   But in any case it is to be remembered that for the last four years of unanswered application Mr. Babbage stood upon the right ground; and also that the rumoured *refusal* to proceed never was made.

The public, we think, has a right to explanation from the Government, and to further explanation from Mr. Babbage.   Sir R. Peel turned it off with a joke in the House of Commons.   He recommended that the machine should be set to calculate the time at which it would be of use.   He ought rather to have advised that it should be set to compute the number of applications which might remain unanswered before a Minister, if the subject were not one which might affect his parliamentary power.   If it had done this, it would have shown that its usefulness had commenced.

From THE ATHENÆUM *of Saturday, Dec. 16th, 1848.*

MR. BABBAGE has reprinted, for private circulation, Mr. Weld's chapter on his *Calculating Machine*, and has appended to it our review [1] of that chapter [see *ante*, p. 1029] with three short foot-notes. The first of these is on a point immaterial to the issue; the second and third contain distinct statements of fact from Mr. Babbage, in reference to our comments upon his proceedings and those of the Government.   Our readers will remember that from September 1834 to November 1842, Mr. Babbage could not procure the attention of the Government to the state of the engine, on which 17,000l. had been spent; and that, about the beginning of that period, Mr. Babbage had invented the new engine, which he called the *Analytical Engine*.   And further, they will remember that all notion of the possibility of blame having been justly incurred by Mr. Babbage rested, in our comment, upon the hypothesis that he had put his wish to abandon the *Difference Engine* and substitute the *Analytical Engine* before the Government in such a form as to give them a right to suppose that he was unwilling to proceed with the former.   On our remark that it is possible that Sir R. Peel and Mr. Goulburn allowed his well-known wish to influence their decision, Mr. Babbage observes:—"It is scarcely possible that this *supposed* wish could have influenced Sir Robert Peel, because he had before him a written disavowal of it from Mr. Babbage himself."

Again, of the first half of the period of unanswered application Mr. Weld gives no account, as to the tenor of the applications therein made to the Government: though he shows by documents that during the second half Mr. Babbage, to repeat our own phrase, "stood upon the right ground." And thereupon we expressed our opinion that the public had a right to explanation from the

---

[3] The two following will sufficiently explain them:—On the 23rd December, 1834, Mr. Babbage addressed a statement to the Duke of Wellington, pointing out the only plans which, in his opinion, could be pursued for terminating the questions relative to the *Difference Engine*, namely:
First, the Government might desire Mr. Babbage to continue the construction of the Engine in the hands of the person who has hitherto been employed in making it.
Secondly, the Government might wish to know whether any other person could be substituted for the engineer at present employed to continue the construction; a course which was possible.
Thirdly, the Government might (although he did not presume that they would) substitute some person to superintend the completion of the Engine instead of Mr. Babbage himself.
Fourthly, the Government might be disposed to give up the undertaking entirely.
A letter to Sir R. Peel from Mr. Babbage, dated 7th April, 1835, and inclosing the above plans, concludes thus:—
"The delays and difficulties of years will, I hope, excuse my expressing a wish that I may at length be relieved from them by an early decision of the Government on the question.—C. B."
[1] We said in that review that Menabrea's Memoir was in Italian:—we should have said French.

Government, and to further explanation from Mr. Babbage.   This further explanation Mr. Babbage now gives, in the following words; among which we insert some bracketed comments:—

"The two following [applications made to the Government] will sufficiently explain them [the undescribed applications of the first half of the period of unanswered applications]:—On the 23rd December, 1834, Mr. Babbage addressed a statement to the Duke of Wellington, pointing out the only [the reader will remark this word *only*] plans which in his opinion could be pursued for terminating the questions relative to the *Difference Engine*, namely—*First*, the Government might desire Mr. Babbage to continue the construction of the engine in the hands of the person who has hitherto been employed in making it. *Secondly*, the Government might wish to know whether any other person could be substituted for the engineer at present employed to continue the construction —a course which was possible. *Thirdly*, the Government might (although he did not presume that they would) substitute some person to superintend the completion of the engine instead of Mr. Babbage himself. *Fourthly*, the Government might be disposed to give up the undertaking entirely." A letter to Sir Robert Peel from Mr. Babbage, dated the 7th of April, 1835, and enclosing the above plans, concludes thus: "The delays and difficulties of years will, I hope, excuse my expressing a wish that I may at length be relieved from them by an early decision of the Government on the question."

From the above it appears that at the end of 1834, Mr. Babbage—though then so full of the *new* engine, that in September he had asked an audience of Lord Melbourne, to communicate the exact state of the case, and to request, of course, his consideration of the question whether the new engine should or should not take the place of the old one—began his applications to the Government with distinct reference to the *old* engine, and to the question of its completion or abandonment. Certainly the first of the two applications was not well timed, for it was made when the Duke of Wellington held all the seals, and a Government courier was hunting Sir Robert Peel all over Italy, to tell him to come home quick and be Prime Minister. But it was repeated to Sir Robert Peel in the April following, when the latter was also in official possession of the previous letter.

Mr. Babbage having thus filled up the only *lacuna* which the public press has brought to his notice, we can but repeat that those who would impute to him the blame of the failure of Government to complete his Calculating Machine must begin by proving his statement to be false or defective. In 1835 he complains *to* the Government of "delays and difficulties," which he implies to be mainly caused *by* the Government, and he gets no answer whatever to repeated applications, until 1843. Those who have propagated the rumours that his conduct was the cause of the delay, and that he compromised his friends in the Royal Society, who had aided in bringing him under the notice of the Government, are bound to abstain in future, or to show cause.

We end by a quotation from Mr. Weld, which we abstained from giving so long as we supposed that the discontinuance of the Calculating Machine might be, in any degree, Mr. Babbage's fault. "Mr. Babbage has shown me letters, by which it appears that he declined offices of great emolument, the acceptance of which would have interfered with his labours upon the *Difference Engine*."

PROCEEDINGS OF THE BRITISH ASSOCIATION, 1878.

Report of the Committee, consisting of Professor Cayley, Dr. Farr, Mr. J. W. L. Glaisher, Dr. Pole, Professor Fuller, Professor A. B. W. Kennedy, Professor Clifford, *and* Mr. C. W. Merrifield, *appointed to consider the advisability and to estimate the expense of constructing Mr.* Babbage's *Analytical Machine, and of printing Tables by its means. Drawn up by Mr.* Merrifield.[*]

We desire in the first place to record our obligations to General Henry Babbage for the frank and liberal manner in which he has assisted the Committee, not only by placing at their disposal all the information within his reach, but by exhibiting and explaining to them, at no small loss of time and sacrifice of personal convenience, the machinery and papers left by his father,[†] the late Mr. Babbage. Without the valuable aid thus kindly rendered to them by General Babbage it would have been simply impossible for the Committee to have come to any definite conclusions, or to present any useful report.

We refer to the chapter in Mr. Babbage's ' Passages from the Life of a Philosopher,'[‡] and to General Menabrea's paper, translated and annotated by Lady Lovelace, in the third volume of Taylor's ' Scientific Memoirs,'[§] for a general description of the Analytical Engine.

I. *The General Principles of Calculating Engines.*

The application of arithmetic to calculating machines differs from ordinary clockwork, and from geometrical construction, in that it is essentially discontinuous. In common clockwork, if two wheels are geared together so as to have a velocity ratio of 10 to 1 (say), when the faster wheel moves through the space of one tooth, the angular space moved through by the slower wheels is one-tenth of a tooth. Now in a calculating machine, which is to work with actual figures and to print them, this is exactly what we *don't* want. We require the second wheel not to move at all until it has to make a complete step, and then we require that step to be taken all at once. The time can be very easily read from the hands of a clock, and so can the gas consumption from an ordinary counter; but a moment's reflection will show what a mess any such machinery would make of an attempt at printing.

This necessity of jumping discontinuously from one figure to another is the fundamental distinction between calculating and numbering machines on the one hand, and millwork or clockwork on the other. A parallel distinction is found in pure mathematics, between the theory of numbers on the one hand, and the doctrine of continuous variation, of which the Differential Calculus is the type, on the other. A calculating machine may exist in either case. The common slide-rule is, in fact, a very powerful calculating machine in which the continuous process is used, and the planimeter is another.

Geometrical construction, being essentially continuous, would be quite out of place in the calculating machine which has to print its results. Linkwork also, for the same reason, is out of place as an

[*] Reprinted by permission of the Council.
[†] The Drawings, Books, Papers, &c., belonging to the Analytical Engine, as well as some Models, Moulds, &c., are now deposited in the Museum at South Kensington.
[‡] *See* page 154.
[§] *See* page 6.

auxiliary in any form to the calculation. It may be of service in simplifying the construction of the machine; but it must not enter into the work as an equivalent for arithmetical computation.

The primary movement of calculating engines is the *discontinuous train*, of which one form is sketched in the accompanying diagram (fig. 1):—B is the follower, an ordinary spur wheel with (say) 10 teeth;

FIG. 1.

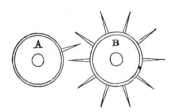

A is its driver, and this has only a single tooth. With a suitable proportion of parts, the single tooth of A only moves B one interval for a whole revolution of A; for it only gears with B by means of this single tooth. When that is not in gear, A simply slips past the teeth of B without moving the latter.

All the other machinery of calculating engines leads up to and makes use of this, or of some transformation of it, as its means of dealing with units of whatever decimal rank, instead of allowing indefinite fractions of units to appear in the result which has to be printed from.

The primary operation of calculation is counting; the secondary operation is addition, with its counterpart, subtraction. The addition and subtraction are in reality effected by means of counting, which still remains the primary operation; but the necessity for economising labour and time forces upon us devices for performing the counting processes in a summary manner, and for allowing several of them to go on simultaneously in the calculating engine. For, if we use simple counting as our only operation, and suppose our engine set to 2312 (say), then, in order to add 3245 to it by mere repetition, we have 3245 unit operations to perform, and this is practically intolerable. If, however, we can separate the counting, so as to count on units to units only, tens to tens only, hundreds to hundreds only, and so forth, we shall only have

$$3 + 2 + 4 + 5 = 14$$

turns of the handle, as against 3245 turns. In general terms the number of operations will be measured by the sum of the digits of the number, instead of by the actual number itself. This is exactly analogous to what we should do ourselves in ordinary arithmetic in working an addition sum, if we had not learnt the addition table, but had to count on our fingers in order to add. This statement of the work is, however, incomplete. In the first place the convenience of machinery obliges us to provide 10 steps for each figure, whatever it may be, and there must be an arrangement by which the setting of the figure to be added shall cause a wheel to *gain ground* by so many steps as the number indicates, and to *mark time* without gaining ground for the other steps up to 10. Thus, in adding 7 our driver must make a complete turn or 10 steps, equivalent to 1 step of the follower; but only 7 of these steps of the driver must be effective steps, the others being skipped steps. There are various devices for this. One of the simplest and most direct is that used in Thomas's Arithmomètre*; another is the Reducing Bar used by Mr. Babbage.† In the second place, the *carrying* has to be provided for just as in ordinary addition of

---

* Let ZO (fig. 2) be a plate with ten ribs of different lengths, Aa, Bb, ... Kk, soldered on it. Let Mm be a square axis on which the wheel N is made to slide by the fork P. Then, supposing N to have teeth which can engage in the ribs Aa &c., when the plate is pushed past the wheel N, the number of teeth through which the wheel N, carrying with it the shaft Mm, is made to rotate, depends upon the number of ribs in which it engages, and this depends upon how far along the axis N is made to slide by means of the fork P. If this fork is set opposite the line marked 3, Mm will turn through a space equivalent to 3 teeth. If a wheel, keyed to the shaft Mm, be geared to other wheels, this enables us to add any digit to any number at a single motion of the plate, by simply changing the position of P to suit the digit required. This is the principle used in Thomas's arithmomètre, only that there the traversing plate is replaced by a rotating cylinder.

† Suppose Aa, Bb, ... Ff (fig. 3) to be a series of racks passing through mortises in a plate xz, and meeting a series of spur-wheels mounted loose on a shaft, so that each wheel gears with one of the racks at the line pq and that all the whole series can be thrown in or out of gear together. Starting with them out of gear, let the racks be drawn out through the plate xz as indicated. Next throw the shaft pq into gear, and then press a plate PQ against the ends of the racks, pushing them back until the plates PQ and xz meet. Then each wheel on pq will turn through the number of teeth corresponding to the original projection of the racks. In this way, if the wheels on pq stood at any given number, say 543243, we should have added 314236 to them, and they would then stand at the sum of these two numbers, namely 857479. This, it will be observed, makes no provision for carrying. PQ is the reducing bar. In practice the arrangement

numbers. Taking account of all this, it follows that by separating the counting on the whole into counting on figure by figure, the number of separate steps is reduced from that expressed by the number itself to eleven times the number of its digits; that is to say, for example, the addition of the number 73592 to any other number is reduced from 73592 to 55 steps, and although of this latter some are slipped, there is no gain of time thereby, except in so far as several of the steps may be made simul-

Fig. 2.

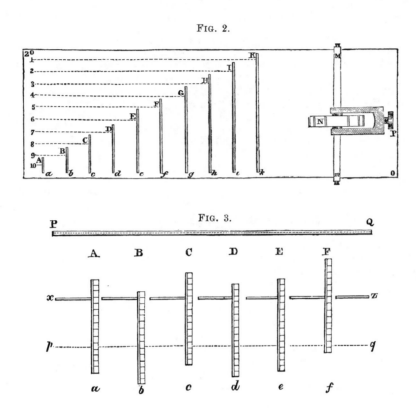

Fig. 3.

taneously. The ordinary engines beat the human calculator in respect of adding all the figures simultaneously; but Mr. Babbage was the first to devise a method of performing all the carrying simultaneously too.

Mechanical invention has not yet gone beyond the reduction of the distinct steps involved in the addition of a number consisting of n digits to less than $11n$; practically, from the necessity of accompanying the carrying with a *warning* step, rather more are required.

In all the calculating machines at present known, including Mr. Babbage's analytical engine, multiplication is really effected by repeated addition. It is true that, by a multiplication of parts, more than one addition may be going on simultaneously; but it yet remains true, as a matter of mechanism, that the process is purely one of iterated addition.

By means of reversing wheels or trains, subtraction is as easily and directly performed as addition, and that without becoming in any degree a tentative process. But it is important to observe that the process can be made tentative, so as to give notice when a minuend is, or is about to become, exhausted. This is the necessary preparation for division, which is thus essentially a tentative process. That does not take it out of the power of the machine, because the machine may be, and is, so devised as to accept and act upon the notice. Nevertheless it is a step *alieni generis* from the direct processes of addition, multiplication, and subtraction. It need hardly be stated that the process of obtaining a quotient consists in counting the number of subtractions employed, up to the machine giving notice of the minuend being exhausted.

is usually circular, the bar PQ revolving about an axis parallel to itself instead of sliding. If the numbers on the wheels pq are placed one way we get addition; if reversed, subtraction. Otherwise we may reverse by introducing an additional set of wheels between the wheels pq and the racks.

This is the bare principle, admitting of many transformations, and making, like the other, no provision for carrying.

Another essentially distinct train is involved in the decimal shift of the unit, in all the four elementary rules. This is most simply and most commonly effected by the sliding of an axis or frame longitudinally, after the manner of a common sliding-scale or rule, so as to bring either the figures, or the teeth which represent them, against those to which their decimal places correspond, and to no others. In multiplication and division, this means a shift for each step of the multiplication and division.

II. *Special Characteristics of Mr. Babbage's Analytical Engine.*

1. *The mill.*—The fundamental operation of Mr. Babbage's analytical engine is simple addition. This and the other elementary rules of subtraction, multiplication, and division, and all combinations of these, are performed in what is called "*the mill.*" All the shifts which have to take place, such as changing addition into subtraction by throwing a reversing train into gear, or the shift of the decimal place, carrying and borrowing, and so forth, are effected by a system of rotating cams acting upon or actuated by bell-cranks, tangs, and other similar devices commonly used in shifting machinery, sometimes under the name of clutches or escapements. These clutches and bell-cranks control the purely additive and carrying processes effected in the additive trains described in the note to § I., and, *being themselves suitably directed*, secure that the proper processes shall be performed upon the proper subject-matter of operation, and duly recorded, or used, as may be required.

2. *The store.*—A series of columns, each containing a series of wheels, constitutes the store. This store, which may be in three or more dimensions, both receives the results of operations performed in the mill, and serves as a store for the numbers which are to be used in the mill, whether as original or as fresh subjects of operation in it. Each column in the store corresponds to a definite number, to which it is set either automatically or by hand, and the number of digits in this number is limited by the number of wheels carried on the shaft of the column. The wheels gear into a series of racks, which can be thrown into or out of gear by means of the cards.

3. *Variable cards.*—All the numbers which are the subject of operation in the mill, whether they are the result of previous operations therein, or new numbers to be operated upon for the first time, are introduced to it in the form of Jacquard* cards, such as are used in weaving. One set of wires or axes transfers the numbers on these cards to the subject of operation in the mill, exactly as similar cards direct which of the warp threads are to be pushed up, and which down, in the Jacquard loom. The mill itself punches such cards when required.

4. *Operation cards.*—A different set of cards selects and prescribes the sequence of operations. These act, not upon the number wheels of the mill or store, but upon the cams and clutches which direct the gearing of these wheels and trains. Thus, in such an operation as $(a\,b + c)\,d$, we should require :—

1st, 4 variable cards with the numbers a, b, c, d.

2nd, an operation card directing the machine to multiply a and b together.

3rd, a record of the result, namely the product $ab = p$, as a fifth variable card.

4th, an operation card directing the addition of p and c.

5th, a record of the result, namely the sum $p + c = q$, as a sixth variable card.

6th, an operation card directing the machine to multiply q and d together.

7th, a record of the result, namely the product $qd = p_2$, either printed as a final result or punched in a seventh variable card.

III. *Capability of the Engine.*

It has already been remarked that the direct work of the engine is a combination and repetition of the processes of addition and subtraction. But in leading up to any given datum by these combinations, there is no difficulty in ascertaining tentatively when this datum is reached, or about to be reached. This is strictly a tentative process, and it appears probable that each such *tentamen* requires to be specially

---

* In a letter written by Mr. Babbage to Arago in December 1839, the following explanation of the use of these cards is given. It probably conveys the idea in the fewest words possible. It is only necessary to add that their twofold employment embodies the separation of the symbols of operation from those of quantity. "You are aware that the system of cards which Jacquard invented are the means by which we can communicate to a very ordinary loom orders to weave any pattern that may be desired. Availing myself of the same beautiful invention, I have by similar means communicated to my calculating engine orders to calculate any formula however complicated; but I have also advanced one stage further, and I have communicated through the same means orders to follow certain laws in the use of those cards, and thus the calculating engine can solve any equations, eliminate between any number of variables, and perform the highest operations of analysis."

provided for, so as to be duly noted in the subsequent operations of the machine. There is, however, no necessary restriction to any particular process, such as division; but any direct combination of arithmetic, such as the formation of a polynomial, can be made to lead up to a given value in such a manner as to yield the solution of the corresponding equation. In any such process, however, it is evident that there can be only (to choose a simile from mechanism) one degree of freedom; otherwise the problem would yield a locus, indeterminate alike in common arithmetic, and as regards the capabilities of the machine. The possibility of several roots would be a difficulty of exactly the same character as that which presents itself in Horner's solution of equations, and the same may be said of imaginary roots differing but little from equality. These, however, are extreme cases, with which it is usually possible to deal specially as they arise, and they need not be considered as detracting materially from the value of the engine. Theoretically, the grasp of the engine appears to include the whole synthesis of arithmetic, together with one degree of freedom tentatively. Its capability thus extends to any system of operations or equations which leads to a single numerical result.

It appears to have been primarily designed with the following general object in view—to be coextensive with numerical synthesis and solution, without any special adaptation to a particular class of work, such as we see in the difference engine. It includes that à majori, and it can either calculate any single result, or tabulate any consecutive series of results just as well. But the absence of any speciality of adaptation is one of the leading features of the design.

Mr. Babbage had also considered the indication of the passage through infinity as well as through zero, and also the approach to imaginary roots. For details upon these points we must refer to his ' Passages from the Life of a Philosopher.'

IV. *Present State of the Design.*

The only part of the analytical engine which has yet been put together is a small portion of " the mill," sufficient to show the methods of addition and subtraction, and of what Mr. Babbage called his " anticipating carriage." It is understood that General Babbage will (independently of this report) publish a full account of this method. No further mention of it will therefore be made here.

A small portion of the work is in gun-metal wheels and cranks, mounted for the most part on steel shafts. But the greater part of the wheels are in a sort of pewter hardened with zinc. This was adopted from motives of economy. They are for the most part not cast, but moulded by pressure, and the moulds of most of them are in existence.

A large number of drawings of the machinery are also in existence. It is supposed that these are complete to the extent of giving an account of every particular movement essential to the design of the engine; but, for the most part, they are not *working drawings*, that is to say, they are not drawings suited to be sent straight to the pattern or fitting shop, to be rendered in metal. There are also drawings for the erection of the engine, and there appears to be a complete set of descriptive notes of it in Mr. Babbage's " mechanical notation." There remains, however, a great deal to be done in the way of calculating quantities and proportions, and in the preparation of working drawings, before any work could actually be set in hand, even if the design be really complete. There is some doubt on this point as the matter stands, and it certainly would be unsafe to rely upon the design being really complete, until the working drawings had been got out. Mechanical engineers are well aware that no complex design can be trusted without this test, at least.

It was Mr. Babbage's rule, in designing mechanism, in the first place to work to his object, in utter disregard of any questions of complexity. This is a good rule in all devising of methods, whether analytical, mechanical, or administrative. But it leaves in doubt, until the design finally leaves the inventor's hands in a finished state, whether it really represents what is meant to be rendered in metal, or whether it is simply a provisional solution, to be afterwards simplified.

V. *Probable Cost.*

It has not been possible for us to form any exact conclusion as to the cost. Nevertheless there are some data in existence which appear to fix a lower limit to the cost. Mr. Babbage, in his published papers, talks of having 1,000 columns of wheels, each containing 50 distinct wheels; this apparently refers to his *store.* Besides the many thousand moulded pewter wheels for these, and the axes on which they are mounted, there is the *mill*, also consisting of a series of columns of wheels and of a vast machinery of cams, clutches, and cranks for their control and connection, so as to bring them within the directing power of the Jacquard systems of variable cards and operation cards. Without attempting any exact estimate we may say that it would surprise us very much if it were found possible to obtain tenders for less than 10,000*l.*, while it would pretty certainly cost a considerable sum to put the design in a fit state for obtaining tenders. On the other hand, it would not surprise us if the cost were to reach three or four times the amount above suggested.

It is understood that towards the close of his life Mr. Babbage had contemplated carrying out the manufacture of the engine on a smaller scale, confining himself to 25 figures instead of 50, and to 200 columns instead of 1000 or more. This would of course reduce the amount of the metal-work proportionately, but we do not think that it would materially reduce the charge which we anticipate for bringing the design into working order.

VI. *Strength and Durability.*

The questions of strength and durability had by no means escaped Mr. Babbage's attention, and a great deal of his detail bears marks of having been designed with especial reference to these two points. That was essential in a large and complex engine with some thousands of wheels, all requiring at some time or other, although not simultaneously, to be driven by the means of one shaft. This necessarily throws a great deal of pressure, and also a great deal of wear and tear, on the main driving shaft and the gear immediately connected with it. We have no means of knowing, in the present state of the design, to what extent Mr. Babbage had succeeded in reducing this, or whether he had always been successful in arranging his cams and cranks so as to secure the best working angles, and to avoid their being jammed at dead points or otherwise. Giving him full credit for being quite aware of the importance of this, we cannot but doubt whether the design was ever in a sufficiently forward state to enable him, or any one else, to speak with certainty on this point. Several of the existing calculating machines show signs of weakness in the driving-pinions.

One of the movements apparently necessary to the tentative processes of the engine is, when the spur-wheels on a given shaft have been brought into certain definite positions depending on previous operations, to bring up a sharp straight-edge against them in a plane passing through the axis of the shaft. This pushes some to right and others to left, according to the position of the crown of the tooth relatively to the straight-edge. This operation is necessary to secure that the clearance of the different parts of the machinery, whether originally provided in order to allow it to work smoothly, or whether afterwards increased by working, shall not introduce a numerical error into the result. The principle of this operation is used generally throughout the analytical engine. Its consequent effect, both in respect of the work which it throws upon the main driving gear, and of the wear of the parts which it pushes, forms an important element in considering the durability of the machine. This bar also serves the purpose of locking part of the machine when required.

On the other hand, it is to be remarked that the use of springs has been wholly discarded by Mr. Babbage, as directors of motion, although he occasionally uses them for return motions.

VII. *Probable Utilization of the Analytical Engine.*

It has been already remarked that one of the main features of the engine is, that its function is coextensive with numerical synthesis and solution, and that there is an absence of any special adaptation. In thus widening the sphere of its capability, it is made to diverge from the general tendency of mechanical design, which is towards the selection and particularization of the work to be performed, and the restriction of the machinery to one particular cycle of operation, usually within close numerical limits, as well as limited in kind. Nevertheless, modern engineering practice finds ample room for "universal" drills, shaping tools, and other machines having very general adjustments and applications. But it remains practically true that each step of freedom of adjustment is also a step in diminution of special aptitude.

While the analytical engine is capable of turning out a single result as the combination of a complex series of numbers and operations performed upon them, it can also yield a series of such results in a consecutive form, and thus give tabulated results. Only it is not restricted, as is the difference engine, to the special method of tabulation by finite differences, nor is tabulation its primary function or intention. If its actual capabilities are found to realize the intentions of its inventor, it will tabulate all functions which are within the reach of numerical synthesis, and those direct inversions of it which are known under the name of solutions. It deals, however, with number, and not with analytical form.

Theoretically it might supersede the difference engine, *à majori*; but for reasons already stated, the specialization of the difference engine would probably give it an advantage over the more powerful engine, when the work was specially suited to finite differences.

There would remain much work, tabular and other, for which differences are not very directly suited. Among these may be mentioned the determination of heavy series of constants and of definite functions of them, such as Bernoulli's numbers, Σx^{-n}, coefficients of various expansions of functions, and inversions of known expansions, solutions of simultaneous equations with large numerical coefficients and many variables, including, as a particular, but important case, the practical correction of observations by the method of least squares. If all sorts of heavy work of this kind could be easily and quickly, as well as certainly, done, by merely selecting or punching a few Jacquard cards and turning a handle, not only much saving of labour would result, but much which is now out of human possibility would be brought within easy reach.

If intelligently directed and saved from wasteful use, such a machine might mark an era in the history of computation, as decided as the introduction of logarithms in the seventeenth century did in trigonometrical and astronomical arithmetic. Care might be required to guard against misuse, especially against the imposition of Sisyphean tasks upon it by influential sciolists. This, however, is no more than has happened in the history of logarithms. Much work has been done with them which could more easily have been done without them, and the old reproach is probably true, that more work has been spent upon making tables than has been saved by their use. Yet, on the whole, there can be no reasonable doubt that the first calculation of logarithmic tables was an expenditure of capital which has repaid itself over and over again. So probably would the analytical engine, whatever its cost, if we could be assured of its success.

VIII. *Possible Modification of the Engine.*

Without prejudging the general question referred to us as to the advisability of completing Mr. Babbage's engine in the exact shape in which it exists in the machinery and designs left by its inventor, it is open to consideration whether some modification of it, to the sacrifice of some portion of its generality, would not reduce the cost, and simplify the machinery, so as to bring it within the range of both commercial and mechanical certainty. The "mill," for example, is an exceedingly good mechanical arrangement for the operations of addition and subtraction, and with a slight modification, with or without store-columns, for multiplication. We have already called attention to the imperfection of the existing machines, which show weakness and occasional uncertainty. It is at least worth consideration whether a portion of the analytical engine might not thus be advantageously specialized, so as to furnish a better multiplying machine than we at present possess. This, we have reason to believe, is a great desideratum both in public and private offices, as well as in aid of mathematical calculators.

Another important desideratum to which the machine might be adapted, without the introduction of any tentative processes (out of which the complications of the machinery chiefly arise) is the solution of simultaneous equations containing many variables. This would include a large part of the calculations involved in the practical application of the method of least squares. The solution of such equations can always be expressed as the quotient of two determinants, and the obtaining this quotient is a final operation, which may be left to the operator to perform by ordinary arithmetic, or which may be the subject of a separate piece of machinery, so that the more direct work of forming the determinant, which is a mere combination of the three direct operations of addition, subtraction, and multiplication, may be entirely freed from the tentative process of division, which would thus be prevented from complicating the direct machinery. In the absence of a special engine for the purpose, the solution of large sets of simultaneous equations is a most laborious task, and a very expensive process indeed, when it has to be paid for, in the cases in which the result is imperatively needed. An engine that would do this work at moderate cost would place a new and most valuable computing power at the disposal of analysts and physicists.

Other special modifications of the engine might also find a fair field for reproductive employment. We do not think it necessary to go into these questions at any great length, because they involve a departure, in the way of restriction and specialization, from Mr. Babbage's idea, of which generality was the leading feature. Nevertheless, we think that we should be guilty of an omission, if we were to fail to suggest them for consideration.

IX. *General Conclusions, and Recommendation.*

1. We are of opinion that the labours of Mr. Babbage, firstly on his Difference Engine, and secondly on his Analytical Engine, are a marvel of mechanical ingenuity and resource.
2. We entertain no doubt as to the utility of such an engine as was in his contemplation when he

undertook the invention of his analytical engine, supposing it to be successfully constructed and maintained in efficiency.

3. We do not consider that the possibilities of its misuse are any serious drawback to its use or value.

4. Apart from the question of its saving labour in operations now possible, we think the existence of such an instrument would place within reach much which, if not actually impossible, has been too close to the limits of human skill and endurance to be practically available.

5. We have come to the conclusion that in the present state of the design of the engine it is not possible for us to form any reasonable estimate of its cost, or of its strength and durability.

6. We are also of opinion that, in the present state of the design, it is not more than a theoretical possibility; that is to say, we do not consider it a certainty that it could be constructed and put together so as to run smoothly and correctly, and to do the work expected of it.

7. We think that there remains much detail to be worked out, and possibly some further invention needed, before the design can be brought into a state in which it would be possible to judge whether it would really so work.

8. We think that a further cost would have to be incurred in order to bring the design to this stage, and that it is just possible that a mechanical failure might cause this expenditure to be lost.

9. While we are unable to frame any exact estimates, we have reason to think that the cost of the engine, after the drawings are completed, would be expressed in tens of thousands of pounds at least.

10. We think there is even less possibility of forming an opinion as to its strength and durability than as to its feasibility or cost.

11. Having regard to all these considerations, we have come, not without reluctance, to the conclusion, that we cannot advise the British Association to take any steps, either by way of recommendation or otherwise, to procure the construction of Mr. Babbage's Analytical Engine and the printing tables by its means.

12. We think it, however, a question for further consideration whether some specialized modification of the engine might not be worth construction, to serve as a simple multiplying machine, and another modification of it arranged for the calculation of determinants, so as to serve for the solution of simultaneous equations. This, however, inasmuch as it involves a departure from the general idea of the inventor, we regard as lying outside the terms of reference, and therefore perhaps rather for the consideration of Mr. Babbage's representatives than ours. We accordingly confine ourselves to the mere mention of it by way of suggestion.

PROCEEDINGS OF THE BRITISH ASSOCIATION, 1888.

Paper read at Bath by MAJOR-GENERAL H. P. BABBAGE, *on 12th September,* 1888.

TEN years have elapsed since a committee of the British Association reported upon the Analytical Engine of my father, Charles Babbage, and I desire now, while offering a few remarks upon that Report, to endeavour to convey some idea of the mechanical arrangements of the engine to those who may be interested in it.

2. I am well assured that a time will come when such an engine will be completed and be a powerful means of enlarging not only pure mathematical science, but other branches of knowledge, and I wish, as far as in me lies, to hasten that time, and to help towards the general appreciation of the labours of my father, so little known or understood by the multitude even of the educated.

3. He considered the Paper by Menabrea, translated with notes by Lady Lovelace, published in volume 3 of Taylor's "Scientific Memoirs," as quite disposing of the mathematical aspect of the invention. My business now is not with that.

4. The idea of the Analytical Engine arose thus : When the fragment of the Difference Engine, now in the South Kensington Museum, was put together early in 1833, it was found that, as had been before anticipated, it possessed powers beyond those for which it was intended, and some of them could be and were demonstrated on that fragment.*

5. It was evident that by interposing a few connecting-wheels, the column of Result can be made to influence the last Difference, or other part of the machine in several ways. Following out this train of thought, he first proposed to arrange the axes of the Difference Engine circularly, so that the Result column should be near that of the last Difference, and thus easily within reach of it.† He called this arrangement "the engine eating its own tail." ‡ But this soon led to the idea of controlling the machine by entirely independent means, and making it perform not only Addition, but all the processes of arithmetic at will in any order and as many times as might be required. Work on the Difference Engine was stopped on 10th April, 1833, and the first drawing of the Analytical Engine is dated in September, 1834.

6. The object may shortly be given thus :—It is a machine to calculate the numerical value or values of any formula or function of which the mathematician can indicate the *method* of solution. It is to perform the ordinary rules of arithmetic in any order as previously settled by the mathematician, and any number of times and on any quantities.

7. It is to be absolutely automatic, the slave of the mathematician, carrying out his orders and relieving him from the drudgery of computing.

8. The Analytical Engine is of course to print the results, or any intermediate result arrived at. He regarded this as an indispensable requisite, without which a calculating machine might indeed be useful for some purposes, but not for those of any scientific value. The perpetual risk of error in copying and transferring lines of figures is most troublesome, and tends to make results, themselves perfectly accurate, unreliable in use.

9. It is at once seen that the necessity of the engine being automatic imposes a gigantic task on the inventor. The first means employed to meet it is the use of cards to govern the engine. These are very similar to those in use in the Jacquard loom, to which we owe the figured patterns in the beautiful fabrics we see everywhere in common use.

10. One set of cards would be used to communicate the " given numbers," or constants of a problem to the machine. I shall call these throughout this paper " Number Cards." §

11. Another set of cards would be used to direct to which particular place or column in the engine these numbers, or any intermediate numbers arising in the course of the calculation, are to be conveyed

---

* See Ninth Bridgewater Treatise, London, 1838 ; pp. 299–304 of this book.   † See p. 154.   ‡ See p. 268.
§ See diagram on general plan, No. 25 of 6 August, 1840.

or transferred; these cards I will call "Directive Cards." There would also be other "Directive Cards" for general purposes of control when necessary.

12. A third sort, called "Operation Cards," * would direct the actual operations to be performed, these would put the engine mechanically into a condition to perform the particular operation required —Addition, Subtraction, &c., &c.

13. I was once asked, "How do you set the question? Do you write it on paper and put it into the machine?" Well, given the problem, the mathematician must first of all settle the operations and the particular quantities each is to be performed on and the time for each operation.

14. Then the superintendent of the engine must make a "Number Card" for each "given number," and settle the particular column in the machine on to which each "given number" is to be first received, and assign columns for every intermediate result expected to arise in the course of the calculation.

15. He will then prepare "Directive Cards" accordingly, and these, together with the necessary "Operation Cards" being placed in the engine, the question will have been set; not exactly, as my friend suggested, written on paper, but in cardboard, and motion being supplied the engine will give the answer.†

16. Now this appears a long process for what may be a simple question, but it is to be noted that the engine is designed for analytical purposes, and it would be like using the steam hammer to crush the nut, to use the Analytical Engine to solve common sums in arithmetic; or, adopting the language of Leibnitz (see page 79): "It is not made for those who sell vegetables or little fishes, but for observatories, or the private rooms of calculators, or for others who can easily bear the expense, and need a good deal of calculation." Moreover, except the "Number Cards," all the cards, once made for any given problem, can be used for the same problem, with any other "given numbers," and it would not be necessary to prepare them a second time—they could be carefully kept for future use. Each formula would require its own set of cards, and by degrees the engine would have a library of its own.‡

17. Thus the values for any number of Life Insurance Policies might be calculated one after the other by merely supplying fresh cards for the age, amount, rate of interest, &c., for each individual case.

18. The separation of "Operation Cards" from the numbers to be operated on is complete. The powers of the engine are the most extended, but each set of cards makes it special for the solution of one particular problem; each individual case of which, again, requires its own "Number Cards."

19. Taking the formula used by Mr. Merrifield $(a\,b + c)\,d$ as an illustration,§ the full detail of the cards of all sorts required, and the order in which they would come into play is this:—

The four cards for the "given numbers" a, b, c and d, strung together are placed by hand on the roller, these numbers have to be placed on the columns assigned to them in a part of the machine called "The Store," where every quantity is first received and kept ready for use as wanted.

Directive Card.	Operation Card.	
1st.	..	Places a on column 1 of Store
2nd.	..	„ b „ 2 „
3rd.	..	„ c „ 3 „
4th.	..	„ d „ 4 „
5th.	..	Brings a from Store to Mill
6th.	..	„ b „ „
..	1	Multiplies a and $b = p$
7th.	..	Takes p to column 5 of Store where it is kept for use and record
8th.	..	Brings p into Mill
9th.	..	Brings c into Mill
..	2	Adds p and $c = q$
10th.	..	Takes q to column 6 of Store
11th.	..	Brings d into Mill
12th.	..	„ q „
..	3	Multiplies $d \times q = p_2$
13th.	..	Takes p_2 to column 7 of Store
14th.	..	Takes p_2 to printing or stereo-moulding apparatus.

20. We have thus besides the "Given Number" Cards, three "Operation Cards" used, and fourteen "Directive Cards;" each set of cards would be strung together and placed on a roller or prism

* See p. 169 and general plan of 6 August, 1840.   † See p. 160.   ‡ See p. 161.   § See p. 326.

of its own; the roller would be suspended and be moved to and fro. Each backward motion would cause the pris.n to move one face, bringing the next card into play, just as on the loom.

21. It is obvious that the rollers must be made to work in harmony, and for this purpose the levers which make the rollers turn would themselves be controlled by suitable means, or by general "Directive Cards," and the beats of the suspended rollers be stopped in the proper intervals.

22. This brings me to the second great distinguishing feature of the engine, the principle of "Chain." This enables us to deal mechanically with any single combination which may occur out of many possible, and thus to be ready for any or every contingency which may arise.

23. Supposing that it is desired to provide for a certain possible combination such as, for example, the concurrence of ten different events, it could be effected mechanically thus*—each event would be represented by an arm turning on its axis, and having at its end a block held loosely and capable of vertical motion independently of the arm which carries it. Now suppose each of these arms to be brought, on the occurrence of the event it represents, into a position so that the blocks should all be in one vertical line, then if the block in the lowest arm was raised by a lever, it would raise all the nine blocks together, and the top one could be made to ring a bell or communicate motion, &c. &c.; but if any one of the ten events had not happened, its block would be out of the "Chain," and the lowest block would be raised in vain and the bell remain silent.

24. This is the simplest form of "Chain," there may be many modifications of it to suit various purposes.

25. In its largest extent it will appear in the Anticipating Carriage† further on, but in its simplest form it appears here as a means of producing intermittent motion at uncertain intervals, which may or may not be previously known either to the mathematician or even to the superintendent of the engine who had prepared the cards.

26. In our illustration the first operation card happens to be multiplication, and the time this would occupy necessarily depends upon the number of digits in the two quantities multiplied. Now when multiplication is directed, the "Chain" to every part of the machine not wanted will be broken, and all motion thus stopped till the multiplication is completed, and whenever the last step of that is reached, the "Chain" will be restored.

27. The Chain for this purpose may be made this way, a link is cut and the two ends are brought side by side, overlapping each other; a part of each link is then cut away exactly the same in both pieces. Now if the piece in communication with the driving power is moved, the other is not, and remains stationary; but if a block is let fall into the space cut away so as to fill the gaps in both parts of the severed link, the chain will be complete and link—block—link will pass on the motion. The block is hung in the end of an arm in which it can slide, and all that is required is to move the arm sideways while the link pieces are at rest in the proper position. This is shown in the diagram on p. 338.

28. In a machine such as the Analytical Engine, consisting of so many distinct trains of motion of which only a few would be in action at a time, it may be easily conceived how useful this application of the principle of Chain is. It helps to realize, too, another important principle largely adopted, viz., to break up every train of motion as far as possible into short courses, the last step of each furnishing a mere guide for a fresh start in the mechanism from the driving power. Of course the link could be broken and restored also by any of the "Directive Cards" mentioned above. The interposition of such links would also be used to save the cards from the wear and tear unavoidable if they were used for the actual transmission of any force.

29. I hope that I have now given some idea of the methods used for the general control of the engine, and shall pass on to illustrate, as well as I am able, some of the details of the machine. It is not to be supposed that I have mastered all these myself; nor, had I done so, would it be possible for me to make them intelligible in the course of a Paper, but some idea of them I hope to convey.

30. The machine consists of many parts. I have found it easier myself to regard these parts as so many separate machines, driven by the same motive power and starting and stopping each other in every possible combination, but otherwise acting independently, though with a settled harmony towards a desired result. When that idea has been reached, it seems easy to imagine them all brought close together, grouped in the positions most convenient.

31. Many general plans have been drawn. Plan No. 25, dated 6 August, 1840 was lithographed, others followed. The fact is, that what suits one part best does not suit another, and the number of possible variations of the parts renders it difficult to settle which combination of the whole may be the best for the general plan.

* See Plate 12 of December 4, 1836.　　　† See p. 157.

2 z 2

32. A part of the machine called " The Store " has already been mentioned. It would consist of a number of vertical columns to receive the " given numbers " and those arising in the course of the calculations from them.

33. In the example already mentioned seven columns would be used ; perhaps for a first machine twenty would do, with twenty-five wheels in each ; each wheel might have a disc with the digits 0 to 9 engraved on its circumference, but this is not absolutely necessary. In fact, the whole machine might be constructed without a figure anywhere except for the printing.

34. The " Directive Cards " would put the column selected to receive a " given number " into gear with a set of racks through which the " given number " (expressed on its card) would be conveyed to the Store column, each wheel being moved as many teeth as in the corresponding place of the number.

35. One revolution of the main shaft would be sufficient to put a " given number " on a column of the Store, or to transfer it from the store to some other part of the engine.

36. The action of the " directive card " in this case is to raise the selected column so that its wheels should be level with a corresponding set of racks, and thus brought into gear with them. Each column of the Store would require its own " directive card."

37. At the top of each column of the Store would be a wheel on which the Algebraic symbol of the quantity could be written,[*] and on the top of those columns assigned to receive the intermediate results this would also be done ; though at first these columns would be at zero.

38. There would also be a wheel for the sign $+$ or $-$. Drawings and notations for such sign wheels exist,[†] and how far the operations of Algebra may be exhibited has been discussed.

39. The Mill is the part of the machine where the quantities are operated on. Two numbers being transferred from the Store to two columns in the Mill the two are put into gear together through racks, and the reduction of one column to zero turns the other the exact equivalent, and thus adds it to the other. Supposing there to be with each wheel a disc with the digits 0 to 9 engraved on its edge, and a screen in front of the column with a hole or window before each disc allowing one figure only to be seen at a time ; during the process of addition the digits will pass before the window just as in counting till the sum is reached : thus if 5 is added to 7, the 7 will disappear and 8, 9, 0, 1, pass before the window in succession till 2 appear. At the moment when 9 passes to 0, a lever will be moved, thus recording the necessity of a carriage to the figure above ; the carriage is made subsequently, and for the Analytical Engine a method of performing the carriages all simultaneously[‡] was invented by my father which he called " Anticipating " Carriage.

40. When two numbers are added, carriages may occur in any or every place except the last ; where the wheels pass as just described from 9 to 0, a carriage arises directly ; in those places where the figure disc comes to rest at 9, no carriage occurs, but one of two things may happen —if there is no carriage to arrive from the next right hand place, the 9 will not be changed ; but should one be due, not only the 9 must be pushed on to 0, but a carriage must be passed on to the next place on the left hand, and if there happen to be a succession of 9's, they will all have to be pushed on. Working with twenty-five places of figures there can be no carriage ever required in the last place, but there may be in any one, any two, any three, &c., up to twenty-four places, where a carriage may arise directly and the same up to twenty-three places where the presence of 9's may indirectly cause it.

41. Now immense as is the number of these two sets of combinations, they can be successfully dealt with mechanically by the principle of " Chain "[§] and indeed every single combination as it arises is presented to the eye. There is a series of blocks two for each place, the lower block is made to serve for the two events. The upper one has a projecting arm, which when moved circularly engages a toothed wheel and moves it on one tooth affecting the figure disc similarly. It rests on the lower block which moves it up and down with itself always. After the addition is completed, should a carriage have become due and the warning lever have been pushed aside, it is made (by motion from the main shaft) to actuate the lower block and throw it into " Chain," when this is raised (again by motion from the main shaft) it raises the upper block which thus effects ordinary carriage ; but supposing no carriage to have become due, there will be at the window either a 9 or some less number. The latter case may be dismissed at once ; as a carriage arriving to it will cause no carriage to be passed on. Should there be a 9 at the window the warning-lever cannot have been pushed aside, but in every place where there is a 9 another lever, again by motion from the main shaft comes into play and pushes the lower block into " Chain " ; not, of course, into " Chain " for ordinary carriage, but into another position for Chain for 9's, so that should there come a carriage from below, the Chain for 9's will be raised as far as it extends

---

[*] See p. 160.   [†] See Plate 5 and Drawings 138 and 139 in List, Page 291, and Notations, p. 286.
       [‡] See p. 157.   [§] See Plate No. 12 and 4 in the List of Drawings, p. 288.

and effect the carriages necessary, be it for a single place, or for several, or for many. Should there be no carriage from below to disturb it the Chain remains passive; it has been made ready for a possible event which has not occurred. A certain time is required for the preparation of "Chain," but that done all the carriages are then effected simultaneously. A piece of mechanism for Anticipating Carriage on this plan to twenty-nine figures exists and works perfectly.

42. When a large number of places of figures is being dealt with, the saving of time is very considerable, especially when it is remembered that multiplication is usually done by successive additions.

43. Another plan for carriage has also been contrived and drawn.* It is obvious that there is no necessity, when there are many successive additions, to make the carriages immediately follow each addition. The additions may be made one after the other, and the carriages having been warned or even actually made on a separate wheel in each place as they arise, can all be made in one lot afterwards; more machinery is required, but the saving of time is very considerable. This plan has been called "Hoarding Carriage," and thoroughly worked out.

44. It is interesting to note that Hoarding Carriage is seen in the little machine of Sir Samuel Moreland invented in 1666, and probably existed in that of Pascal still earlier.

45. Now it may happen in addition that two or more numbers being added together, there may not be room at the top of the column for the left hand figure of the result. This would usually happen from an oversight in preparing or arranging the cards when space should be left; but it might so happen† that the calculation led, as mathematical problems sometimes do, through infinity. In either case a bell would be rung and the engine stopped; or if the contingency had been anticipated by the mathematician, fresh directive cards previously prepared might be brought into play.

46. One addition would be executed in each turn of the main axis, the intermittent motions required being produced by cams on the main axis. These would be flat discs with projecting parts on them acting on arms with friction rollers at the end. Each cam would be double, *i.e.* have two discs; the projections on the one corresponding with depressions on the other. Such cams are easily made and fixed and adjusted, about six or seven are sufficient for addition. The illustration shows such a double cam, together with a "Chain" for throwing into and out of gear, as explained on page 333. This may or may not be wanted.

47. Subtraction is performed by the interposition of an additional pinion, which turns the figure discs the reverse way; the figures decreasing in succession as they pass the window, and the carriage arises whenever the 0 passes and a 9 appears. The same arrangement is applied to the 0's in subtraction as to the 9's in addition, and the same principle of "Chain"; indeed, the same actual mechanism serves for both, the change being made by the movement of a single lever. See page 338.

48. In subtraction, when a larger is subtracted from a smaller number, there will be a warning made in the highest place for a carriage to a place above which does not exist, zero has been passed; the warning lever will ring a bell and stop the engine, unless the contingency has been anticipated and provided for by the mathematician.‡

49. Several ways have been worked out§ and drawn for multiplication. A skilled computer dealing with many places of figures having to multiply would make by successive additions a table of the first ten multiples of the multiplicand; if he has done this correctly the tenth multiple will be the same as the first, only all moved one place to the left and with a 0 in the units place. Using this table he picks out in succession the multiples required and puts them in the proper places, then adding all up gets his result, dispensing altogether with the multiplication table and doing nothing beyond addition. For the machine this way has been worked out and many drawings and notations exist for it.

50. Another way by the use of barrels‖ has also been drawn. The way by succession additions and stepping is perhaps the simplest.

51. When two numbers each of any number of places from one up to twenty or thirty have to be multiplied, it becomes necessary, in order to save time, to ascertain which has the fewest significant digits; special apparatus has been designed for this, called "Digit Counting Apparatus."¶ The smaller of the two is made the multiplier. Both are brought into the Mill and put on the proper columns. As the successive additions are made, the figure wheels of the multiplier are successively reduced to zero; when this happens for any one figure of the multiplier a cam on its wheel pushes out a lever which breaks the link or "chain" for addition, and completes that for stepping; so that the next revolution of the main axis causes stepping instead of addition, and that being done the links are changed back and the successive additions go on.

52. By this process the multiplier column is all reduced to zero; but if need be there can be

* See Drawings 3, Fig. 1. 57, 180** and 183* in List, p. 288, and 18 of Nov. 1857, p. 293.   † See p. 176.
‡ See p. 176.   § See p. 275.   ‖ See List of Drawings, 87 and 88.   ¶ See List of Drawings, 36 and 52.

another column alongside to which it may step by step be worked on—anyhow, it stands on record in it own column in the Store.

53. Multiplication would ordinarily be performed from the highest place downwards, and from the decimal point onwards; so that the result would be complete to the lowest place of decimals; but, of course, would contain the accumulated error due to cutting off the figures beyond in each quantity operated on. As, however, there would be a counting apparatus recording the successive additions, which at the end of the multiplication, would give the sum of all the digits of the multiplier, the maximum possible error would be known, and if thought advisable, a correction ordered to be made for it; for instance, the machine might be directed to halve the total of the digits, and add it to the last figures of the result wherever cut off.

54. Division is a more troublesome operation by far than multiplication. Mr. Merrifield in his report has called this "essentially a tentative process," and so it is as regards the computer with the pen; but as regards the Analytical Engine, I do not assent to it.

55. As, however, he considered the striking part of a clock a tentative process, while I do not, the difference might be one of definition of a word, and I only notice it as maintaining that the processes of the engine are only so far tentative as the guiding spirit of the mathematician leads him to make them.

56. Division by Table and also by barrels has been thoroughly drawn and worked out—the process by successive subtraction also. The divisor and dividend being brought into the Mill, the successive subtractions proceed and their number is recorded; when the correct figure has been reached, the subtraction is thrown out by "Chain" and stepping caused.

57. There is no trial and error whatever. Some additional machinery is required for this and more time is occupied, but the result is certain; the engine stops of necessity at the right figure of the quotient, and gives the order for stepping and so on to the end.

58. For the extraction of the square root a barrel would be used, but the ordinary process might be followed step by step without a barrel.

59. Counting machines of sorts would perform important functions in the general directive. Some would be mere records of the progress of the different steps going on, but others, of which the multiplier column may serve as a sort of example, would themselves act at appointed times as might be previously arranged.

60. This principle of "Chain" is used also to govern the engine in those cases where the mathematician himself is not able to say beforehand what may happen, and what course is to be pursued, but has to let it depend on the intermediate result of the calculation arrived at. He may wish to shape it in different ways according as one or several events may occur, and "Chain" gives him the power to do it mechanically. By this contrivance machines to play simple games of skill such as "tit tat too" have been designed.

61. Take a simpler case where the mathematician desired to deal with the largest of two numbers arrived at, not knowing beforehand which it might be.* The numbers would have been placed on the two columns of the Store previously assigned for their reception, and cards would have been arranged to direct the two numbers to be subtracted each from the other. In the one case there would be a remainder, but in the other the carriage warning lever would have been moved in the highest place, which would be made to bring an alternative set of cards, previously prepared, into play.

62. I have not been quite able to accept Mr. Merrifield's opinion as to the "capability of the engine." He says: "Its capability thus extends to any system of operations or equations which leads to a single numerical result." Now it could furnish not only one root, but every root of an equation, if there were more than one, capable of arithmetical expression, and there are many such equations. It could follow the processes of the mathematician be they tentative or direct,† wherever he could show the way to any number of numerical results. It is only a question of cards and time. Fabrics have been woven requiring several thousand cards. I possess one‡ made by the aid of over twenty thousand cards, and there is no reason why an equal number of cards should not be used if necessary, in an Analytical Engine for the purposes of the mathematician.

63. There exist over two hundred drawings, in full detail, to scale, of the engine and its parts. These were beautifully executed by a highly skilled draughtsman and were very costly.§

64. There are over four hundred notations of different parts.‖ These are, in my father's system of mechanical notation, an outline of which I had the honour to submit to this Association at Glasgow in 1855.¶ Not many years ago I was looking over one of my own drawings with a very intelligent mechanical engineer in Clerkenwell. I wished to get motion for some particular purpose. He suggested to get it from an axis which he pointed to on the drawing. I answered, " No, that will not

---

* See p. 176.   † See p. 173.   ‡ A portrait of Jacquard woven in silk, see p. 169 of 'Passages from the Life
          of a Philosopher.'   § See p. 288.   ‖ See p. 271.   ¶ See pp. 242 to 257.

do; I see by the drawing that it is a 'dead centre.'" He replied, "You have some means of knowing which I have not." I certainly had, for I used this mechanical notation. The system in whole or part should be taught in our Art and Technical Schools.

65. In addition to the above things there will shortly be available, as I have said before, the reprint of the various papers published relating to these machines, and a full list of the drawings and notations will be included.

66. I believe that the present state of the design would admit of the engine being executed in metal; nor do I think that, as suggested by Mr. Merrifield, quantities and proportions would have to be calculated. Of course working drawings would have to be made—it would not be wise to commence any work without such drawings—but they would mostly be simply copies from the originals with such details as workmen want added. It would also be wise to make models of particular parts. The shapes of the cams, for instance, might be tried in wood which would afterwards suit as patterns for castings.

67. Mr. Merrifield doubts "whether the drawings really represent what is meant to be rendered in metal, or whether it is simply a provisional solution to be afterwards simplified." I have no doubt that the drawings do represent what at the time was intended to be put into metal; but as certainly were intended to be superseded when anything better could be found. Very few machines indeed are invented which do not undergo modification. It is almost invariably the case that the second machine made has improvements on the first design. In such a machine as an Analytical Engine, this would be important; but no one ever stops a useful invention for fear of improvements. A gun or an ironclad, for example, has been scarcely made before superseded by something better, and so it must be with all inventions, though fortunately not at the same speed.

68. As to the possible modification of the engine (Chapter VIII. of the Report) I may say that I am myself of opinion that the general design might with practical advantage be restricted. The engine would be still very useful indeed, if made not quite so automatic; even the Mill by itself would, I believe, be extensively useful, if a printing or stereo-moulding apparatus were joined to it. Perhaps, if that existed, the wants of further parts such as the Store would be felt and supplied.

69. As to the general conclusion and recommendation (Chapter IX. of the Report) there is little for me to notice. I see no hope of any Analytical Engine, however useful it might be, bringing any profit to its constructor, and beyond the preparation of this Paper, and the publication of the volume I have mentioned as shortly to follow, there is little or no temptation to do more. Those who wish for such an engine would, I think, give it a helping hand if they could show what pecuniary benefit it would bring. The History of Babbage's Calculating Machines is sufficient to damp the ardour of a dozen enthusiasts.

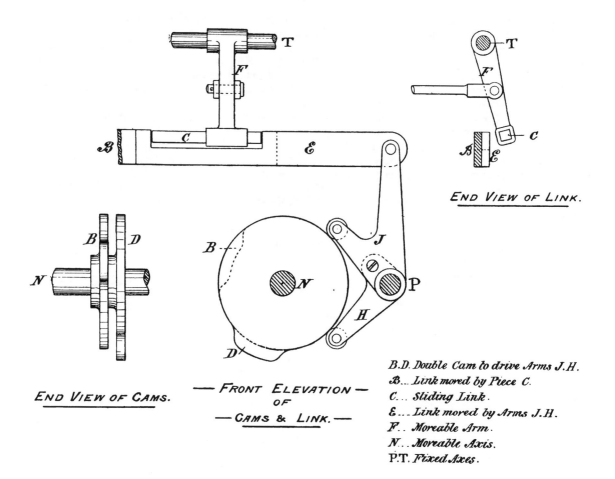

T

F

B   C   E

T

F

B E   C

END VIEW OF LINK.

B   D

N

END VIEW OF CAMS.

B

N

J

P

H

D

— FRONT ELEVATION —
OF
— CAMS & LINK. —

B.D. Double Cam to drive Arms J.H.
B... Link moved by Piece C.
C... Sliding Link.
E... Link moved by Arms J.H.
F.. Moveable Arm.
N... Moveable Axis.
P.T. Fixed Axes.

CONCLUSION.

CHARLES BABBAGE was born on 26th December, 1791, at Totnes in Devonshire. He was entered at Trinity College, Cambridge, on 21st April, 1810, and resided there from October. He had at an early age shown a taste for mathematics, and was already acquainted with the works of La Croix and other continental mathematicians when he went to Cambridge. The first idea of a calculating machine occurred to him when a student at Cambridge in 1812 or 1813. He was familiar with the method of Differences, and knew that by that method Logarithmic and many other Tables could be produced by simple addition, and he saw that if a machine could be constructed to perform only simple addition, such tables could be produced mechanically. The idea seems to have lain dormant for a time, but some years afterwards, when he went to Paris, he studied the details of the arrangement by which the celebrated French Tables had been computed under the direction of Prony, and he came in contact with Didot, from whom he purchased at a high price a copy of his stereotyped Table of Natural Sines and their Differences, to twenty places of figures. By permission of the French Board of Longitude, he copied the Logarithms to fourteen places of figures of every 500th number from 10,000 to 100,000 from the Tables deposited in the observatory at Paris. In 1819 he was working at the subject (p. 42), and on 14th June, 1822, he announced to the Astronomical Society that he had completed a machine of two orders of differences, capable of producing tables of squares, triangular numbers, values of $x^2 + x + 41$, &c. (p. 211). On the 3rd July of that year, he addressed a letter on the subject to Sir H. Davy, then President of the Royal Society. This letter was published and advertised for sale in the newspapers of the day. It is reprinted at page 212 of this book. It was followed by a letter to Dr. Brewster, dated 6th November, 1822, published in "Brewster's Journal of Science" (p. 216), "On the theoretical principles of the machinery for calculating Tables;" and in December 1822 by "Observations on the application of machinery to the computation of mathematical Tables," published in the memoirs of the Astronomical Society (p. 220). These four papers show how fully, even at that time, the invention was developed; little, if anything, was subsequently added to the general design. It was intended to produce Tables by the method of Differences, requiring mechanism for addition only, and to print the computed results without a chance of error. The very essence of the design was accuracy; it is doubtful if he would have proceeded with it otherwise. It was to compute Tables with both positive and negative Differences (such as Natural Sines, &c.), and even those with no constant Differences (pp. 212, 224), and to extract the roots of equations and numbers (p. 212). He had actually then constructed a machine which produced any Table whose second differences are constant (such as square numbers, values of $x^2 + x + 41$, &c.). He had also designed a machine for multiplying.

We learn further from the Paper by F. Baily (p. 225) that it was intended to produce in the machine impressions on a soft substance, of the figures computed by the machine which were afterwards to be stereotyped, and that the last figure was to be automatically increased by one if the next following figure should happen to be five or upwards, thus making the printed result correct to the nearest figure. In this paper by Mr. Baily, the powers of the Difference Engine and its possible uses are very fully set forth, and more in the same way will be found in Dr. Lardner's article at pp. 53 to 56, and 73 to 74. The mechanism of the engine is described at some length in the same Article, the whole of which is worth close attention.

On the 1st April, 1823, the Lords of the Treasury referred the letter to Sir H. Davy to the Royal Society (p. 111), requesting their opinion on the " merits and utility of this invention." On the 1st

3 A

May, the Royal Society reported to the Treasury in its favour (see Parliamentary Paper, No. 370, printed 22 May, 1823), and on 21st July, 1823, £1,500 was advanced by the Treasury, and very shortly after the actual construction of the Difference Engine was commenced (p. 113) in the workshops of Mr. Clement in Lambeth. Work was stopped on 9th May, 1829. It was resumed in March 1830 (p. 121) and continued till 10 April, 1833, when Clement, the engineer, stopped the work and discharged the workmen employed (p. 124).

Work was never resumed on it, and the design was abandoned by the Government in 1842. The history of the connection of the Government with the machine is given in pages 110–138.

Up to 9 May, 1829, £6,628 had been expended, and up to April, 1833, probably the total expense did not exceed £11,000 or £12,000, all paid to the engineer for the actual construction of the machine.

The purchase by the Government of the lease of some property (p. 123), and the construction on it of fireproof buildings, brought the total sum expended by Government up to about £17,000 (p. 145), but for these last items it is believed that the Government have received a good equivalent in rents.

Among the workmen discharged by Clement on 10th April, 1833, was J. Whitworth (afterwards Sir J. Whitworth, Bart.). He and others saw in the Difference Engine on which they were daily employed the most exact workmanship; they saw many pieces made identically the same size to gauge: surfaces planed and turned accurately flat, screws (made from about 1827) with a fixed number of threads to the inch), and tools which produced work not to be excelled in accuracy even at the present day. A lathe made to order in 1823–4 by Clement is on deposit loan in the South Kensington Museum. Its guide screws have divided circles on the handles, so that the tool can be set to the fraction of a thread; they are pivoted on a large divided circle, so that a cone of any angle can be turned, and two cutting tools can be set at once, so that with a little management two different diameters may be cut, one after the other, on piece after piece without shifting the tool, or touching the guide screw. Many other improved tools were also in use (see the Report of the committee of the Royal Society, February, 1829, p. 233). It cannot be denied that the requirements of the Difference Engine led to a considerable advance in the art of mechanical construction generally, which was a distinct gain to the country.

When the work was stopped by Clement in April, 1833, a part of the machine had been put together and was found to do its work perfectly; most if not all of the pieces, both of the printing part and of the calculating part existed, and many of these last had been actually fitted and marked ready to be put together (some of these pieces so marked came into my own possession in 1871). These all remained in the control of the Government till 1842. The Report of the committee of the Royal Society, dated 12 February, 1829, states that even then "the actual work of the calculating part is in a great measure constructed, though not put together." On the 19th November, 1829, the Duke of Wellington, then Prime Minister, accompanied by Mr. Goulburn, the Chancellor of the Exchequer, had visited Clement's workshop, and inspected the progress of the engine. In the letter written shortly afterwards (3rd December), Mr. Goulburn suggested that the calculating should be separated from the printing part, "so that if any failure should take place in the attempt to print, the calculating part should nevertheless be perfect." No doubt whatever of the success of the calculating part was ever expressed. A reference to the Royal Society had been proposed on 16th January, 1836 (pp. 130 and 132), by Mr. Spring Rice, Chancellor of the Exchequer, which unfortunately never took place, and apparently without any further enquiry or report of any sort, Sir Robert Peel, the Prime Minister, with his Chancellor of the Exchequer, Mr. Goulburn, summarily decided in 1842 to abandon the construction of the engine.

Now it is amazing that in this state of things, with which Mr. Goulburn at least must have been acquainted, Sir Robert Peel did not see the advisability of having the calculating part completed. A few hundred pounds would at that time probably have been sufficient for this. Can any one doubt that if that part had been at any time between 1833 and 1842 completed, the fate of the Difference Engine would have been changed? There would have been something to show to the public to justify the expense incurred; the House of Commons would have been satisfied, and in all probability there would have been no difficulty about funds for the completion of the printing part. Sir Robert Peel certainly made no reference to my father in this direction, nor consulted him in the matter. It is probable indeed that their relations had then become strained, but Sir Robert Peel does not appear to have taken the advice of anybody, nor is it known that he made any enquiries whatever regarding the state of the work. After communicating with Mr. Goulburn, his Chancellor of the Exchequer, the latter announced his decision to my father (p. 136). Had another course been taken, a Difference Engine might have existed fifty years ago. Many Tables of unerring accuracy and of many places of figures might have been by this time in existence and in use over the world, and perhaps even an Analytical Engine might have been now in active work, enlarging the sphere of our knowledge. Nor would the inventor himself have remained unhonoured and unrewarded, nor he and his family have had cause to regret the day on which he became entangled with the English Government.

The full history of this entanglement is set forth in the statement (pp. 110–138), and in Chapter XI. of Weld's History of the Royal Society (p. 311), and in Professor de Morgan's review of it (page 318). The question has there been dealt with in another aspect; and I am not aware that any one has handled it in the same aspect as myself, but from my own personal knowledge it seems to me to rise beyond the other, and to dwarf it entirely.

The construction being abandoned, the completed portion was placed for some years in Somerset House. It is now in the museum at South Kensington, where some photographic copies of drawings of the complete engine are also exhibited. A representation of this piece is given at the end of this book. When completed there would have been seven columns of about eighteen figures each.

The rest of the machinery was disposed of at something like the price of old metal. My father bought some (p. 196, line seven) which he cut up and used as he found occasion during his lifetime. What was left came to me after his death, in 1871, and from it a few sample pieces were put together. One of these I sent to Cambridge University, one to University College, London, and another to Owens College, Manchester. One I sent to America, and two more I have myself. All the rest was melted up some years ago, and about the same time some of the work which had remained in Clement's workshop was, I heard, melted up on the works being closed by his nephew and successor. The fire-proof rooms and premises were rented for other purposes, and I believe remained so till the lease expired, yielding to the Government a fair return on the expenditure.

The circumstances connected with the Difference Engine No. 2 are set forth at pp. 139 to 153 of this book, and it is not necessary to say more about it here.

The publication of Dr. Lardner's article in the *Edinburgh Review* of July, 1834, led to the construction by the Messrs. Schentz of Stockholm, of a Difference Engine, which they brought to England in 1854, and publicly exhibited.

They had not anticipated a friendly welcome from my father, and were equally surprised and gratified when they found him ready and pleased to forward their objects as far as lay in his power. Mr. Gravatt, F.R.S., and M. Inst. C.E., showed great interest in their invention, and took active steps to aid the inventors. A committee of the Royal Society reported favourably on it on 21st June, 1855 (p. 264), and my father at the anniversary meeting of the Royal Society on 30th November, 1855, spoke in its behalf, gave a short history of it, and recommended the inventors for one of the Society's medals (p. 260).

Having closely studied this machine, I made a description of it in the system of mechanical notation invented by my father, and exhibited it at the meeting of the British Association in that year at Glasgow (p. 246). The diagrams were subsequently shown at a meeting of the Institution of Civil Engineers in London, and the mechanical notation was more fully explained (p. 248).

The original Paper by my father on the subject was published in vol. ii. p. 250 of the "Philosophical Transactions" (p. 236), and in 1851 a short paper was drawn up and circulated for consideration (p. 242). The Swedish machine was exhibited at the Paris Exhibition in 1855, where it obtained the gold medal of its class (p. 261), and was finally acquired for the Dudley Observatory at Albany, U. S. America (p. 270). A small volume of specimen tables produced by the machine was published in 1857, dedicated by George and Edward Schentz to my father.

A second machine, a copy of the first, was made and placed in the Registrar General's Office, London, where it was used in the computations for the English Life Tables, by W. Farr, London, 1864. The reader is referred to p. cxxxix. of the Appendix of that work, where the author, after mentioning the use of the machine, says: "This volume is the result, and thus—if I may use the expression—the soul of the machine is exhibited in a series of Tables which are submitted to the criticism of the consummate judges of this kind of work in England and in the world."

Several other Difference Engines have been constructed, but it is foreign to the object of this volume to describe them, nor would Schentz's engine have here been mentioned were it not for the active part my father took in its favour, and the fact of my having taken their machine on which to illustrate his system of Mechanical Notation.

Before I leave the Difference Engine I may mention that about 1856 my father said in conversation with me, that he was by no means sure that if he had unlimited wealth he should not first construct a Difference Engine. Work had been stopped on the Difference Engine on 10th of April, 1833, and it was not till July 1834 that the parts of the machine and the drawings were removed to the fireproof buildings near Dorset Street. Meanwhile the idea of the Analytical Engine had arisen (p. 154). Some insight into the various steps of the idea may be gained from reading the extracts from the "Ninth Bridgewater Treatise," printed at pp. 299 to 304. The idea first was to arrange the axes of the

Difference Engine round a central wheel, to facilitate the control of the last Difference by the column of result; a process which he called "the engine eating its own tail" (p. 268). Then came the idea of controlling it in other ways and by other means, and gradually the general design of the Analytical Engine was developed. Pages 307 to 310 and 154 to 180 give some of the main divisions of the machine. I have added copies of the drawings mentioned at p. 6, and of the general plan No. 25 of 6th August, 1840. They will give some idea of the machinery, though they have been more or less superseded by others following. I have endeavoured in pages 331 to 337 to give some further insight into the general arrangements of the engine.

From 1834 till the day of his death, my father employed his faculties on the design of the Analytical Engine, varied occasionally by studies of other subjects; many of which appear in the last of his published works at the end of this volume. He spent much of his time and some thousands of pounds on this, the great work of his life. Any one competent to judge, who examines the mere list of Drawings or pages 288 to 293, and peruses the list of Notations given at pages 273 to 287 may see how thoroughly every detail of the design was considered and worked out, and may obtain some faint idea of his gigantic task and the labour he devoted to it.

Should any ardent student wish to pursue the subject further, he should first master the system of mechanical notation explained in pp. 242 to 259 of this book. He should then read through the Scribbling Books, mentioned on p. 294, and afterwards study at leisure the drawings and notations of which lists are given at pp. 271 to 293. For the benefit, however, of such students, I will close with a quotation from the statement drawn up by the late Sir H. Nicolas (p. 137):—

"In this statement the heavy sacrifices both pecuniary and personal which the invention of these machines has entailed upon their author have been alluded to as slightly as possible. Few can imagine and none will ever know their full extent. Some idea of those sacrifices must nevertheless have occurred to every one who has read this statement."

HENRY P. BABBAGE.

October 1888.

FINIS

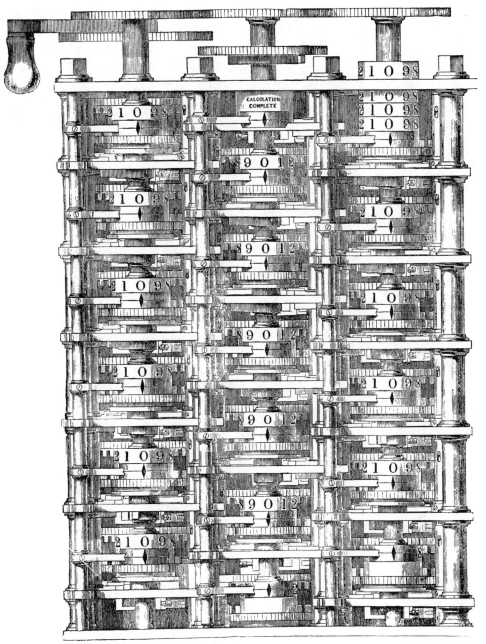

CALCULATION COMPLETE

B. H. Babbage del.

Impression from a woodcut of a small portion of Mr. Babbage's Difference Engine, No. 1, the property of Government, at present deposited in the Museum at South Kensington.

It was commenced 1823.
This portion put together 1833.
The construction abandoned 1842.
This plate was printed 1889.

PLATE I.

PLAN OF THE FIGURE WHEELS FOR ONE METHOD OF ADDING NUMBERS.

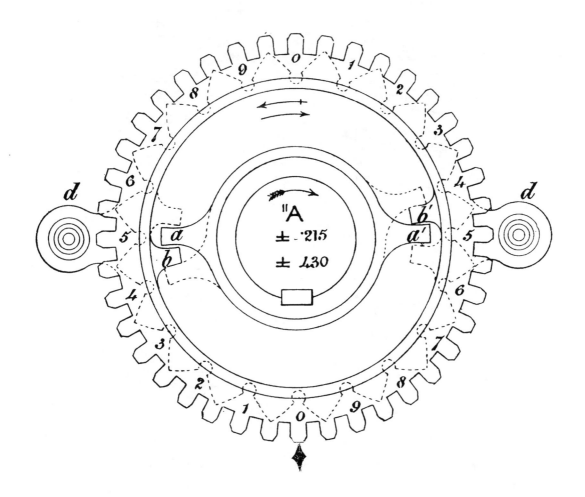

PLATE II.

ELEVATION OF THE WHEELS AND AXIS FOR ONE METHOD OF ADDING NUMBERS.

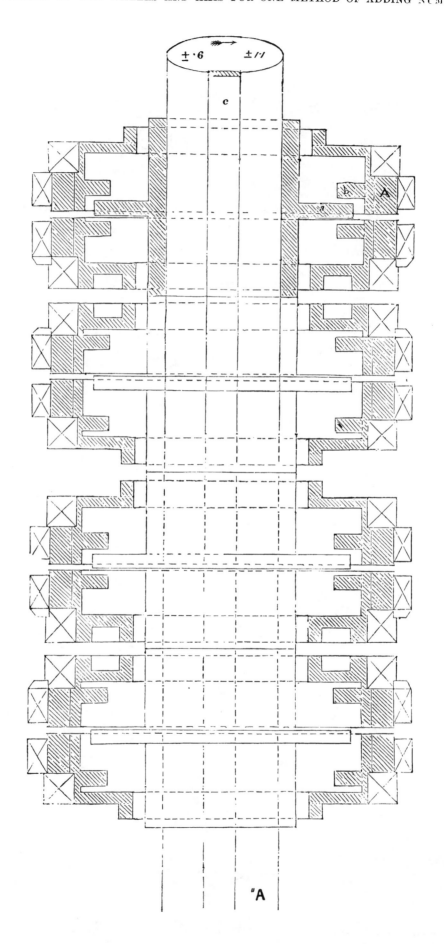

PLATE III.

ELEVATION OF FRAMING ONLY FOR ONE METHOD OF ADDING NUMBERS.

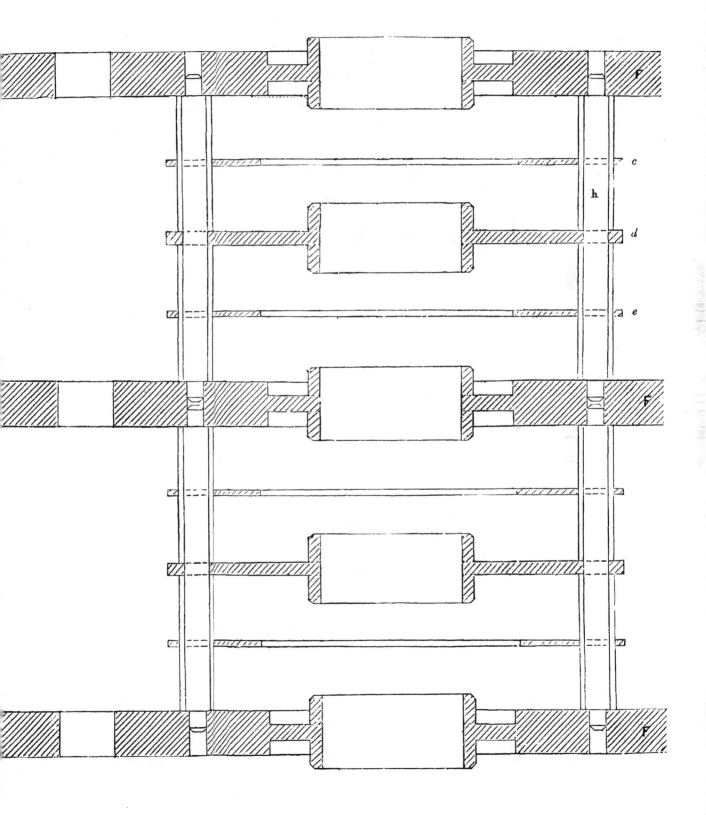

PLATE IV.

SETION OF ADDING WHEELS AND FRAMING TOGETHER.

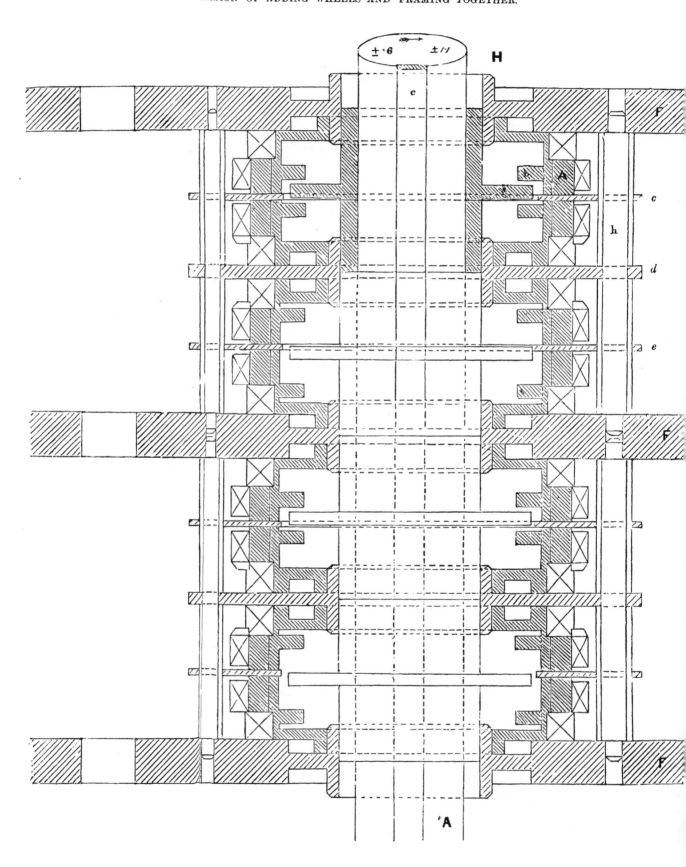

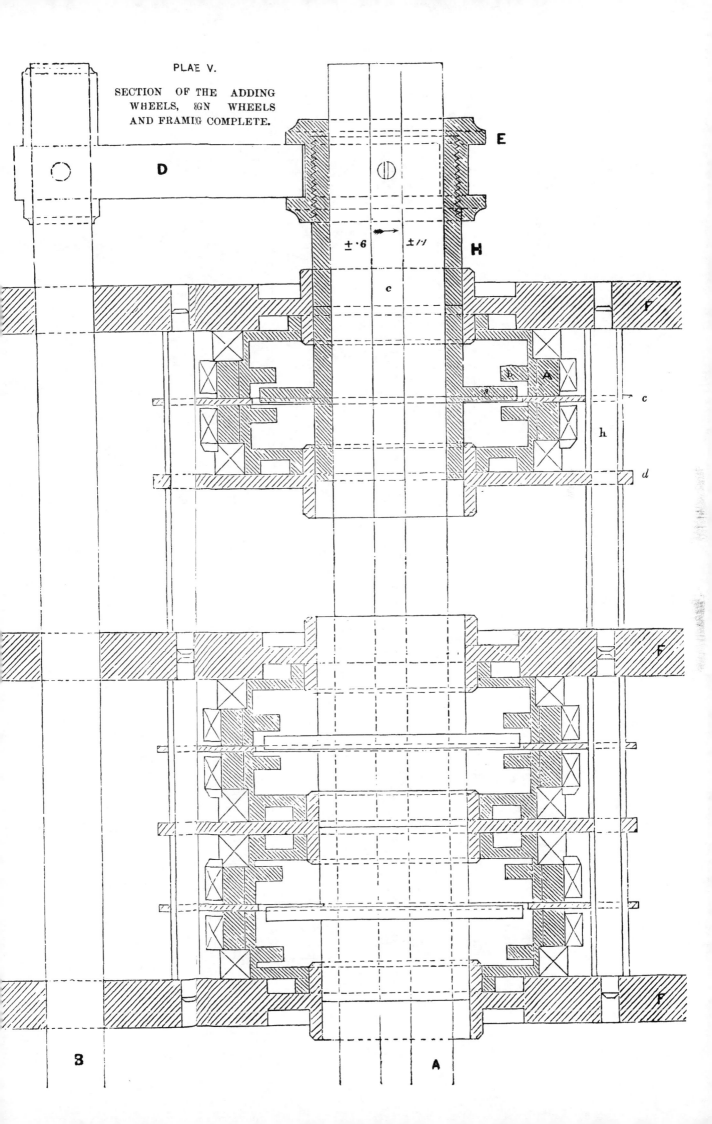

PLAE V.

SECTION OF THE ADDING
WHEELS, ŁGN WHEELS
AND FRAMIG COMPLETE.

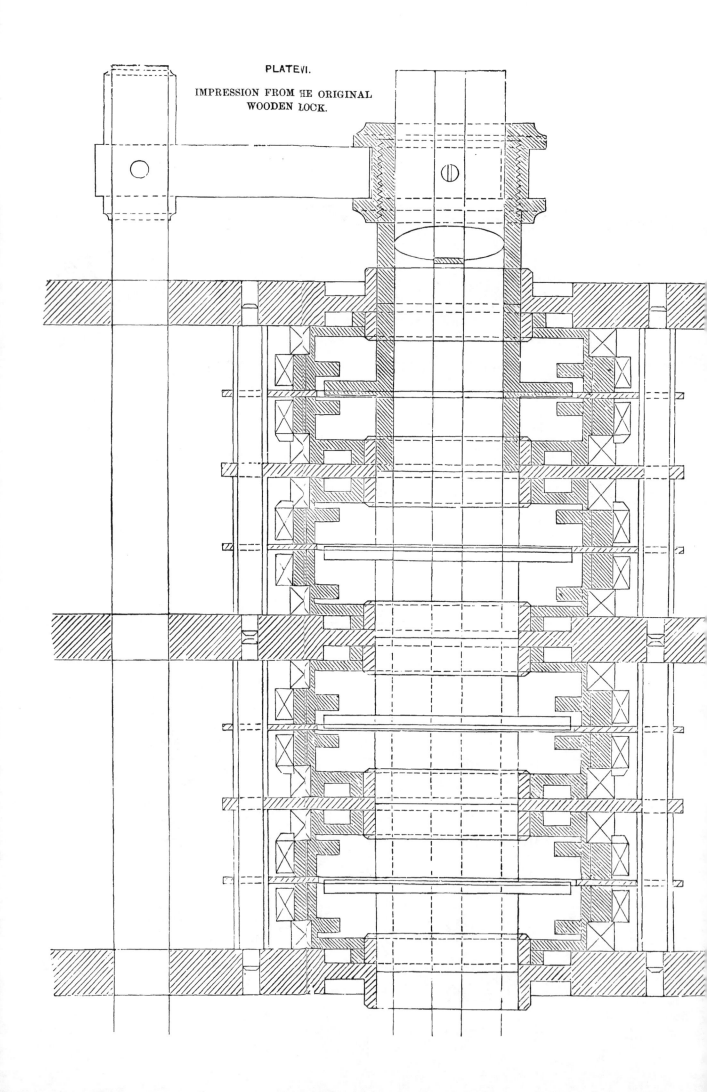

PLATE VI.

IMPRESSION FROM HE ORIGINAL
WOODEN LOCK.

PLATE VII.

IMPRESSIONS FROM A STEREOTYPE CAST OF No. VI., WITH THE LETTERS AND SIGNS INSERTED. Nos. II., III., IV., AND V. WERE STEREOTYPES TAKEN FROM THIS.

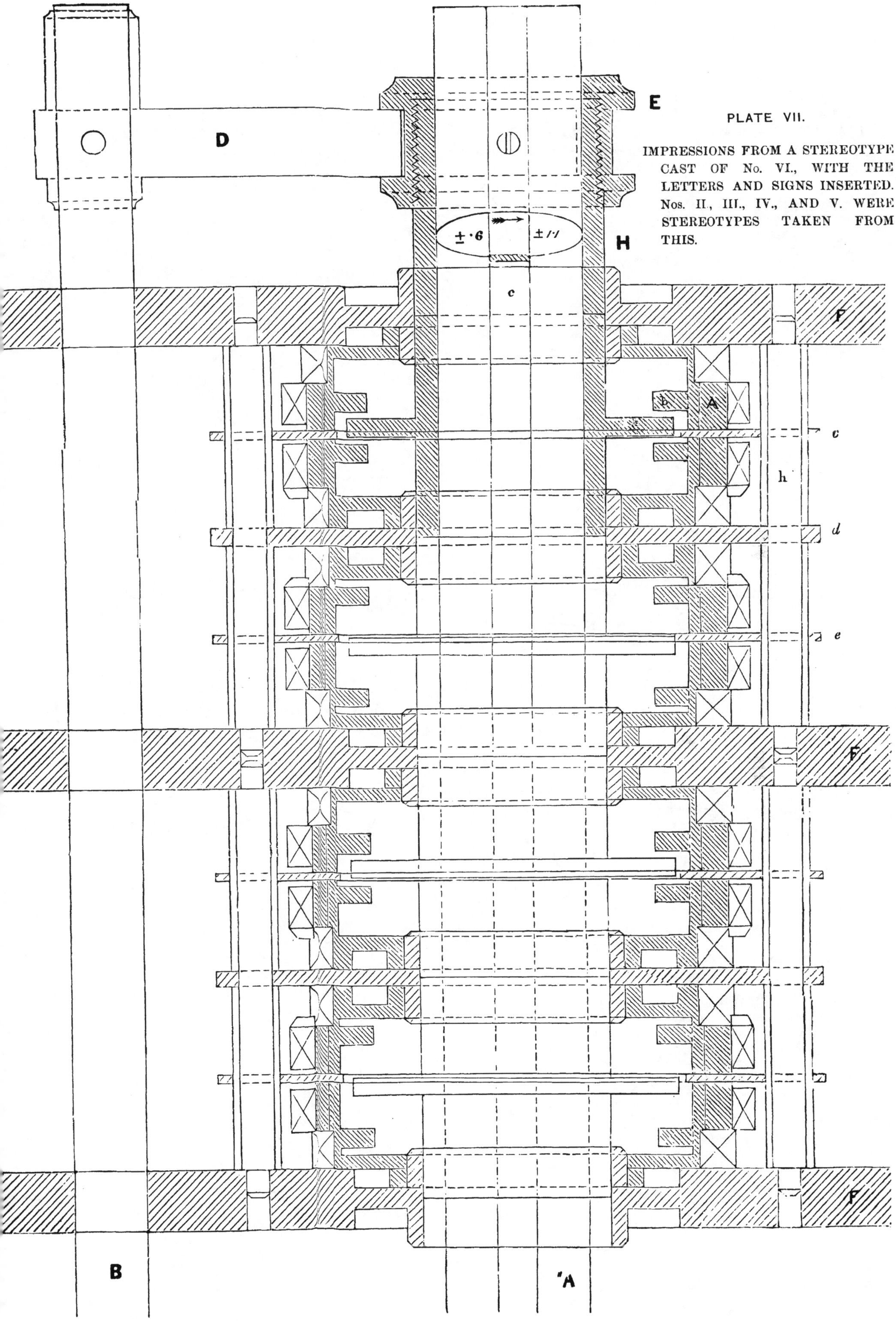

$\pm \cdot 6$ $\pm \prime\prime$

D

E

H

F

F

F

B

'A

PLATE VIII.

PLAN OF ADDING WHEELS AND OF LONG AND SHORT PINIONS, BY MEANS OF WHICH *STEPPING* IS ACCOMPLISHED.

N.B.—This process performs the operation of multiplying or dividing a number by any power of ten.

PLATE IX.

ELEVATION OF LONG PINIONS IN THE POSITION FOR ADDITION.

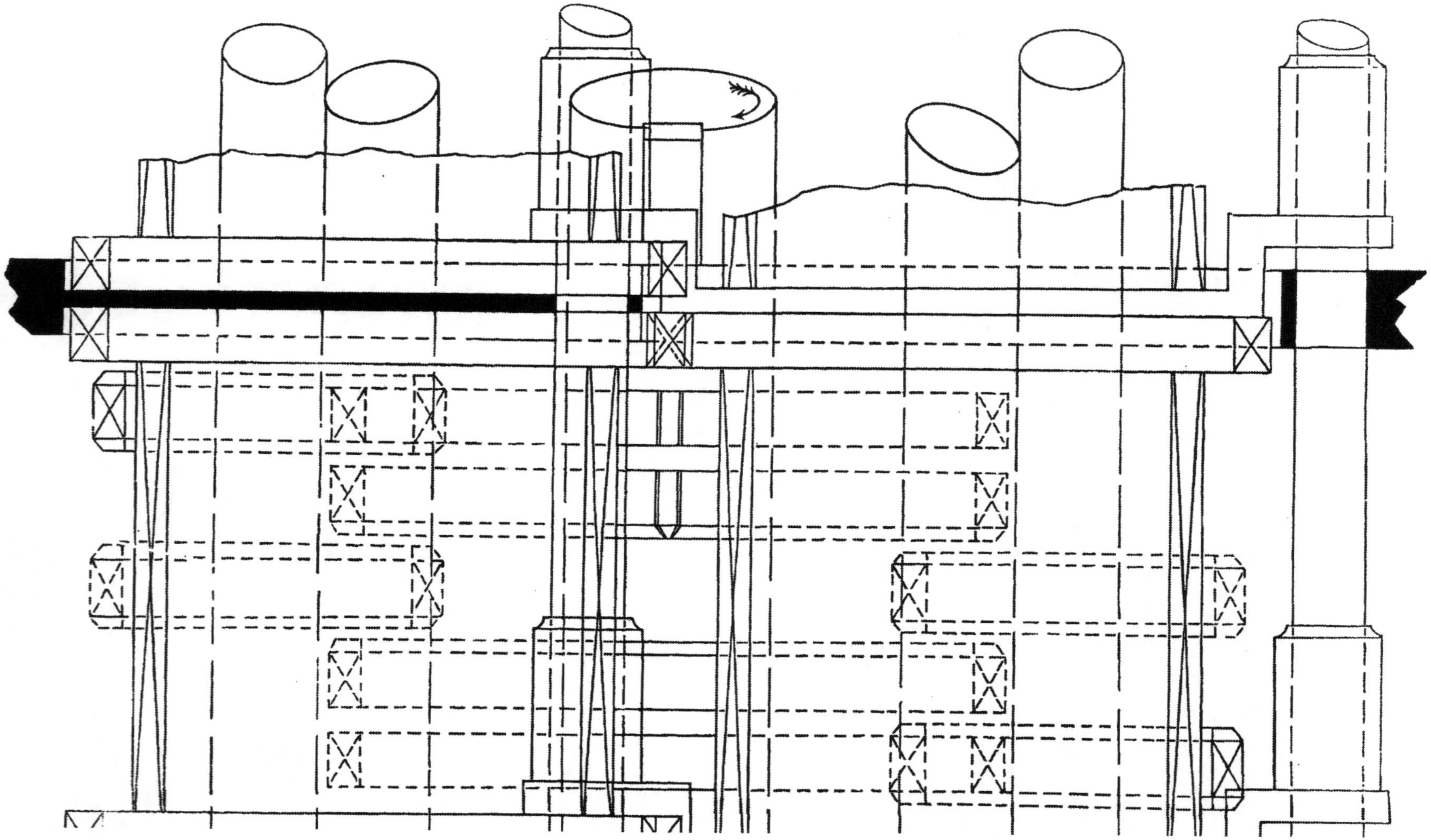

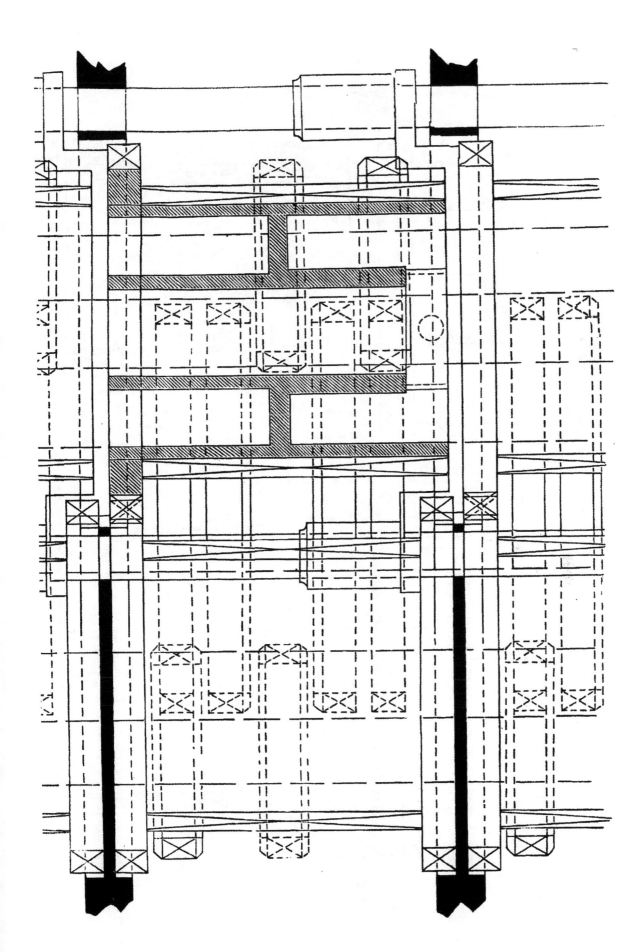

PLATE X.

ELEVATION OF LONG PINIONS IN THE POSITION FOR STEPPING.

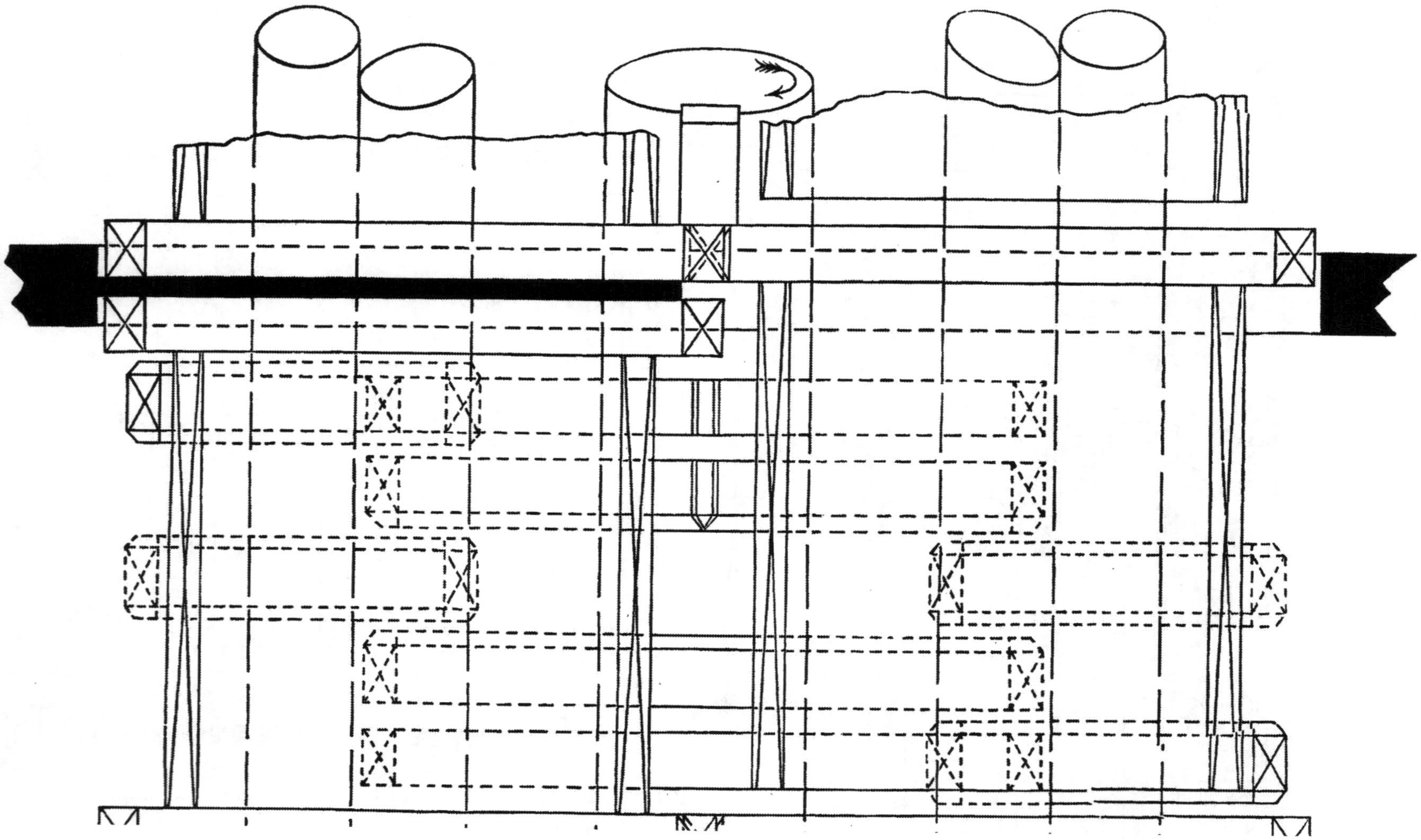

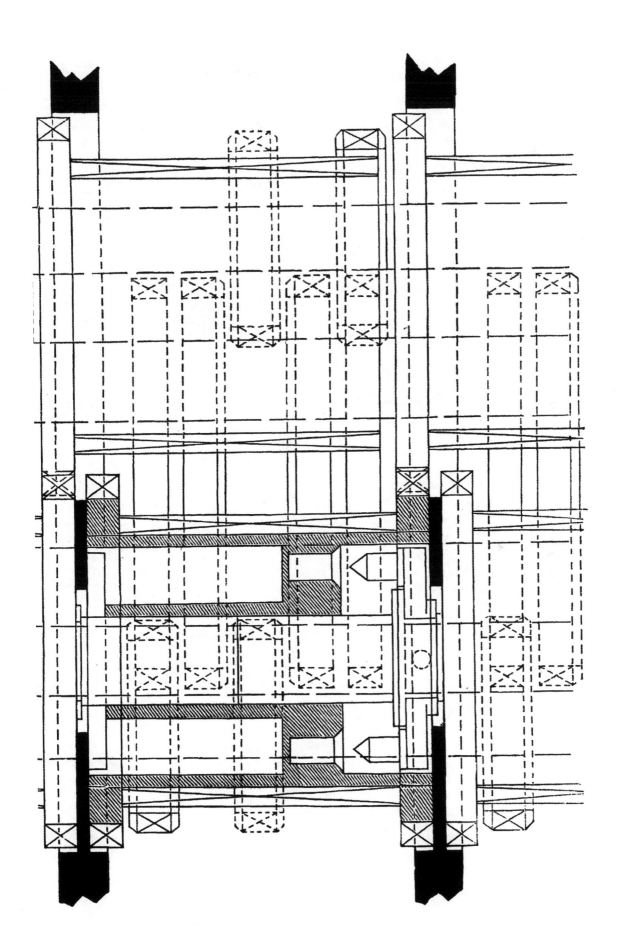

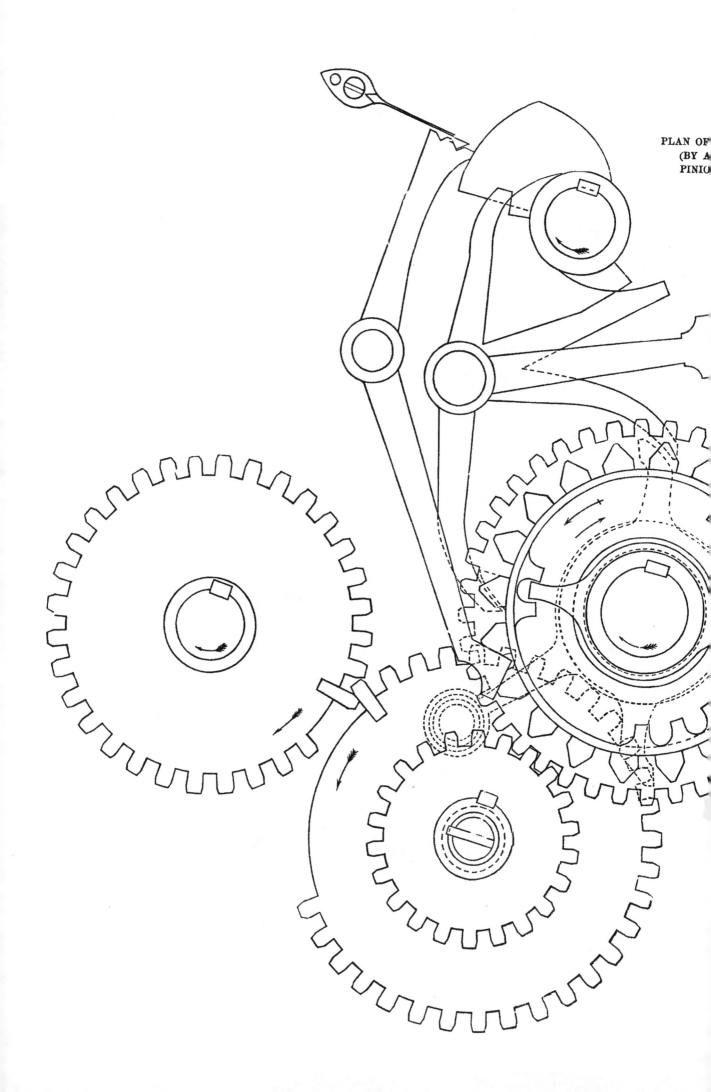

PLAN OF
(BY A
PINIO

PLATE XII.

SECTION OF THE CHAIN OF WIRES FOR ANTICIPATING CARRIAGE.

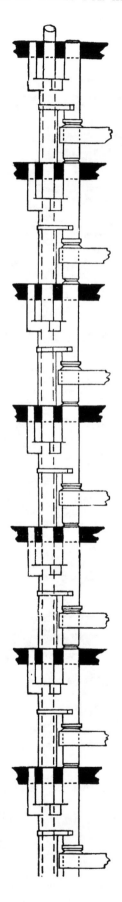

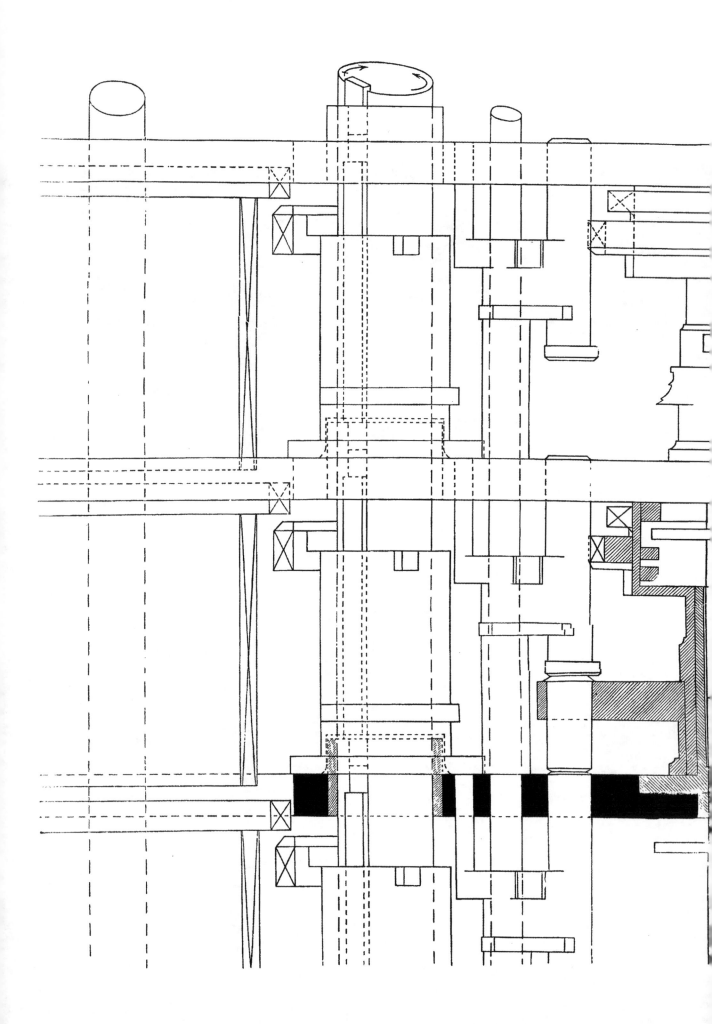

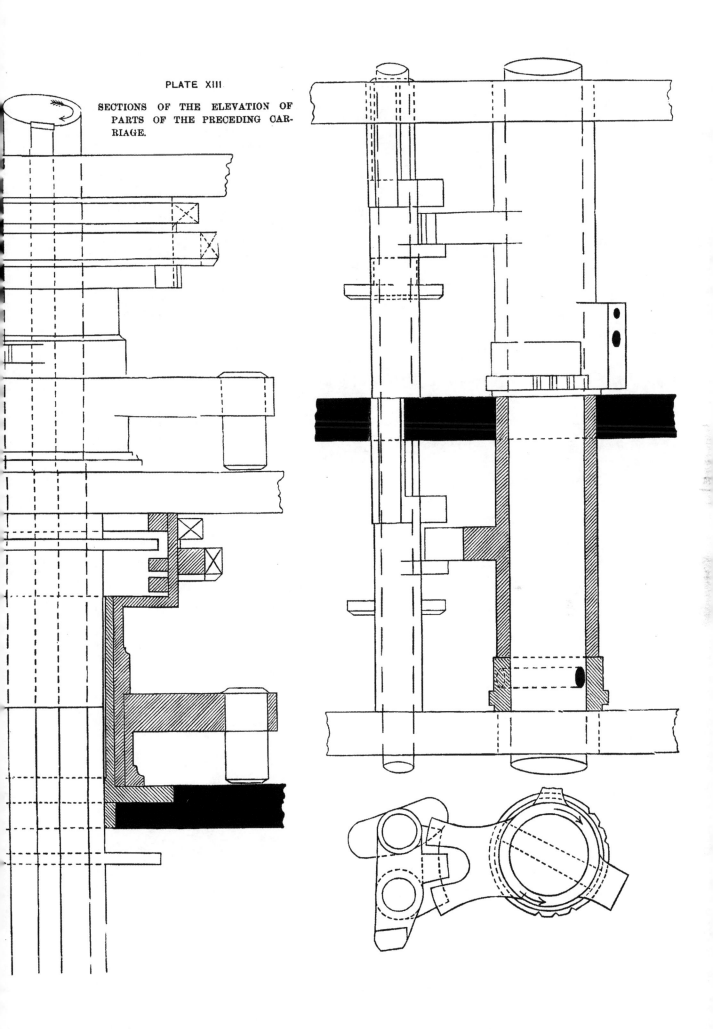

PLATE XIII

SECTIONS OF THE ELEVATION OF
PARTS OF THE PRECEDING CAR-
RIAGE.

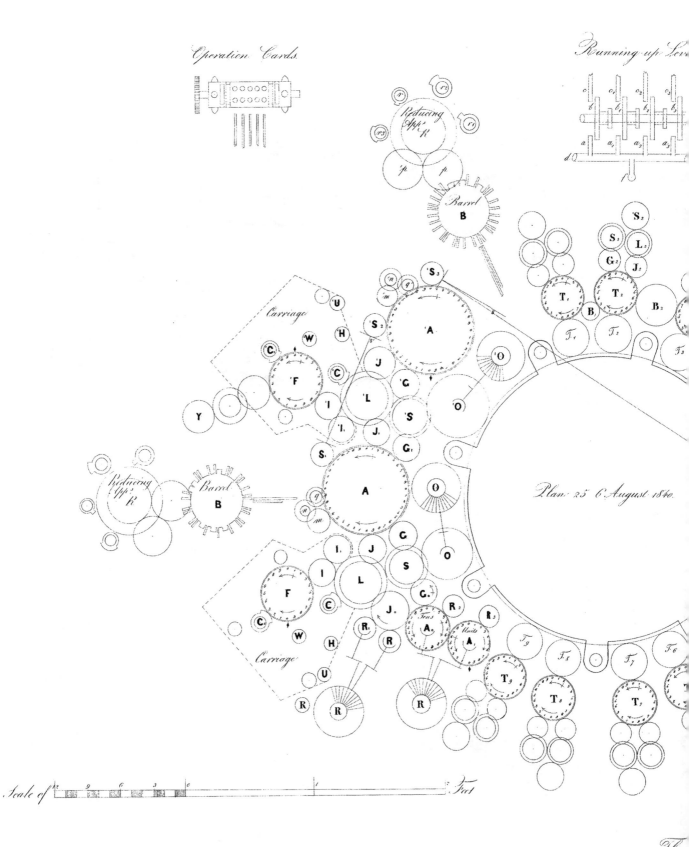

Operation Cards

Running up Lev

Reducing
App^r R

Barrel
B

Carriage

Reducing
App^r R

Barrel
B

Plan 25 6 August 1840.

Scale of   Feet

The
Mr Babbage's

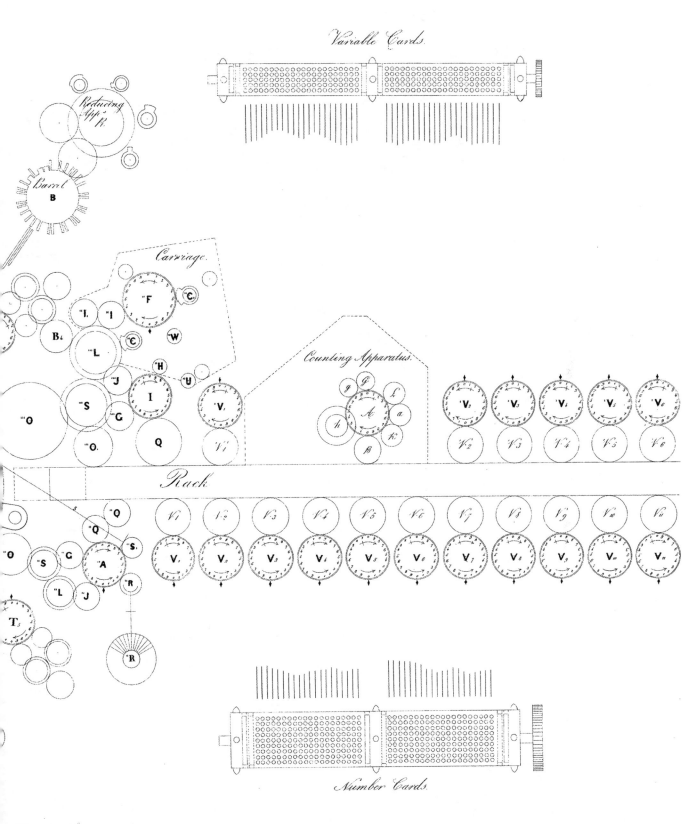

Variable Cards.

Reducing Appᵗ R.

Barrel
B

Carriage.

"F" "C"

"I." "I"

B₄

"C"

"W"

"'L"

"H"

"J"

"U"

Counting Apparatus.

"S"

"I"

"'V₁"

g   G

"O"

"G"

h   A   a

"O."

Q

V₇

B   K

'V₁ 'V₃ 'V₄ 'V₅ 'V₆

V₂ V₃ V₄ V₅ V₆

Rack

"Q"

"Q"

V₁ V₂ V₃ V₄ V₅ V₆ V₇ V₈ V₉ V₁₀ V₁₁

"O" "S" "G" "A" "S"

"L" "J" "R"

V₁ V₂ V₃ V₄ V₅ V₆ V₇ V₈ V₉ V₁₀ V₁₁

T₅

"R"

Number Cards.

Plan of
Calculating Engine

LIST OF MR. BABBAGE'S PRINTED PAPERS.

Many applications having been made to the Author and to his Publishers, for detached Papers which he has from time to time printed, he takes this opportunity of giving a list of those Papers, with references to the Works in which they may be found.

---

1. The Preface; jointly with Sir John Herschel. ⎱ *Memoirs of the Analytical Society.* 4to. Cam-
2. On Continued Products. ⎰ *bridge*, 1813.

3. An Essay towards the Calculus of Functions.—*Phil. Trans.* 1815.

4. An Essay towards the Calculus of Functions, Part 2.—*Phil. Trans.* 1816. P. 179.

5. Demonstrations of some of Dr. Matthew Stewart's General Theorems, to which is added an Account of some New Properties of the Circle.—*Roy. Inst. Jour.* 1816, Vol. i. p. 6.

6. Observations on the Analogy which subsists between the Calculus of Functions and other branches of Analysis.—*Phil. Trans.* 1817. P. 179.

7. Solution of some Problems by means of the Calculus of Functions.—*Roy. Inst. Jour.* 1817. P. 371.

8. Note respecting Elimination.—*Roy. Inst. Jour*, 1817. P. 355.

9. An Account of Euler's Method of Solving a Problem relating to the Knight's Move at Chess.— *Roy. Inst. Jour.* 1817. P. 72.

10. On some new Methods of Investigating the Sums of several Classes of Infinite Series.—*Phil. Trans.* 1819. P. 245.

11. Demonstration of a Theorem relating to Prime Numbers.—*Edin. Phil. Jour.* 1819. P. 46.

12. An Examination of some Questions connected with Games of Chance.—*Trans. of Roy. Soc. of Edin.* 1820. Vol. ix. p. 153.

13. Observations on the Notation employed in the Calculus of Functions.—*Trans. of Cam. Phil. Soc.* 1820. Vol. i. p. 63.

14. On the Application of Analysis, &c. to the Discovery of Local Theorems and Porisms.—*Trans. of Roy. Soc. of Edin.* Vol. ix. p. 337. 1820.

15. Translation of the Differential and Integral Calculus of La Croix, ⎫ These two works were exe-
1 vol. 1816.   ⎪ cuted in conjunction with
16. Examples to the Differential and Integral Calculus. 2 vols. 8vo. ⎬ the Rev. G. Peacock (Dean
1820.   ⎪ of -Ely) and Sir John Her-
   ⎭ schel. Bart.

17. Examples of the Solution of Functional Equations. Extracted from the preceding. 8vo. 1820.

18. Note respecting the Application of Machinery to the Calculation of Mathematical Tables.— *Memoirs of the Astron. Soc. June,* 1822. Vol. i. p. 309.

19. A Letter to Sir H. Davy, F.R.S., on the Application of Machinery to the purpose of calculating and printing Mathematical Tables. 4to. *July,* 1822.

20. On the Theoretical Principles of the Machinery for calculating Tables.—*Brewster's Edin. Jour. of Science.* Vol. viii. p. 122. 1822.

21. Observations on the application of Machinery to the Computation of Mathematical Tables, Dec. 1822.—*Memoirs of Astron. Soc.* 1824. Vol. i. p. 311.

22. On the Determination of the General Term of a new Class of Infinite Series.—*Trans. Cam. Phil. Soc.* 1824. Vol. ii. p. 218.

23. Observations on the Measurement of Heights by the Barometer.—*Brewster's Edin. Jour. of Science,* 1824. P. 85.

24. On a New Zenith Micrometer.—*Mem. Astron. Soc.* March, 1825.

25. Account of the repetition of M. Arago's Experiments on the Magnetism manifested by various substances during Rotation. By C. Babbage, Esq., and Sir John Herschel.—*Phil. Trans.* 1825. P. 467

26. On the Diving Bell.—*Ency. Metrop.* 4to. 1826.

27. On Electric and Magnetic Rotation.—*Phil. Trans.* 1826. Vol. ii. p. 494.

28. On a method of Expressing by Signs the Action of Machinery.—*Phil. Trans.* 1826. Vol. ii. p. 250.

29. On the Influence of Signs in Mathematical Reasoning.—*Trans. Cam. Phil. Soc.* 1826. Vol. ii. p. 218.

LIST OF MR. BABBAGE'S PRINTED PAPERS.

30. A Comparative View of the different Institutions for the Assurance of Life. 1 vol. 8vo. 1826. German Translation. Weimar, 1827.

31. On Notation.—*Edinburgh Encyclopedia.* 4to.

32. On Porisms.—*Edinburgh Encyclopedia.* 4to.

33. A Table of the Logarithms of the Natural Numbers, from 1 to 108,000, Stereotyped. 1 vol. 8vo. 1826.

34. Three editions on coloured paper, with the Preface and Instructions translated into German and Hungarian, by Mr. Chas. Nagy, have been published at Pesth and Vienna. 1834.

35. Notice respecting some Errors common to many Tables of Logarithms.—*Mem. Astron. Soc.* 4to. 1827. Vol. iii. p. 65.

Evidence on Savings-Banks, before a Committee of the House of Commons, 1827.

36. Essay on the general Principles which regulate the Application of Machinery.—*Ency. Metrop.* 4to. 1829.

37. Letter to T. P. Courtenay on the Proportion of Births of the two Sexes amongst Legitimate and Illegitimate Children.—*Brewster's Edin. Jour. of Science.* Vol. ii. p. 85. 1829. This letter was translated into French and published by M. Villermé, Member of the Institute of France.

38. Account of the great Congress of Philosophers at Berlin, on 18 Sept. 1828.—Communicated by a Correspondent [C. B.]. *Edin. Journ. of Science by David Brewster.* Vol. x. p. 225. 1829.

39. Note on the Description of Mammalia.—*Edin. Jour. of Science*, 1829. Vol. i. p. 187. *Ferussac Bull*, vol. xxv. p. 296.

40. Reflections on the Decline of Science in England, and on some of its causes. 4to. and 8vo. 1830.

41. Sketch of the Philosophical Characters of Dr. Wollaston and Sir H. Davy. Exttacted from the *Decline of Science.* 1830.

42. On the Proportion of Letters occurring in Various Languages, in a letter to M. Quételet.— *Correspondence Mathematique et Physique.* Tom. vi. p. 136.

43. Specimen of Logarithmic Tables, printed with different Coloured inks and on variously-coloured papers, in twenty-one volumes 8vo. London. 1831.

The object of this Work, of which *one single copy only* was printed, is to ascertain by experiment the tints of the paper and colours of the inks least fatiguing to the eye.

One hundred and fifty-one variously-coloured papers were chosen, and the same two pages of my stereotype Table of Logarithms were printed upon them in inks of the following colours: light blue, dark blue, light green, dark green, olive, yellow, light red, dark red, purple, and black.

Each of these twenty volumes contains papers of the same colour, numbered in the same order, and there are two volumes printed with each kind of ink.

The twenty-first volume contains metallic printing of the same specimen in gold, silver, and copper, upon vellum and on variously-coloured papers.

For the same purpose, about thirty-five copies of the complete table of logarithms were printed on thick drawing paper of various tints.

An account of this work may be found in the *Edin. Journ. of Science (Brewster's)*, 1832. Vol. vi. p. 144.

44. Economy of Manufactures and Machinery. 8vo. 1832.

There are many editions and also American reprints, and several Translations of this Work into German, French, Italian, Spanish, &c.

45. Letter to Sir David Brewster, on the Advantage of a Collection of the Constants of Nature and Art.—*Brewster's Edin. Jour. of Science.* 1832. Vol. vi. p. 334. Reprinted by order of the British Association for the Promotion of Science. Cambridge, 1833. See also pp. 484, 490, Report of the Third Meeting of the British Association. Reprinted in Compte Rendu des Traveaux du Congres Général de Statistique, Bruxelles, Sept. 1853.

46. Barometrical Observations, made at the Fall of the Staubbach, by Sir John Herschel, Bart., and C. Babbage, Esq.—*Brewster's Edin. Jour. of Science.* Vol. vi. p. 224. 1832.

47. Abstract of a Paper, entitled Observations on the Temple of Serapis, at Pozzuoli, near Naples ; with an attempt to explain the causes of the frequent elevation and depression of large portions of the earth's surface in remote periods, and to prove that those causes continue in action at the present time. Read at Geological Society, 12 March, 1834. See *Abstract of Proceedings of Geol. Soc.* Vol. ii. p. 72.

This was the first *printed* publication of Mr. Babbage's Geological Theory of the Isothermal Surfaces of the Earth.

48. The Paper itself was published in the *Proceedings of the Geological Soc.* 1846.

49. Reprint of the same, with Supplemental Conjectures on the Physical State of the Surface of the Moon. 1847.

50. Letter from Mr. Abraham Sharpe to Mr. J. Crosthwait, Hoxton, 2 Feb. 1721–22. Deciphered by Mr. Babbage. See *Life of Flamsteed*, by Mr. F. Baily. Appendix, pp. 348, 390. 1835.

51. The Ninth Bridgewater Treatise. 8vo. May, 1837 ; Second Edition, Jan. 1838.

52. On some Impressions in Sandstone.—*Proceedings of Geological Society.* Vol. ii. p. 439. Ditto, *Phil. Mag.* Ser. 3. Vol. x. p. 474. 1837.

52*. Short account of a method by which Engraving on Wood may be rendered more useful for the Illustration and Description of Machinery.—*Report of Meeting of British Association at Newcastle.* 1838. P. 154.

53. Letter to the Members of the British Association. 8vo. 1839.

54. General Plan, No. 25, of Mr. Babbage's Great Calculating or Analytical Engine, lithographed at Paris. 24 by 36 inches. 1840.

LIST OF MR. BABBAGE'S PRINTED PAPERS.

55. Statement of the circumstances respecting Mr. Babbage's Calculating Engines. 8vo. 1843.

56. Note on the Boracic Acid Works in Tuscany. *Murray's Handbook of Central Italy.* First Edition, p. 178. 1843.

57. On the Principles of Tools for Turning and Planing Metals, by Charles Babbage. Printed in the Appendix of Vol. ii. Holtzapffel Turning and Mechanical Manipulation. 1846.

58. On the Planet Neptune.—*The Times*, 15th March, 1847.

59. Thoughts on the Principles of Taxation, with reference to a Property Tax and its Exceptions. 8vo. 1848. Second Edition, 1851. Third Edition, 1852.

An Italian translation of the first edition, with notes, was published at Turin, in 1851.

60. Note respecting the pink projections from the Sun's disc observed during the total solar eclipse in 1851.—*Proceedings of the Astron. Soc.*, vol. xii., No. 7.

61. Laws of Mechanical Notation, with Lithographic Plate. Privately printed for distribution. 4to. July, 1851.

62. Note respecting Lighthouse (Occulting Lights). 8vo. Nov. 1851.
Communicated to the Trinity House, 30 Nov. 1851.
Reprinted in the Appendix to the Report on Lighthouses presented to the Senate of the United States, Feb. 1852.
Reprinted in the *Mechanics' Magazine,* and in various other periodicals and newspapers. 1852-3.
It was reprinted in various parts of the Report of Commissioners appointed to examine into the state of Lighthouses. Parliamentary Paper, 1861.

63. The Exposition of 1851; or, Views of the Industry, the Science, and the Government of England. 6s. 6d. Second edition, 1851.

64. On the Statistics of Lighthouses. Compte Rendu des Traveaux du Congres Général, Bruxelles, Sept. 1853.

65. A short description of Mr. Babbage's Ophthalmoscope is contained in the Report on the Ophthalmoscope by T. Wharton Jones, F.R.S.—*British and Foreign Medical Review.* Oct. 1854. Vol. xiv. p. 551.

66. On Secret or Cipher Writing. Mr. T.'s Cipher Deciphered by C.—*Jour. Soc. Arts*, July, 1854, p. 707.

67. On Mr. T's Second Inscrutable Cipher Deciphered by C.—*Jour. Soc. Arts*, p. 777, Aug. 1854.

68. On Submarine Navigation.—*Illustrated News*, 23rd June, 1855.

69. Letter to the Editor of the *Times*, on Occulting Lights for Lighthouses and Night Signals. Flashing Lights at Sebastopol. 16th July, 1855.

72. Sur la Machine Suédoise de M. Scheutz pour Calculer les Tables Mathématiques. 4to. *Comptes Rendus et l'Académie des Sciences.* Paris, Oct. 8, 1855.

73. On the Action of Ocean-currents in the Formation of the Strata of the Earth.—*Quarterly Journal Geological Society*, Nov. 1856.

74. Observations by Charles Babbage on the Mechanical Notation of Scheutz's Difference Engine, prepared and drawn up by his Son, Major Henry Prevost Babbage, addressed to the Institution of Civil Engineers. *Minutes of Proceedings*, vol. xv. 1856.

75. Statistics of the Clearing-House. Reprinted from *Trans. of Statistical Soc.* 8vo. 1856.

76. Observations on Peerage for Life. July, 1833. Reprinted, 1856.

77. Observations addressed to the President and Fellows of the Royal Society on the Award of their Medals for 1856. 8vo.

78. Table of the Relative Frequency of Occurrence of the Causes of Breaking Plate-Glass Windows.—*Mech. Mag.* 24th Jan. 1857.

79. On Remains of Human Art, mixed with the Bones of Extinct Races of Animals. *Proceedings Roy. Soc.* 26th May, 1859.

80. Passages from the Life of a Philosopher. 8vo. 1864.

Printed in the United States
By Bookmasters